Methods in Molecular Biology

Series Editor
John M. Walker
School of Life and Medical Sciences
University of Hertfordshire
Hatfield, Hertfordshire, UK

For further volumes:
http://www.springer.com/series/7651

For over 35 years, biological scientists have come to rely on the research protocols and methodologies in the critically acclaimed Methods in Molecular Biology series. The series was the first to introduce the step-by-step protocols approach that has become the standard in all biomedical protocol publishing. Each protocol is provided in readily-reproducible step-by step fashion, opening with an introductory overview, a list of the materials and reagents needed to complete the experiment, and followed by a detailed procedure that is supported with a helpful notes section offering tips and tricks of the trade as well as troubleshooting advice. These hallmark features were introduced by series editor Dr. John Walker and constitute the key ingredient in each and every volume of the Methods in Molecular Biology series. Tested and trusted, comprehensive and reliable, all protocols from the series are indexed in PubMed.

Manuscript Instructions https://drive.google.com/file/d/1K4Az5vpAFVzYadJ3jKx2-QSo7Dqa_SNA/view?usp=sharing

Nanopore Technology

Methods and Protocols

Second Edition

Edited by

Fanjun Li

Federale de Lausanne, Ecole Polytechnique, Vaud, Switzerland

Humana Press

Editor
Fanjun Li
Federale de Lausanne
Ecole Polytechnique
Vaud, Switzerland

ISSN 1064-3745 ISSN 1940-6029 (electronic)
Methods in Molecular Biology
ISBN 978-1-0716-5526-9 ISBN 978-1-0716-5527-6 (eBook)
https://doi.org/10.1007/978-1-0716-5527-6

This Humana imprint is published by the registered company Springer Science+Business Media, LLC, part of Springer Nature.
The registered company address is: 1 New York Plaza, New York, NY 10004, U.S.A.

Preface

The ability to observe and manipulate matter at the single-molecule level has transformed our understanding of the molecular world. At the heart of this transformation lies nanopore technology, a deceptively simple concept that has evolved into one of the most versatile analytical platforms in modern science. By monitoring the ionic current passing through a single nanometer-scale opening, researchers can now "see" the size, shape, charge, internal dynamics of individual biomolecules, as well as interactions between different molecules (e.g., protein–ligand, protein–protein) in real-time.

This book is intended as a comprehensive methodological guide for the modern nanopore laboratory. As the field moves beyond its initial focus on DNA sequencing and into the complex realms of proteomics, enzymology, and environmental monitoring, the technical requirements for success have become increasingly multidisciplinary. This volume bridges the gap between protein engineering and materials science, providing researchers with the protocols necessary to navigate this expanding landscape. With its focus on nanopore technology and biomolecule characterization, this book will hold the interest of biophysicists, biochemists, bioengineers, and molecular biologists.

Unlike purely theoretical texts, this book is designed as a practical, bench-side manual. It is the result of a global collaboration between 64 scientists working at the forefront of the field. Our collective goal is to provide step-by-step procedures to build, characterize, and utilize nanopore sensors, reducing the "trial-and-error" phase that often hinders new research. This book includes 17 chapters, which are organized into three parts.

Part I consists of nine chapters discussing various protein nanopores including their preparation, engineering, characterization, and applications in single-molecule sensing, as well as using computational methods to study intrinsic nanopore behavior.

Part II consists of six chapters presenting various solid-state nanopores. In particular, nanopore tomography, which focuses on extracting quantitative geometrical parameters rather than genomic sequence information, are outlined.

Part III consists of two chapters specifically detailing the use of the technique droplet interface bilayer in nanopore sensing and membrane biophysical studies.

Whether you are a nanopore expert looking to transition to solid-state architectures or a newcomer attempting your first protein nanopore insertion, these chapters provide the technical "know-how" to accelerate your work. We hope these protocols empower you to build your own platforms, move past the technical hurdles of fabrication, and focus on the most vital part of the process: the discovery.

We are deeply grateful to the contributors who shared their hard-earned protocols and insights. Their dedication to open scientific exchange is what allows this field to move forward.

Vaud, Switzerland *Fanjun Li*
2026

Contents

Part II Solid-State Nanopores

Part III Droplet Interface Bilayer System

Contributors

ISHITA AGRAWAL • *Department of Physics, National University of Singapore, Singapore, Singapore*
ANDREA BONINI • *Department of Chemistry "Ugo Schiff", University of Florence, Sesto Fiorentino Florence, Italy*
DANIEL L. BURDEN • *Wheaton College, Department of Chemistry, Wheaton, IL, USA*
SAURABH CHAUBEY • *Department of Chemistry, Indian Institute of Technology (Banaras Hindu University), Varanasi, UP, India*
KAIKAI CHEN • *School of Nanoscience and Nanotechnology, University of Chinese Academy of Sciences, Beijing, China*
KUO CHEN • *Department of Polymer Science and Engineering, University of Massachusetts, Amherst, MA, USA*
MIN CHEN • *Molecular and Cellular Biology Program, University of Massachusetts Amherst, Amherst, MA, USA; Department of Chemistry, University of Massachusetts Amherst, Amherst, MA, USA*
WEIDI CHEN • *School of Nanoscience and Nanotechnology, University of Chinese Academy of Sciences, Beijing, China*
XIAOYU CHEN • *School of Nanoscience and Nanotechnology, University of Chinese Academy of Sciences, Beijing, China*
CHRISTINA M. CHISHOLM • *Molecular and Cellular Biology Program, University of Massachusetts Amherst, Amherst, MA, USA*
JAEON CHO • *Department of Biomedical Engineering, University of Massachusetts Amherst, Amherst, MA, USA*
MATTEO DAL PERARO • *Institute of Bioengineering, School of Life Sciences, Swiss Federal Institute of Technology Lausanne, Lausanne, Switzerland*
THEO EMMERICH • *Laboratoire de Physique, UMR CNRS 5672, ENS de Lyon, Université de Lyon, Lyon, France*
ANNIKA E. EVENSON • *Wheaton College, Department of Chemistry, Wheaton, IL, USA*
MONIFA A. V. FAHIE • *Molecular and Cellular Biology Program, University of Massachusetts Amherst, Amherst, MA, USA*
JOSHUA C. FOSTER • *Department of Chemistry, University of Massachusetts Amherst, Amherst, MA, USA*
SHOKO FUJITA • *Department of Biotechnology and Life Science, Tokyo University of Agriculture and Technology, Koganei, Tokyo, Japan*
SLAVEN GARAJ • *Department of Physics, National University of Singapore, Singapore, Singapore; Department of Bioengineering, National University of Singapore, Singapore, Singapore; Department of Materials Science and Engineering, National University of Singapore, Singapore, Singapore*
MADISON A. HANDYSIDE • *Wheaton College, Department of Chemistry, Wheaton, IL, USA*
JOSEPH S. HARVEY • *Wheaton College, Department of Chemistry, Wheaton, IL, USA*
NIECK JORDY VAN DER HEIDE • *Chemical Biology, Groningen Biomolecular Sciences & Biotechnology Institute, University of Groningen, Groningen, The Netherlands*

RUI HU • *School of Nanoscience and Nanotechnology, University of Chinese Academy of Sciences, Beijing, China; School of Physics, Central South University, Changsha, China*
JING HUANG • *Department of Chemistry, University of Massachusetts Amherst, Amherst, MA, USA*
RAN JI • *School of Nanoscience and Nanotechnology, University of Chinese Academy of Sciences, Beijing, China*
RYUJI KAWANO • *Department of Biotechnology and Life Science, Tokyo University of Agriculture and Technology, Koganei, Tokyo, Japan*
LISA M. KERANEN-BURDEN • *Wheaton College, Department of Chemistry, Wheaton, IL, USA*
SARAH E. KOENEMAN • *Wheaton College, Department of Chemistry, Wheaton, IL, USA*
SEEEUN LEE • *Wheaton College, Department of Chemistry, Wheaton, IL, USA*
AIYUAN LI • *School of Nanoscience and Nanotechnology, University of Chinese Academy of Sciences, Beijing, China*
FANJUN LI • *Department of Chemistry, University of Massachusetts Amherst, Amherst, MA, USA*
KUN LI • *Department of Physics, National University of Singapore, Singapore, Singapore*
XIN LI • *Department of Chemistry, University of Massachusetts Amherst, Amherst, MA, USA*
YUFEI LI • *School of Nanoscience and Nanotechnology, University of Chinese Academy of Sciences, Beijing, China*
JIE LIANG • *Department of Bioengineering, University of Illinois at Chicago, Chicago, IL, USA*
CHANG LIU • *Department of Biomedical Engineering, University of Massachusetts Amherst, Amherst, MA, USA*
YAXIAN LIU • *School of Nanoscience and Nanotechnology, University of Chinese Academy of Sciences, Beijing, China*
ZHONGFEI LIU • *School of Nanoscience and Nanotechnology, University of Chinese Academy of Sciences, Beijing, China*
CHUNZHE LU • *Chemical Biology, Groningen Biomolecular Sciences & Biotechnology Institute, University of Groningen, Groningen, The Netherlands*
CHENGWEI LUO • *School of Nanoscience and Nanotechnology, University of Chinese Academy of Sciences, Beijing, China*
YANMEI MA • *School of Nanoscience and Nanotechnology, University of Chinese Academy of Sciences, Beijing, China*
GIOVANNI MAGLIA • *Chemical Biology, Groningen Biomolecular Sciences & Biotechnology Institute, University of Groningen, Groningen, The Netherlands*
MARIA JOSE MARCAIDA • *Institute of Bioengineering, School of Life Sciences, Swiss Federal Institute of Technology Lausanne, Lausanne, Switzerland*
TIHOMIR MARJANOVIĆ • *Department of Physics, National University of Singapore, Singapore, Singapore*
SIMON FINN MAYER • *Institute of Bioengineering, School of Engineering, Swiss Federal Institute of Technology Lausanne, Lausanne, Switzerland; Institute of Bioengineering, School of Life Sciences, Swiss Federal Institute of Technology Lausanne, Lausanne, Switzerland*
KATHRYN A. MICHAEL • *Wheaton College, Department of Chemistry, Wheaton, IL, USA*

HELENA MILJKOVIC • *Laboratory of Nanoscale Biology (LBEN), Institute of Bioengineering, School of Engineering, Swiss Federal Institute of Technology Lausanne (EPFL), Lausanne, Switzerland*

MARIANNA FANOURIA MITSIONI • *Institute of Bioengineering, School of Engineering, Swiss Federal Institute of Technology Lausanne, Lausanne, Switzerland; Institute of Bioengineering, School of Life Sciences, Swiss Federal Institute of Technology Lausanne, Lausanne, Switzerland*

MURUGAPPAN MUTHUKUMAR • *Department of Polymer Science and Engineering, University of Massachusetts, Amherst, MA, USA*

YUXIA NIE • *School of Nanoscience and Nanotechnology, University of Chinese Academy of Sciences, Beijing, China*

MARCO VAN DEN NOORT • *Chemical Biology, Groningen Biomolecular Sciences & Biotechnology Institute, University of Groningen, Groningen, The Netherlands*

KM MAMATA PATEL • *Department of Chemistry, Indian Institute of Technology (Banaras Hindu University), Varanasi, UP, India*

LIJUN PENG • *Department of Biomedical Engineering, University of Massachusetts Amherst, Amherst, MA, USA*

ZUGUI PENG • *Department of Biotechnology and Life Science, Tokyo University of Agriculture and Technology, Koganei, Tokyo, Japan*

ALAN PEREZ-RATHKE • *Department of Bioengineering, University of Illinois at Chicago, Chicago, IL, USA*

BACH PHAM • *Department of Chemistry, University of Massachusetts Amherst, Amherst, MA, USA*

ALEKSANDRA RADENOVIC • *Laboratory of Nanoscale Biology (LBEN), Institute of Bioengineering, School of Engineering, Swiss Federal Institute of Technology Lausanne (EPFL), Lausanne, Switzerland; Institute of Bioengineering, School of Engineering, Swiss Federal Institute of Technology Lausanne, Lausanne, Switzerland*

MANA SATO • *Department of Biotechnology and Life Science, Tokyo University of Agriculture and Technology, Koganei, Tokyo, Japan*

KAYLA R. SENNESE • *Wheaton College, Department of Chemistry, Wheaton, IL, USA*

IZUMI SHIBAYAMA • *Department of Biotechnology and Life Science, Tokyo University of Agriculture and Technology, Koganei, Tokyo, Japan*

SABINE STRAATHOF • *Chemical Biology, Groningen Biomolecular Sciences & Biotechnology Institute, University of Groningen, Groningen, The Netherlands*

NANAMI TAKEUCHI • *Department of Biotechnology and Life Science, Tokyo University of Agriculture and Technology, Koganei, Tokyo, Japan*

PRABHAT TRIPATHI • *Department of Chemistry, Indian Institute of Technology (Banaras Hindu University), Varanasi, UP, India*

JAKOB H. VIEL • *Chemical Biology, Groningen Biomolecular Sciences & Biotechnology Institute, University of Groningen, Groningen, The Netherlands*

BIB YANG • *Department of Chemistry, University of Massachusetts Amherst, Amherst, MA, USA*

WAYNE YANG • *Laboratory of Nanoscale Biology (LBEN), Institute of Bioengineering, School of Engineering, Swiss Federal Institute of Technology Lausanne (EPFL), Lausanne, Switzerland; Biomedical Sciences Division, King Abdullah University of Science and Technology (KAUST), Thuwal, Saudi Arabia*

KITTIPITCH YOOPRASERTCHUTI • *Department of Physics, National University of Singapore, Singapore, Singapore*
ZEMIN ZHOU • *School of Nanoscience and Nanotechnology, University of Chinese Academy of Sciences, Beijing, China*

Part I

Protein Nanopores

Check for updates

Chapter 1

Purification and Oligomerization of YaxAB Nanopores

Marco van den Noort, Nieck Jordy van der Heide, Jakob H. Viel, Sabine Straathof and Giovanni Maglia

Abstract

Large nanopores are powerful tools for studying conformational dynamics of folded proteins on a single-molecule level. Additionally, they can be used as real-time metabolite sensors. YaxAB is a large funnel-shaped, hetero-oligomeric nanopore of variable size. It has a unique shape that enables the detection of a broad range of folded proteins and it can even provide insight into protein dimerization. Here, we present a straightforward protocol for the expression and purification of the monomeric components of YaxAB, as well as for their assembly into functional nanopores. Furthermore, we describe how specific oligomeric sizes of the nanopore can be enriched.

Keywords Nanopore, Pore-forming toxin, YaxAB, Protein purification, Oligomerization, Electrophysiology, Single-molecule, Protein trapping, Biosensor

1 Introduction

Nanopores are best known for their role in DNA sequencing [1–5]. However, their potential extends far beyond, enabling applications such as protein sequencing [6–8], metabolite sensing [9–11], single-molecule enzymology [12–14], and folded protein identification [15, 16]. Nanopores are nanometer-sized apertures embedded within an insulating membrane that separates two electrolyte-filled chambers. When a voltage bias is applied across the membrane, it generates a measurable current and an electrophoretic force on charged analytes, and induces and electro-osmotic flow through the nanopore, driving analytes through the pore or trapping them inside. Analytes of sufficient size partially block the ionic current, providing a measurable signal that serves as the basis for detection and analysis.

For nanopore-based analysis of folded proteins and single-molecule enzymology, a protein has to be trapped for sufficient amount of time, preferably seconds or longer. Solid-state nanopores generally lack the structural features required to stably trap proteins, whereas

Fanjun Li (ed.), *Nanopore Technology*, Methods in Molecular Biology, vol. 3072,
https://doi.org/10.1007/978-1-0716-5527-6_1,

most biological nanopores are too narrow for proteins to enter. To overcome this, DNA-origami spheres can act as molecular "corks" in solid-state nanopores, creating a nanocavity that enables reversible protein trapping for minutes to hours [17, 18]. In the case of biological nanopores, cytolysin A (ClyA) has been widely used for single-molecule studies of folded proteins [9, 10, 12–14, 19, 20]. However, its cylindrical geometry restricts protein capture to a size range of approximately 10–40 kDa.

As an alternative to ClyA, the YaxAB toxin from *Yersinia enterocolitica* has been employed [15, 16, 21, 22]. YaxAB belongs to the same family of α-helical pore-forming toxins (α-PFTs) as ClyA. The protein forms pores of different oligomeric sizes, most commonly forming 16-, 18-, and 20-mers [22, 23]. Unlike ClyA, YaxAB assembles into hetero-oligomeric pores of YaxA and YaxB subunits, with YaxB subunits lining the interior of the pore and inserting into the membrane [23]. Moreover, YaxAB has a conical shape. The pore diameter ranges from approximately 3.5 nm at its narrowest point to 15 nm at its widest. This particular shape allows for capturing a much wider range of protein size. To date, proteins between 12 and 125 kDa have been studied using YaxAB, and even larger proteins could potentially be accommodated [15, 16, 21, 22].

Herein, we describe a method for recombinant expression and purification of YaxA and YaxB monomers, subsequent oligomerization into YaxAB nanopores, and separation of the different oligomeric sizes of YaxAB. The methods are similar to the purification-oligomerization-separation method, we described for ClyA [20]. However, some major differences exist that pressed us to write a separate method for YaxAB. First of all, YaxAB consists of two different subunits of which YaxA expresses significantly less well in a normal expression strain. Second, YaxA and YaxB monomers form nanopores once they are mixed, even in the absence of detergent. In this case, the mild detergent CYMAL-6 has to be added to destroy the formation of multimers of oligomers. Altogether, YaxAB nanopores can be obtained within 1 week using basic biochemical lab equipment, but attention has to be paid to some caveats presented in this chapter.

2 Materials

All solutions are prepared using MilliQ water (Millipore Merck). All reagents are prepared and stored at room temperature, unless otherwise indicated.

2.1 Transformation and Expression of YaxA and YaxB Monomers

1. 1 L Lysogeny Broth (LB). Weigh 25 g of premixed LB-powder (10 g tryptone, 5 g yeast extract, 10 g NaCl, pH 7.0) and transfer to a liter bottle. Fill the bottle up to 1 L with water to dissolve the LB powder. Sterilize the LB-medium by autoclaving for 15 min at 121°C.

2. 1 L 2×YT broth. Weigh 31 g of premixed 2×YT-powder (16 g tryptone, 10 g yeast extract, 5 g NaCl, pH 7.0) and transfer to a liter bottle. Fill the bottle up to 1 L with water to dissolve the 2×YT powder. Sterilize the 2×YT-medium by autoclaving for 15 min at 121°C.
3. 100 mL 40% (w/v) glucose. Weigh 40 g of glucose and fill a 100 mL bottle with 50 mL water. While stirring, slowly add the glucose to the bottle with water. When the glucose is dissolved, fill the bottle up to 100 mL with water. Sterilize the glucose solution by autoclaving for 15 min at 121°C.
4. 10 mL of recovery-medium (LB-medium with 1% glucose). Add 250 μL of 40% glucose solution to 9.75 mL LB-medium. Aliquot the recovery-medium (1 mL) and store at −20°C.
5. 10 mL 100 mg/mL ampicillin. Dissolve 1 g of ampicillin in 10 mL of water. Aliquot the ampicillin stocks (500 μL) and store at −20°C.
6. 10 mL 1 M Isopropyl-β-D-thiogalactopyranoside (IPTG). Dissolve 2.38 g of IPTG in 10 mL of water. Aliquot the IPTG stocks (500 μL) and store at−20°C.
7. LB agar plates (LB-agar with 1% glucose). Weigh 12.5 g LB-powder and 7.5 g Agar-Agar powder and transfer to a liter bottle. Fill the bottle up to 500 mL with water and sterilize the LB-agar by autoclaving for 15 min at 121°C. Cool down to ~60°C, and add 12.5 mL of 40% glucose solution and 500 μL of 100 mg/mL ampicillin stock. Pour the LB-agar plates (~15 mL per plate) next to a Bunsen burner. Store the solidified LB-agar plates at 4°C.
8. Autoclave system.
9. Bunsen burner.
10. Toothpicks for inoculation. Place toothpicks in a glass bottle and sterilize by autoclaving for 15 min at 121°C.
11. 2×50 μL aliquot of E. cloni® EXPRESS BL21(DE3) or SoluBL21™ competent cells stored at −80°C (*see* **Note 1**).
12. Plasmid DNA (~100 ng/μL) of YaxA and YaxB in the pRSET-A vector, containing an N-terminal 6× histidine purification tag and TEV-cleavage site.
13. Heath block capable of maintaining 42°C, 37°C, and 30°C.
14. Stove capable of maintaining 60°C and 37°C.
15. Shaking incubator capable of maintaining 37°C, 25°C, and 200 rpm.
16. Centrifuge capable of maintaining 4°C and 8000 ×*g*.
17. Spectrophotometer (Nanodrop) for measuring OD_{600} and protein concentrations.

18. Petri dishes and Drigalski spatulas.
19. 3×1 L glass bottle and 1×100 mL glass bottle.
20. 2×100 mL Erlenmeyer flasks (starter culture) and 2×1 L Erlenmeyer flasks (Expression culture).
21. Ice box with crushed ice.

2.2 Purification of YaxA and YaxB Monomers

1. 100 mL lysis buffer A: 50 mM Tris (121.14 g/mol), 300 mM NaCl (58.44 g/mol), 10 mM Imidazole (68.08 g/mol), 2 M Urea (60.06 g/mol), and 1 mM $MgCl_2$ (95.21 g/mol). Add the components to a 100 mL bottle and fill up to 90 mL with water (*see* **Note 2**). Adjust the pH to 8.0 using HCl and fill up to 100 mL with water, store at 4°C.
2. 100 mL lysis buffer B: 50 mM Tris (121.14 g/mol), 300 mM NaCl (58.44 g/mol), 10 mM Imidazole (68.08 g/mol), and 1 mM $MgCl_2$ (95.21 g/mol). Add the components to a 100 mL bottle and fill up to 90 mL with water. Adjust the pH to 8.0 using HCl and fill up to 100 mL with water, store at 4°C.
3. 100 mL wash buffer: 50 mM Tris (121.14 g/mol), 300 mM NaCl (58.44 g/mol), and 10 mM Imidazole (68.08 g/mol). Add the components to a 100 mL bottle and fill up to 90 mL with water. Adjust the pH to 8.0 using HCl and fill up to 100 mL with water, store at 4°C.
4. 100 mL Elution buffer: 50 mM Tris (121.14 g/mol), 300 mM NaCl (58.44 g/mol), and 100 mM Imidazole (68.08 g/mol). Add the components to a 100 mL bottle and fill up to 90 mL with water. Adjust the pH to 7.5 and fill up to 100 mL with water, store at 4°C.
5. Pierce Protease Inhibitor Mini Tablets (EDTA free).
6. Probe sonicator compatible with 50 mL tubes.
7. Benchtop microtube centrifuge capable of maintaining $10{,}000 \times g$.
8. Tube rotator/shaker compatible with 50 mL tubes.
9. Plate reader and 96-well microplates.
10. Bradford dye.
11. DNase I (1 U/μL).
12. Ni-NTA affinity resin.
13. TEV-protease (10 U/μL).
14. 100 mL of 500 mM EDTA disodium dihydrate (372.24 g/mol) pH 8. Dissolve 18.6 g EDTA in ~80 mL MQ, while stirring. Add slowly NaOH pellets to facilitate the dissolving process, until the pH is set to 8 and EDTA is fully dissolved.
15. 1 mL of 100 mM dithiothreitol (DTT; 154.25 g/mol).

16. Gravity flow columns (2 mL, 5 mL, or 10 mL columns).
17. 4 × 100 mL glass bottles.

2.3 Hemolytic Activity of YaxAB

1. Plate reader and transparent, flat-bottom, 96-well microplates.
2. Defibrinated sheep blood.
3. 1% (v/v) Triton X-100.
4. Multichannel pipette (200 μL).

2.4 Oligomerization of YaxAB

1. 10 mL 10% (w/v) 6-Cyclohexylhexyl β-D-maltoside (CYMAL-6). Dissolve 1 g of CYMAL-6 in 10 mL of water. Aliquot the CYMAL-6 stocks (1 mL) and store at −20°C.
2. 500 mL SEC buffer (filtered and degassed): 25 mM HEPES (238.31 g/mol), 150 mM NaCl (58.44 g/mol), and 0.05% (w/v) CYMAL-6 (*see* **Note 3**). Add the components to a 1 L bottle and fill up to 400 mL with water. Adjust the pH to 7.0 and fill up to 500 mL with water, store at 4°C.
3. FPLC (ÄKTA pure™) with 500 μL sample loop.
4. Superose 6 10/300 GL Increase size exclusion chromatography column.
5. 1 × 1 L glass bottle.
6. (Optional) Native-PAGE (Criterion™ Cell; Bio-Rad) running system and loading dye.
7. (Optional) 3%–8% Criterion™ XT Tris-Acetate Protein Gel.
8. (Optional) Invitrogen™ NativePAGE™ Cathode Buffer Additive (20×) and NativePAGE™ Running Buffer (20×).
9. (Optional) Scalpel blade.

3 Methods

3.1 Transformation and Expression of YaxA and YaxB Monomers (Day 1–3)

1. During transformation and expression, work sterile next to a Bunsen burner.
2. Thaw the aliquots of chemical competent E. cloni® EXPRESS BL21(DE3) cells on ice (*see* **Note 4**).
3. Add 0.5 μL plasmid DNA (~100 ng/μL) of YaxA or YaxB (in the pRSET-A vector) to 50 μL competent cells and incubate for 20 min on ice.
4. To transform the cells with the DNA, heat-shock the cells for 1 min at 42°C followed by a cold-shock on ice for 5 min.
5. Add 200 μL recovery-medium to the cells and incubate at 37°C for 45 min (*see* **Note 5**).

6. Plate out 50 μL of the transformed cells on LB-agar plates and incubate at 37°C overnight.
7. On day 2, pick one colony with a sterile toothpick and inoculate it into a 100 mL Erlenmeyer flasks containing 10 mL LB-medium with 100 μg/mL ampicillin (starter culture). Incubate the starter culture overnight at 37°C and 200 rpm.
8. On day 3, inoculate 2 mL starter culture into a 1 L Erlenmeyer flasks containing 200 mL 2×YT-medium with 100 μg/mL ampicillin (expression culture). Incubate the expression culture at 37°C and 200 rpm for 2–3 h until the optical density at 600 nm wavelength (OD_{600}) is between 0.6–0.8.
9. Cool down the expression culture on ice for 5 min.
10. To induce protein expression, add 100 μL IPTG (1 M) to 200 mL expression culture for a final concentration of 0.5 mM IPTG.
11. Incubate the expression culture overnight at 20°C and 200 rpm.

3.2 Purification of YaxA and YaxB Monomers (Day 4)

1. Harvest the bacterial cells in 50 mL plastic tubes by centrifugation at 7000 × *g* for 5 min at 4°C. Spin down 50–400 mL cells in one tube by repeated centrifugation.
2. Discard the supernatant and store the cell pellets at −20°C, −80°C or directly continue with the protocol.
3. Resuspend the cell pellets in lysis buffer (20–40 mL lysis buffer per pellet of 50–400 mL culture). Use Lysis buffer A for YaxA and lysis buffer B for YaxB (*see* **Note 2**).
4. Add one protease inhibitor tablet per 10–20 mL of resuspended cell pellet and 0.05 units/mL of DNaseI. Incubate the cell suspension for 15 min at ~4°C–6°C (or at room temperature with ice-cold lysis buffer) while rotating at 10 rpm to dissolve the protease inhibitor tablet.
5. Place the cell suspension in an ice-water bath and lyse the cells by sonication for 10 min at 20% output power (cycles with 5 s on/5 s off time, 5 min total on time).
6. Pellet the cell debris by centrifugation at 7500 × *g* for 30 min at 4°C. The target protein is in the supernatant.
7. Transfer 200 μL of Ni-NTA resin (400 μL of Ni-NTA suspension) to a gravity-flow column, and wash two times with 5 mL MQ and one time with 1 mL wash buffer. Between each step, wait for the solution to completely flow through.
8. Transfer the resin from the tube to a 50 mL tube containing the supernatant from Step 6 using 0.5 mL of wash buffer to resuspend the resin. Incubate for 15 min at ~4°C–6°C while

rotating at 10 rpm or at room temperature after cooling the supernatant on ice.

9. Transfer the supernatant with the incubated Ni-NTA resin to the gravity flow column and wait for the solution to completely flow through.
10. Wash the Ni-NTA resin in at least two steps, using 5 mL of wash buffer in each step.
11. Elute the monomers from the Ni-NTA resin in 6× elution steps of 120 μL Elution buffer. During each elution step incubate 1–2 min with Elution buffer and collect the flowthrough fractions separately.
12. Check the protein concentration using a Bradford assay or by measuring the absorbance with a Nanodrop at 280 nm (*see* **Note 6 and** 7). Combine the protein-containing fractions.
13. Add 1 mM DTT, 0.5 mM EDTA and the TEV proteases [1:100 TEV:protein (w/w)] to cleave off the N-terminal histidine-tag. Incubate for 2 h at 30°C or overnight at 4°C (*see* **Note 8**).
14. Proceed with the hemolytic activity test of the monomers or directly with the oligomerization of YaxAB. The monomers can also be stored at 4°C up to 1 week for later use.

3.3 Hemolytic Activity of YaxAB (Day 4)

1. Prepare the defibrinated sheep blood by washing 500 μL blood ~3× times with 500 μL wash buffer. During each wash step add 500 μL wash buffer, mix the cell suspension by pipetting up and down, centrifuge at 10,000 × *g* for 30 s, discard the supernatant and repeat until the supernatant is clear.
2. Dilute the washed blood in ~20 mL wash buffer to reach an optical density of 0.6–0.8 at 650 nm wavelength. The OD_{650} can be measured with a plate-reader by adding 100 μL of diluted blood to a 96-well microplate.
3. Add the controls to the microplate: 10 μL of 1% Triton X-100 (100% lysis) and 10 μL of elution buffer (0% lysis).
4. First add 10 μL of YaxA (~0.1–1 mg/mL protein concentration) to the microplate.
5. Add 100 μL diluted blood to the controls and YaxA quickly (with a multichannel pipette or stepper pipette) and incubate for 20 min (*see* **Note 9**).
6. Add 10 μL YaxB in a 1:1 ratio to YaxA quickly and incubate for ~30 min, while measuring the OD_{650}.

3.4 Oligomerization of YaxAB (Day 5)

1. Setup the FPLC system by connecting the Superose 6 column and the 500 μL sample loop (*see* **Note 10**).

2. Wash the column first with two column volumes of water, followed by equilibration of at least one column volume SEC-buffer (*see* **Note 11**).
3. Spin down the monomers at 20,000×g and 4°C for 10 min to remove large aggregates, which otherwise will reduce the performance of the Superose 6 column.
4. Combine 250 µL YaxA and 250 µL YaxB monomers together in a 1:1 ratio (~1–10 mg/mL protein concentration) (*see* **Note 12**). Incubate the monomer mixture for 30 min at room temperature.
5. Add 1.5% (w/v) CYMAL-6 to the protein mixture, mix well by pipetting, and incubate for 30 min at 4°C (*see* **Note 13**).
6. Directly load 500 µL of the protein mixture into the sample loop of the FPLC.
7. Load the protein mixture onto the Superose 6 column with a flowrate of 0.5 mL/min and continuously measure the absorbance at 280 nm wavelength in the eluent. Collect the protein peaks in fractions of 200-500 µL.
8. A normal FPLC-run of YaxAB on the Superose 6 column should result in four different absorbance peaks at 280 nm wavelength. Peak 1 at 8–9 mL contains unseparated multimers of YaxAB oligomers, peak 2 at 10–13 mL contains different oligomeric forms of YaxAB, and peak 3 and 4 at 16–18 mL contain non-oligomerized monomers of YaxA and YaxB, respectively (Fig. 1A).
9. Aliquot per 20 µL and flash freeze the YaxAB oligomers in liquid nitrogen and store for long-term at −80°C.

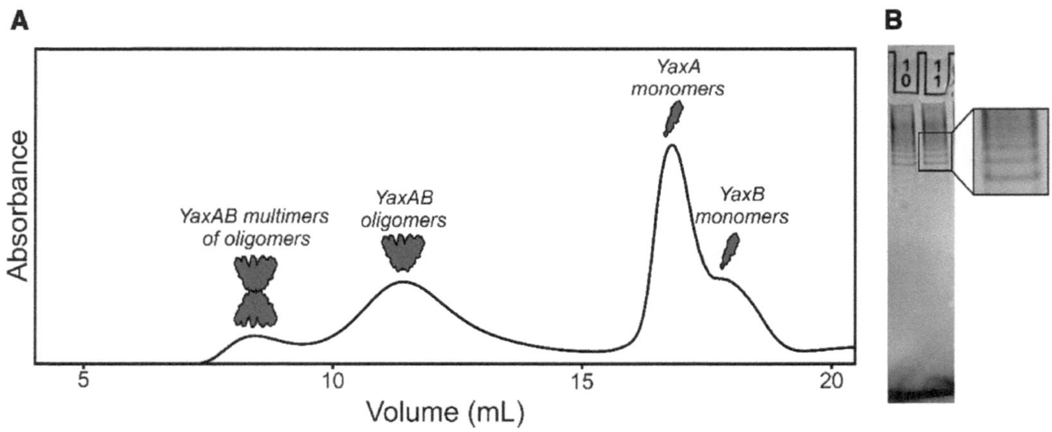

Fig. 1 Purification of YaxAB oligomers. (**A**) A typical size-exclusion chromatography profile of YaxAB after oligomerization. (**B**) Isolation of specific YaxAB oligomer sizes separated on a 3%–8% Tris-Acetate Native-PAGE gel. The three bands represent three oligomeric sizes of YaxAB. From bottom to top, the bands most likely correspond to 16-, 18-, and 20-mers, respectively

10. YaxAB nanopores can be used in electrophysiology experiments as described before [16, 21].

3.5 Enrichment of Specific Oligomer Sizes (Day 5) (Optional)

1. After Step 8 of the oligomerization of YaxAB, collect the fractions belonging to absorbance peak 2 and add Native-PAGE loading dye to the protein mixture. Load the protein mixture with loading dye directly onto a Tris-acetate Native-PAGE gel (3%–8% gradient).
2. Run the Native-PAGE gel for 10 min at 120 V with 1× NATIVE running buffer in the chamber and 1× NATIVE running buffer supplemented with 2 mL 20× cathode buffer additive in the top compartment. Replace the cathode buffer additive with 1× native running buffer in the top compartment and continue running the gel for 80 min at 120 V.
3. Three to four clear bands should appear in the top part of the gel, reflecting different oligomeric sizes of YaxAB (Fig 1B) [15, 22]. Cut the protein bands from the Native-PAGE gel with a scalpel blade and store them separately in microtubes. Store the YaxAB protein bands for long-term at −80°C.
4. For electrophysiology experiments, redissolve the YaxAB oligomers by incubating the protein band for at least 30 min in 25 μL of SEC buffer on the day of the experiment.

4 Notes

1. It is advisable to use *E. coli*SoluBL21™ competent cells for a better YaxA expression. YaxB works equally well with both E. cloni® EXPRESS BL21(DE3) and SoluBL21™ competent cells.
2. YaxA aggregates during expression and partly ends up in inclusion bodies. Lysis buffer A contains 2 M urea to redissolve some YaxA aggregates. Higher urea concentrations do not increase the amount of functional YaxA monomers.
3. Before adding the CYMAL-6 detergent, first adjust the pH, filter and degas the SEC-buffer to prevent the formation of foam.
4. Thaw the chemical competent cells on top of the ice and not pressed into the ice. When pressed into the ice the temperature increases quickly. This temperature shock can reduce the competency of the cells.
5. This incubation step helps cells recover from the transformation process, allows the cells to divide and overcome the antibiotic selection.

6. Expect a protein yield of about 0.25–1 mg YaxA and 1.5–3 mg YaxB per 100 mL of culture.
7. The protein concentration can also be determined by Nanodrop. However, the concentration estimation of YaxB can be less reliable because the protein only contains one tryptophan.
8. Make sure the imidazole concentration in the protein sample does not exceed 150 mM. Higher concentrations of imidazole inhibit the activity of the TEV-protease. Optionally the imidazole can be removed by desalting-columns or dialysis to increase protease activity.
9. No hemolytic activity should be observed when only YaxA is present.
10. When connecting the Superose 6 column use the drop-to-drop method to prevent air from entering the column. Air bubbles can damage the Superose 6 column.
11. Make sure that the buffers and water are filtered and degassed to prevent the formation of air bubbles inside the FPLC system.
12. To obtain higher concentrations of YaxA and/or YaxB, the protein can be concentrated prior oligomerization using an Amicon 10 K filtration spin-column. Spin down in steps of 2 min at 14,000 × *g* and 4°C.
13. YaxAB oligomers are formed once the monomers are mixed (Fig. 1). In the absence of detergent, the oligomers from multimers of oligomers. To destroy those multimers, the sample is incubated with 1.5% CYMAL-6. Do not incubate for longer than 30 min. Otherwise, a significant fraction of YaxAB oligomers fall apart to monomers again.

References

1. Maglia G, Heron AJ, Stoddart D et al (2010) Chapter 22—analysis of single nucleic acid molecules with protein nanopores. In: Walter NG (ed) Methods in enzymology. Academic Press, pp 591–623
2. Clarke J, Wu H-C, Jayasinghe L et al. (2009) Continuous base identification for single-molecule nanopore DNA sequencing. Nature Nanotech 4:265–270
3. Wang Y, Zhao Y, Bollas A et al. (2021) Nanopore sequencing technology, bioinformatics and applications. Nat Biotechnol 39:1348–1365
4. Manrao EA, Derrington IM, Laszlo AH et al. (2012) Reading DNA at single-nucleotide resolution with a mutant MspA nanopore and phi29 DNA polymerase. Nat Biotechnol 30:349–353
5. Deamer D, Akeson M, Branton D (2016) Three decades of nanopore sequencing. Nat Biotechnol 34:518–524
6. Lu C, Bonini A, Viel JH et al. (2025) Toward single-molecule protein sequencing using nanopores. Nat Biotechnol 43:312–322
7. Brinkerhoff H, Kang ASW, Liu J et al. (2021) Multiple rereads of single proteins at single–amino acid resolution using nanopores. Science 374:1509–1513
8. Zhang M, Tang C, Wang Z et al. (2024) Real-time detection of 20 amino acids and

discrimination of pathologically relevant peptides with functionalized nanopore. Nat Methods 21:609–618
9. Galenkamp NS, Soskine M, Hermans J et al. (2018) Direct electrical quantification of glucose and asparagine from bodily fluids using nanopores. Nat Commun 9: 4085
10. Zernia S, Heide NJ, van der, Galenkamp NS et al. (2020) Current Blockades of Proteins inside Nanopores for Real-Time Metabolome Analysis. ACS Nano 14:2296–2307
11. Rauf S, Zhang L, Ali A et al. (2017) Label-free nanopore biosensor for rapid and highly sensitive cocaine detection in complex biological fluids. ACS Sens 2:227–234
12. Galenkamp NS, Zernia S, Van Oppen YB et al. (2024) Allostery can convert binding free energies into concerted domain motions in enzymes. Nat Commun 15: 10109
13. Li F, Fahie MA, Gilliam KM et al. (2022) Mapping the conformational energy landscape of Abl kinase using ClyA nanopore tweezers. Nat Commun 13: 3541
14. Galenkamp NS, Biesemans A, Maglia G (2020) Directional conformer exchange in dihydrofolate reductase revealed by single-molecule nanopore recordings. Nat Chem 12:481–488
15. Ryu M, Oh S, Jeong K-B et al. (2024) Single-molecule-based, label-free monitoring of molecular glue efficacies for promoting protein–protein interactions using YaxAB nanopores. ACS Nano 18:31451–31465
16. Straathof S, Di Muccio G, Yelleswarapu M et al. (2023) Protein sizing with 15 nm conical biological nanopore YaxAB. ACS Nano 17:13685–13699
17. Schmid S, Stömmer P, Dietz H et al. (2021) Nanopore electro-osmotic trap for the label-free study of single proteins and their conformations. Nat Nanotechnol 16:1244–1250
18. Schmid S, Dekker C (2021) The NEOtrap—en route with a new single-molecule technique. iScience 24:103007
19. Soskine M, Biesemans A, De Maeyer M et al. (2013) Tuning the size and properties of ClyA nanopores assisted by directed evolution. J Am Chem Soc 135:13456–13463
20. Galenkamp NS, Meervelt V, Mutter NL et al (2021) Preparation of cytolysin A (ClyA) nanopores. In: Fahie MAV (ed) Nanopore technology: methods and protocols. Springer, New York, NY, USA, pp 11–18
21. Straathof S, Di Muccio G, Maglia G (2025) Nanopores with an engineered selective entropic gate detect proteins at nanomolar concentration in complex biological sample. J Am Chem Soc 147:15050–15065
22 Jeong K-B, Ryu M, Kim J-S et al. (2023) Single-molecule fingerprinting of protein-drug interaction using a funneled biological nanopore. Nat Commun 14: 1461
23. Bräuning B, Bertosin E, Praetorius F et al. (2018) Structure and mechanism of the two-component α-helical pore-forming toxin YaxAB. Nat Commun 9: 1806

Chapter 2

Expression and Purification of CytK Nanopores

Nieck Jordy van der Heide, Andrea Bonini, Giovanni Maglia and Chunzhe Lu

Abstract

Nanopores are emerging as powerful tools for peptide and protein sensing and sequencing. Cytotoxin K (CytK), a highly stable β-barrel pore, shows strong potential for both applications. The incorporation of an acidic-aromatic recognition site significantly enhances its ability to resolve different peptides. Other CytK variants with enhanced electroosmotic flow enable the measurement of unfolded native proteins. Together, these properties position CytK as an effective tunable nanopore for label-free detection of peptides and for analyzing native full-length proteins under denaturing conditions. This chapter provides a step-by-step protocol for the expression and purification of CytK nanopores.

Keywords Nanopore, Protein identification, Peptide sequencing, Cytotoxin K, Protein purification, Label-free detection, Electrophysiology, Single-molecule, Electroosmotic flow

1 Introduction

Biological nanopores are multimeric membrane proteins that regulate the transport of ions, water, small molecules, and other substances across cellular membranes. Owing to their precisely defined size, shape, and geometry, many of these pores have been repurposed as powerful tools for single molecule sensing and sequencing. The most notable successes of nanopore technology have been its application to DNA and RNA sequencing [1–3], which have revolutionized life science research, enabled rapid disease diagnostics including COVID-19 studies, and accelerated vaccine development. Building on these achievements, researchers are now exploring nanopores for protein identification and sequencing at the single-molecule level.

Proteins, however, pose greater challenges than nucleic acids. DNA consists of only four uniformly negatively charged nucleotides, whereas proteins are built from 20 amino acids with diverse charges and chemical properties, giving rise to far more complex structures. The forces that drive their transport through a nanopore also differ.

Fanjun Li (ed.), *Nanopore Technology*, Methods in Molecular Biology, vol. 3072,
https://doi.org/10.1007/978-1-0716-5527-6_2,

Single-stranded DNA is electrophoretically drawn through a nanopore by an externally applied electric field, while proteins are often transported by electroosmotic flow (EOF), the movement of water induced by net ion accumulation within the nanopore.

Two principal strategies have emerged for analyzing full-length proteins with nanopores: (1) chop-and-drop and (2) strand sequencing [4]. In the chop-and-drop approach, proteins are first enzymatically digested into short peptides, typically with trypsin, and the resulting fragments are identified using nanopores such as Fragaceatoxin C (FraC) [5], aerolysin (AeL) [6], α-hemolysin (α-HL) [7], cytotoxin K (CytK) [6], or engineered proteasome activator 28α (REG) [8]. The generated peptide "spectra" provide fingerprints that can be matched to protein databases, a method termed nanopore peptide spectrometry. Strand sequencing, in contrast, involves chemically or enzymatically unfolding a native protein into a linearized polypeptide chain that can be threaded through nanopores. Several pores have been engineered and used for this purpose, including *Mycobacterium smegmatis* porin A (MspA) [9–11], α-HL [12, 13], bacterial curli transport lipoprotein (CsgG) [14], and CytK [15].

CytK (PDB ID 8RJ8) is a β-barrel pore-forming toxin from *Bacillus cereus*. It features a ~1.1 nm *cis*-side entry constriction and a ~1.2 nm barrel [16], dimensions well suited for detecting both peptides and unfolded proteins. Although wild-type CytK cannot efficiently detect peptides, engineered variants such as $CytK^{K128D}$, $CytK^{K128F}$, and $CytK^{S126F\text{-}K128D}$ can capture and detect peptides from tyrosinated lysozyme, driven primarily by electrophoretic forces rather than EOF [6]. As for protein identification, the EOF enhanced variant CytK-2E-4D was demonstrated to drive unidirectional transport of unstructured natural polypeptides across the nanopore [15]. At −100 mV this variant generates forces of roughly 20 pN, comparable to those acting on single-stranded DNA and sufficient to pull highly charged polypeptides even against opposing electric fields. The concurrent EOF helps stretch the translocating chains, producing characteristic ionic-current signatures that enable protein identification and, potentially, sequence analysis, thereby addressing a fundamental challenge in nanopore-based protein studies.

This chapter describes the methods for expressing and purifying the CytK nanopore. CytK is notably stable, retaining good membrane-insertion activity even after several months of refrigerated storage. The following protocol highlights the essential steps and key considerations for reliable nanopore preparation.

2 Materials

All solutions are prepared using deionized water (Millipore Merck). All reagents are prepared and stored at room temperature, unless otherwise indicated.

2.1 Transformation and Expression of CytK

1. 1 L Lysogeny Broth (LB). Weigh 25 g of premixed LB-powder (10 g tryptone, 5 g yeast extract, 10 g NaCl, pH 7.0) and transfer to a liter bottle. Fill the bottle up to 1 L with water to dissolve the LB powder. Sterilize the LB-medium by autoclaving for 15 min at 121°C.
2. 1 L 2xYT broth. Weigh 31 g of premixed 2xYT-powder (16 g tryptone, 10 g yeast extract, 5 g NaCl, pH 7.0) and transfer to a liter bottle. Fill the bottle up to 1 L with water to dissolve the 2xYT powder. Sterilize the 2xYT-medium by autoclaving for 15 min at 121°C.
3. 100 mL 40% w/v glucose. Weigh 40 g of glucose and fill a 100 mL bottle with 50 mL water. While stirring, slowly add the glucose to the bottle with water. When the glucose is dissolved, fill the bottle up to 100 mL with water. Sterilize the glucose solution by autoclaving for 15 min at 121°C.
4. 10 mL of recovery-medium (LB-medium with 1% glucose). Add 250 μL of 40% glucose solution to 9.75 mL LB-medium. Aliquot the recovery-medium (1 mL) and store at −20°C.
5. 10 mL 100 mg/mL Ampicillin. Dissolve 1 g of ampicillin in 10 mL of water. Aliquot the ampicillin stocks (500 μL) and store at −20°C.
6. 10 mL 1 M Isopropyl-β-D-thiogalactopyranoside (IPTG). Dissolve 2.38 g of IPTG in 10 mL of water. Aliquot the IPTG stocks (500 μL) and store at −20°C.
7. LB agar plates (LB-agar with 1% glucose). Weigh 12.5 g LB-powder and 7.5 g Agar-Agar powder and transfer to a liter bottle. Fill the bottle up to 500 mL with water and sterilize the LB-agar by autoclaving for 15 min at 121 °C. Cool down to ~60°C, and add 12.5 mL of 40% glucose solution and 500 μL of 100 mg/mL ampicillin stock. Pour the LB-agar plates (~15 mL per plate) next to a Bunsen burner. Store the solidified LB-agar plates at 4°C.
8. Autoclave system.
9. Bunsen burner.
10. Toothpicks for inoculation. Place toothpicks in a glass bottle and sterilize by autoclaving for 15 min at 121°C.
11. 2 × 50 μL aliquot of *E. cloni*® EXPRESS BL21(DE3) competent cells stored at−80°C.
12. Plasmid DNA (~100 ng/μL) of CytK in the pT7-SC1 vector, containing a C-terminal 6× histidine purification tag.
13. Heath block capable of maintaining 42°C and 37°C.
14. Stove capable of maintaining 60°C and 37°C.
15. Shaking incubator capable of maintaining 37°C, 25°C, and 200 rpm.
16. Centrifuge capable of maintaining 4°C and 8000 × *g*.

17. Spectrophotometer (Nanodrop) for measuring OD_{600} and protein concentrations.
18. Petri dishes and Drigalski spatulas.
19. 3 × 1 L glass bottle and 1 × 100 mL glass bottle.
20. 2 × 100 mL Erlenmeyer flasks (starter culture) and 2 × 1 L Erlenmeyer flasks (Expression culture).
21. Ice box with crushed ice.

2.2 Purification of CytK

1. 100 mL Lysis buffer: 50 mM Tris (121.14 g/mol), 150 mM NaCl (58.44 g/mol), 10 mM Imidazole (68.08 g/mol), 1 mM MgCl2 (95.21 g/mol), and 0.02% DDM. Add the components to a 100 mL bottle and fill up to 90 mL with water (*see* **Notes 1 and 2**). Adjust the pH to 7.5 using HCl and fill up to 100 mL with water, store at 4°C.
2. 100 mL Wash buffer: 50 mM Tris (121.14 g/mol), 150 mM NaCl (58.44 g/mol), 10 mM Imidazole (68.08 g/mol), and 0.02% DDM. Add the components to a 100 mL bottle and fill up to 90 mL with water. Adjust the pH to 7.5 using HCl and fill up to 100 mL with water, store at 4°C.
3. 100 mL Elution buffer: 50 mM Tris (121.14 g/mol), 150 mM NaCl (58.44 g/mol), 250 mM Imidazole (68.08 g/mol), and 0.02% DDM. Add the components to a 100 mL bottle and fill up to 90 mL with water. Adjust the pH to 7.5 and fill up to 100 mL with water, store at 4°C.
4. 10 mL 10% (w/v) Dodecyl-β-D-maltoside (DDM). Dissolve 1 g of DDM in 10 mL of water. Aliquot the DDM stocks (1 mL) and store at −20°C.
5. Pierce Protease Inhibitor Mini Tablets (EDTA free).
6. DNase I (1 U/μL).
7. Probe sonicator compatible with 50 mL tubes.
8. Benchtop microtube centrifuge capable of maintaining 10,000 rpm.
9. Tube rotator/shaker compatible with 50 mL tubes.
10. Plate reader and 96-well microplates.
11. Bradford dye.
12. Ni-NTA affinity resin.
13. Gravity flow columns (2 mL, 5 mL, or 10 mL columns).
14. 3 × 100 mL glass bottles.

2.3 Hemolytic Activity of CytK Monomers

1. Plate reader and transparent, flat-bottom, 96-well microplates.
2. Defibrinated sheep blood.
3. 1 mL 1% v/v Triton X-100.
4. Multichannel pipette (200 μL).

3 Methods

3.1 Transformation and Expression of CytK (Day 1–3)

1. During transformation and expression, work sterile next to a Bunsen burner.
2. Thaw the aliquots of chemical competent *E. cloni*® EXPRESS BL21(DE3) cells on ice (*see* **Note 3**).
3. Add 0.5 μL plasmid DNA (~100 ng/μL) of CytK (in the pT7-SC1 vector) to 50 μL competent cells and incubate for 20 min on ice.
4. To transform the cells with the DNA, heat-shock the cells for 1 min at 42°C followed by a cold-shock on ice for 5 min.
5. Add 200 μL recovery-medium to the cells and incubate at 37°C for 45 min (*see* **Note 4**).
6. Plate out 50 μL of the transformed cells on LB-agar plates and incubate at 37°C overnight.
7. On day 2, pick one colony with a sterile toothpick and inoculate it into a 100 mL Erlenmeyer flask containing 10 mL LB-medium with 100 μg/mL ampicillin (starter culture). Incubate the starter culture overnight at 37°C and 200 rpm.
8. On day 3, inoculate 2 mL starter culture into a 1 L Erlenmeyer flasks containing 200 mL 2xYT-medium with 100 μg/mL ampicillin (expression culture). Incubate the expression culture at 37°C and 200 rpm for 2–3 h until the optical density at 600 nm wavelength (OD600) is between 0.6–0.8.
9. Cool down the expression culture on ice for 5 min.
10. To induce protein expression, add 100 μL IPTG (1 M) to 200 mL expression culture for a final concentration of 0.5 mM IPTG.
11. Incubate the expression culture overnight at 25°C and 200 rpm.

3.2 Purification of CytK (Day 4)

1. Harvest the bacterial cells in 50 mL plastic tubes by centrifugation at 7000 × *g* for 5 min at 4°C. Spin down 50–400 mL cells in one tube by repeated centrifugation.
2. Discard the supernatant and store the cell pellets at −80°C or directly continue with the protocol.
3. Resuspend the cell pellets in lysis buffer (20–40 mL buffer per pellet of 50–400 mL culture) (*see* **Note 1**).
4. Add one protease inhibitor tablet per 10–20 mL of resuspended cell pellet and 0.05 units/mL of DNaseI. Incubate the cell suspension for 15 min at ~48°C–6°C (or at room temperature with ice-cold lysis buffer) while rotating at 10 rpm to dissolve the protease inhibitor tablet.

5. Place the cell suspension in an ice-water bath and lyse the cells by sonication for 10 min at 20% output power (cycles with 5 s on/5 s off time, 5 min total on time).
6. Pellet the cell debris by centrifugation at 7500 ×*g* for 30 min at 4°C. The target protein is inside the supernatant.
7. Transfer 200 μL of Ni-NTA resin (400 μL of Ni-NTA suspension) to a gravity-flow column, and wash two times with 5 mL MQ and one time with 1 mL wash buffer. Between each step, wait for the solution to completely flow through.
8. Transfer the resin from the column to a 50 mL tube containing the supernatant from Step 6 using 0.5 mL of wash buffer to resuspend the resin. Incubate for 15 min at ~4°C–6°C while rotating at 10 rpm or at room temperature after cooling the supernatant on ice.
9. Transfer the supernatant with the incubated Ni-NTA resin back to the gravity flow column and wait for the solution to completely flow through.
10. Wash the incubated Ni-NTA resin in at least two steps, using 5 mL of wash buffer in each step.
11. Elute CytK from the Ni-NTA resin in 6× elution steps of 150 μL elution buffer. During each elution step, incubate 1–2 min with elution buffer and collect the flowthrough fractions separately.
12. Check the protein concentration using a Bradford assay or by measuring the absorbance with a nanodrop at 280 nm (*see* **Notes 5 and 6**). Combine the protein-containing fractions.
13. Flash freeze the CytK oligomers (Fig. 1) in liquid nitrogen and store for long-term at −80°C.
14. CytK nanopores can be used in electrophysiology experiments as described before [6, 15].

3.3 Hemolytic Activity of CytK Monomers (Day 4)

1. The hemolytic activity of CytK can only be tested with CytK monomers. CytK monomers can be obtained by purifying without 0.02% DDM in all the buffers (*see* **Note 7**).
2. Prepare the defibrinated sheep blood by washing 500 μL blood ~3× times with 500 μL wash buffer. During each wash step add 500 μL wash buffer, mix the cell suspension by pipetting up and down, centrifuge at 10,000 ×*g* for 30 s, discard the supernatant and repeat until the supernatant is clear.
3. Dilute the washed blood in ~20 mL wash buffer to reach an optical density of 0.6–0.8 at 650 nm wavelength. The OD_{650} can be measured with a plate-reader by adding 100 μL of diluted blood to a 96-well microplate.

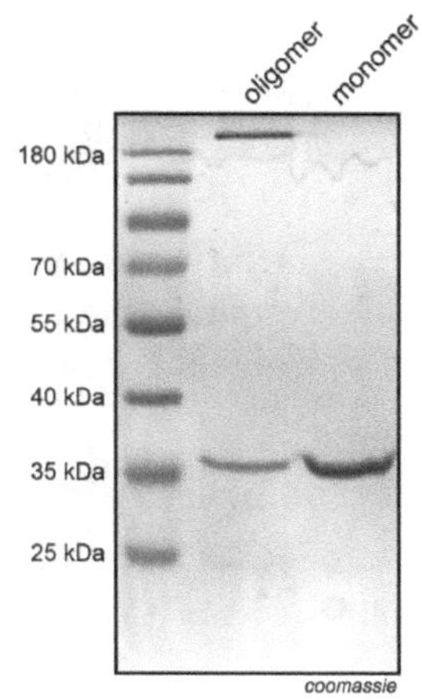

Fig. 1 SDS-PAGE of CytK nanopore and its monomer. The sample of CytK oligomer was directly loaded by omitting the heating step

4. Add the controls to the microplate: 10 μL of 1% Triton X-100 (100% lysis) and 10 μL of elution buffer (0% lysis).
5. Add 10 μL of CytK monomers (~0.1–0.5 mg/mL protein concentration) to the microplate.
6. Quickly Add 100 μL diluted blood to the controls and CytK (with a multichannel pipette or stepper pipette) and incubate for 15 min, while measuring the OD_{650}.

4 Notes

1. The 0.02% DDM in the buffers, initiates the oligomerization of the CytK monomers. If you want to measure the Hemolytic activity of CytK, you need CytK monomers. Purify without 0.02% DDM in the buffers to obtain CytK monomers.
2. Before adding the detergent DDM, first dissolve the other components and adjust the pH to prevent the formation of foam.
3. Thaw the chemical competent cells on top of the ice and not pressed into the ice. When pressed into the ice the temperature increases quickly. This temperature shock can reduce the competency of the cells.
4. This incubation step helps cells recover from the transformation process, allows the cells to divide and overcome the antibiotic selection.
5. The protein concentration might be slightly off, since the DDM detergent can affect the Bradford dye.

6. Expect a protein yield of about 1–3 mg/mL for CytK oligomers per 100 mL of culture. Expect a less pure sample with protein yield of about 0.2–0.5 mg/mL for CytK monomers per 100 mL of culture when purifying without DDM.
7. Detergents like DDM completely lyse all red blood cells and initiates oligomerization of CytK monomers. The CytK oligomers are not hemolytically active.

References

1. Wang Y, Zhao Y, Bollas A et al. (2021) Nanopore sequencing technology, bioinformatics and applications. Nat Biotechnol 39:1348–1365. https://doi.org/10.1038/s41587-021-01108-x
2. Ying YL, Hu ZL, Zhang S et al. (2022) Nanopore-based technologies beyond DNA sequencing. Nat Nanotechnol 17:1136–1146. https://doi.org/10.1038/s41565-022-01193-2
3. Dorey A, Howorka S (2024) Nanopore DNA sequencing technologies and their applications towards single-molecule proteomics. Nat Chem 16:314–334. https://doi.org/10.1038/s41557-023-01322-x
4. Lu C, Bonini A, Viel JH, Maglia G (2025) Toward single-molecule protein sequencing using nanopores. Nat Biotechnol 43(3):312–322. https://doi.org/10.1038/s41587-025-02587-y
5. Huang G, Voet A, Maglia G (2019) FraC nanopores with adjustable diameter identify the mass of opposite-charge peptides with 44 dalton resolution. Nat Commun 10:1–10. https://doi.org/10.1038/s41467-019-08761-6
6. Versloot RCA, Straathof SAP, Stouwie G, et al. (2021) β-barrel nanopores with an acidic-aromatic sensing region identify proteinogenic peptides at low pH. ACS Nano 16(5):7258–7268. https://doi.org/10.1021/acsnano.1c11455
7. Zhang Y, Yi Y, Li Z et al. (2024) Peptide sequencing based on host–guest interaction-assisted nanopore sensing. Nat Methods 21:102–109. https://doi.org/10.1038/s41592-023-02095-4
8. Zhang S, Huang G, Versloot RCA et al. (2021) Bottom-up fabrication of a proteasome–nanopore that unravels and processes single proteins. Nat Chem 13:1192–1199. https://doi.org/10.1038/s41557-021-00824-w
9. Brinkerhoff H, Kang ASW, Liu J et al. (2021) Multiple rereads of single proteins at single-amino acid resolution using nanopores. Science (80-) 374:1509–1513. https://doi.org/10.1126/science.abl4381
10. Chen Z, Wang Z, Xu Y et al. (2021) Controlled movement of ssDNA conjugated peptide through: Mycobacterium smegmatis porin A (MspA) nanopore by a helicase motor for peptide sequencing application. Chem Sci 12:15750–15756. https://doi.org/10.1039/d1sc04342k
11. Yan S, Zhang J, Wang Y et al. (2021) Single molecule ratcheting motion of peptides in a mycobacterium smegmatis porin A (MspA) nanopore. Nano Lett 21:6703–6710. https://doi.org/10.1021/acs.nanolett.1c02371
12. Yu L, Kang X, Li F et al. (2023) Unidirectional single-file transport of full-length proteins through a nanopore. Nat Biotechnol 41:1130–1139. https://doi.org/10.1038/s41587-022-01598-3
13. Martin-Baniandres P, Lan WH, Board S et al. (2023) Enzyme-less nanopore detection of post-translational modifications within long polypeptides. Nat Nanotechnol 18:1335–1340. https://doi.org/10.1038/s41565-023-01462-8
14. Motone K, Kontogiorgos-Heintz D, Wee J et al (2024) Multi-pass, single-molecule nanopore reading of long protein strands. Nature 633:662–669. https://doi.org/10.1038/s41586-024-07935-7
15. Sauciuc A, Morozzo della Rocca B, Tadema MJ, et al. (2023) Translocation of linearized full-length proteins through an engineered nanopore under opposing electrophoretic force. Nat Biotechnol 42(8):1275–1281. https://doi.org/10.1038/s41587-023-01954-x
16. Jiang HJ, Underwood TC, Bell JG et al (2024) Blobs form during the single- file transport of proteins across nanopores. Proc Natl Acad Sci 120:e2405018121. https://doi.org/10.1073/pnas

Chapter 3

Tuning Rectification and Gating in Biological Nanopores Through Lumen Charge Mutation

Simon Finn Mayer, Marianna Fanouria Mitsioni, Maria Jose Marcaida, Theo Emmerich, Aleksandra Radenovic and Matteo Dal Peraro

Abstract

β-barrel nanopores are involved in crucial biological processes, from ATP export in mitochondria to antibiotic resistance in Gram-negative bacteria, and represent a promising platform for emerging sequencing technologies. However, in contrast to ion channels, the understanding of the fundamental principles governing ion transport through these nanopores remains in its early stages. In this chapter, we describe the production and mutation of three major biological nanopores—aerolysin, MspA, and α-HL—as well as experimental approaches to elucidate ion transport mechanisms in these biological nanopores. Specifically, we provide tools to characterize two distinct nonlinear phenomena: open-pore rectification and gating, as well as guidelines to tune these phenomena to achieve a certain rectification or gating behavior.

Keywords Gating, Beta barrel, Rectification, Lumen charge, Ion transport, Mutant, Biological nanopore

1 Introduction

Biological nanopores are proteins that form channels or holes in membranes on the scale of nanometers, allowing for the flux of ions, water, and small molecules. In living organisms, they achieve various functions all associated with cross-membrane transport. In a technological context, they are used as label-free analysis tools, famously for nanopore sequencing of DNA as well as detecting small analytes [1–3] (Fig. 1a). Efforts are underway to employ this technique to sequence proteins [4, 5]. There are several challenges associated with reaching this goal; one of them is the lack of fundamental understanding of the underlying transport physics through these pores. Across bacteria, viruses, and eukaryotes, biological nanopores mediate diverse functions. Thus, they are diverse in their structural manifestations. There are monomeric and multimeric nanopores.

Fanjun Li (ed.), *Nanopore Technology*, Methods in Molecular Biology, vol. 3072,
https://doi.org/10.1007/978-1-0716-5527-6_3,

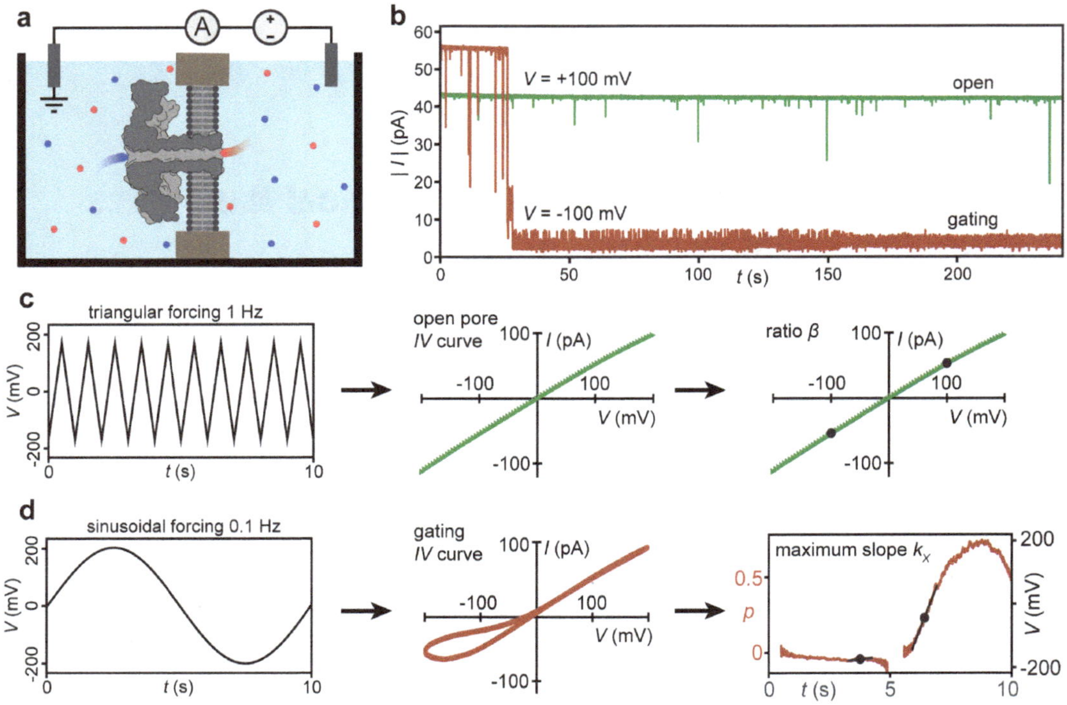

Fig. 1 Quantification of nonlinear ion transport in biological nanopores. (**a**) Not-to-scale schematic of the experimental set-up. A lipidic membrane separating two reservoirs filled with electrolyte (cations are shown in blue; anions are shown in red) with a single aerolysin wild-type pore incorporated; the current through the pore is measured via Ag/AgCl electrodes. (**b**) Absolute current over time traces at constant potential of a single aerolysin wild-type nanopore. The same pore is measured first at 100 mV (green) and then at −100 mV (red) for 240 s. Rectification is clearly visible as the absolute current is higher at −100 mV. At 100 mV, the measured current is constant, while at −100 mV, the current decreases spontaneously; this behavior is referred to as gating. (**c**) Triangular forcing at 1 Hz, 200 mV, was used to extract the single open-pore *IV* curve of aerolysin wild-type. Rectification can be quantified using a ratio β (see Eq. 1) of the current at ±100 mV (black dots). (**d**) Sinusoidal forcing at 0.1 Hz, 200 mV, was used to extract the gating *IV* curve of aerolysin wild type. Gating can be quantified at each polarity using the maximum slope (black dots and lines) of the closed-state probability (p) corresponding to k_X, the maximum closing rate. All experiments were conducted in 1 M KCl buffered with 10 mM phosphate to pH 6.2. (Part of this Figure was adapted from [6])

They can be classified as beta barrel and alpha helical nanopores depending on the main structural motif that spans the membrane and forms the lumen itself. Many, if not all, of these nanopores have charged residues inside their conductive lumen. In monomeric ones, the charge distribution is less regular than in multimeric ones, where the center symmetry leads to rings of charges in the lumen. These lumen charges have been under investigation as they govern ion transport and resulting electro-osmotic flow.

Lumen charges can be modified selectively by mutating the pores. This approach has been employed to improve the pores' signal-to-noise ratio, molecule capture rate, or retention time [7–9].

Recently, lumen charges have been linked to the nonlinear behavior of nanopores. In particular, rectification, where the pore's conductance at positive and negative voltage is different, and gating, where pores spontaneously close under the application of an external voltage (Fig. 1b) [6].

In this chapter, we detail experimental procedures to study and quantify the aforementioned nonlinear behaviors of biological nanopores. We present a benchmarking and validation nanopore library comprising three major β-barrel nanopores: aerolysin, Mycobacterium smegmatis porin A (MspA), and α-hemolysin (α-HL), and their respective mutants. This chapter is a detailed protocol of some of the methods employed in [6], in which we tested and refined our gating hypothesis using the herein detailed mutation strategies, quantitative rectification, and gating measurements.

2 Materials

All solutions are prepared using deionized water (given a resistance value of 18 MΩ). Special attention should be paid to the final pH of all solutions, in particular when using pH-altering substances at large concentrations that may overwhelm a buffers' capacity, such as high-molarity imidazole-containing buffers. All reagents can be prepared and stored at room temperature (~25°C), unless otherwise noted.

2.1 Expression and Purification of Three Major β-Barrel Nanopores

1. Host: *E. coli* BL21 (DE3) or BL21 (DE3) pLys.
2. Vectors and tags
   - Aerolysin: pET22b, NcoI/XhoI insertion, C-terminal His-tag.
   - α-HL: pET28-b, NcoI/XhoI insertion, C-terminal His-tag.
   - MspA: pET28-b, NcoI/XhoI insertion, C-terminal His-tag.
3. Media/antibiotics: LB medium, Kanamycin (MspA and a-HL), Ampicillin & chloramphenicol (Aerolysin).
4. Inducer: IPTG.
5. Protease inhibitor: EDTA-free cocktail and TurboNuclease.
6. Chromatography: HisTrap HP (Ni-NTA) 5 mL, ÄKTA FPLC, desalting (HiPrep 26/10 or PD-10 where noted).
7. Filtration: 0.45 μm syringe filters.
8. Detergents/additives:
   - Triton X-100 (α-HL lysis).
   - Genapol X-80 (MspA lysis/wash/elution).
   - OTOE (MspA final buffer).

9. Buffers:

   - Aerolysin Buffer A (lysis/equilibration): 500 mM NaCl, 50 mM Tris–HCl pH 7.4.
   - Aerolysin Buffer B (elution): 500 mM NaCl, 50 mM Tris–HCl pH 7.4, 500 mM imidazole.
   - Aerolysin Buffer F (final): 150 mM NaCl, 20 mM Tris–HCl pH 7.4.
   - α-HL lysis buffer: 500 mM NaCl, 20 mM HEPES pH 8.0, 1% (v/v) Triton X-100.
   - α-HL wash buffer: 500 mM NaCl, 20 mM HEPES pH 8.0, 5 mM imidazole.
   - α-HL elution buffer: 500 mM NaCl, 20 mM HEPES pH 8.0, 500 mM imidazole.
   - MspA lysis buffer: 100 mM sodium phosphate, 0.1 mM EDTA, 150 mM NaCl, 0.5% (w/v) Genapol X-80, pH 6.5.
   - MspA wash buffer: 500 mM NaCl, 20 mM HEPES pH 8.0, 5 mM imidazole, 0.5% (w/v) Genapol X-80.
   - MspA elution buffer: 500 mM NaCl, 20 mM HEPES pH 8.0, 500 mM imidazole, 0.5% (w/v) Genapol X-80.
   - MspA final buffer: 500 mM NaCl, 20 mM HEPES pH 8.0, 0.6% (v/v) OTOE.

3 Methods

3.1 Preparation of Aerolysin, α-HL, and MspA Mutants

3.1.1 Guidelines for Creating Mutant Libraries Targeting Lumen Charges

1. Rationale: rings of charge in multimeric β-barrel lumens.

   β-barrel nanopores, such as aerolysin, α-hemolysin, and MspA, because of their structure, have multimericity feature rings of charges that modulate ion selectivity and gating. Upon oligomerization into an n-mer, each monomer contributes one residue to any given ring of charges. Along the lumen, there are often multiple, sometimes alternating rings of charge, creating an electrostatic potential landscape for ions that governs both open-pore rectification and voltage-dependent gating (Fig. 2). In the case of the roughly cylindrical heptameric aerolysin, there are 13 rings of charge throughout the lumen (Fig. 2). Mutating one residue effectively alters the lumen charge by ±7

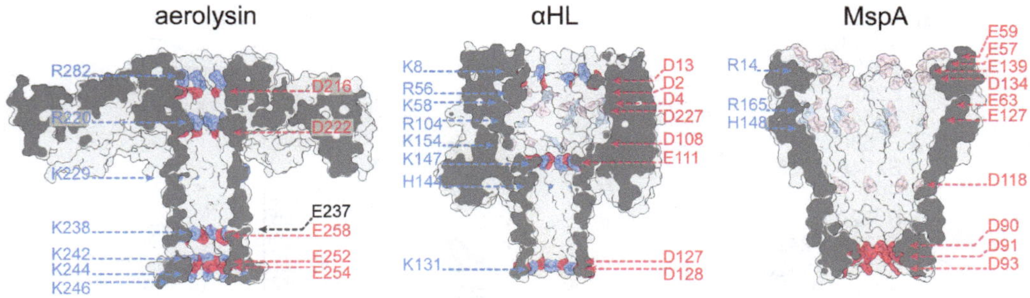

Fig. 2 β-barrel pores and their lumen charges. From left to right, the wild type of aerolysin, α-HL, and MspA. All charges in the lumen are colored in the structures with positive residues in blue and negative residues in red. Amino acid type and position in the peptide chain are indicated. Charges in regions that exhibit diameters larger than 2.5 nm carry a lighter shade of their respective color. Charges outside of the lumen or stem regions are not shown. (Figure reproduced from [6])

charges. The focus of this chapter is mostly charged-to-neutral amino acid substitutions.

In any channel, ion transport depends strongly on the distribution and amount of charge that is present. Thus, by modulating the charge, the ion transport and selectivity can be modulated. Diode-like, strongly, and weakly gating and potentially highly ion selective biological nanopores can be created for different applications (Fig. 3) (*see* **Note 1**).

2. Which lumen charges influence rectification and gating

- Locate bottlenecks of ion transport: In a cylindrical pore, such as aerolysin rings of charges that are located near, but not exactly at, the mouth of the pore, most strongly bias open-pore rectification (Fig. 3a) and gating (Fig. 3b). The reason is that these positions influence the ion concentration inside the pore the most; they are the largest bottlenecks for ions. In less cylindrical pores, this rule is more complex, but the same rationale can be applied. The higher the charge at these locations, the stronger the selectivity and the gating are.
- Enhancing conductivity by increasing rectification: A charge at a bottleneck will ease the transport of ions at one voltage polarity (direction of ion flow) and oppose it at the opposite voltage polarity. Such a rectification can be enhanced by increasing the charge at that location or when an appropriate charge is placed at another location, e.g., an opposite charge at the opposite end of the lumen. The adverse effect is that the more the pore rectifies, the more gating takes place.

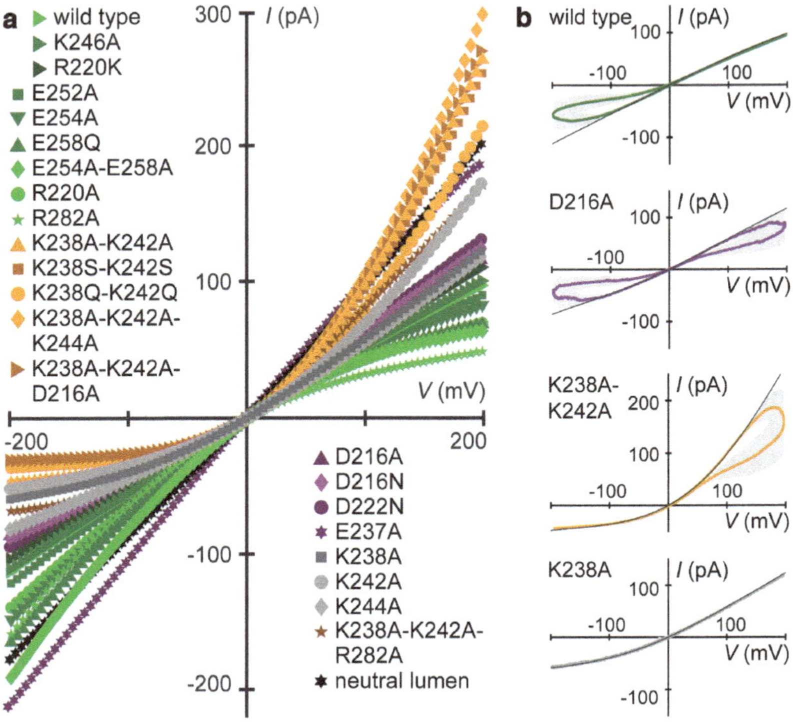

Fig. 3 Tunability of rectification and gating in aerolysin nanopores.(**a**) Open-pore *IV* curves of several mutants of aerolysin showcasing the different degrees of rectification tuned by the mutation of lumen charges (1 M KCl, 10 mM phosphate pH 6.2, 1 Hz). (**b**) Gating behavior was measured with different aerolysin mutants, showcasing the great tunability of gating by changing lumen charge. Mutants gating at negative potentials are shown in green, at positive potentials in orange, mutants showing low levels of gating in grey, and mutants that gate at both polarities in purple. Each plot shows the open-pore *IV* curve in black, showing the differences in rectification. Further, each plot shows ±1 standard deviation of the mean over the number of experiments *n* plotted as a grey area. All experiments were conducted in 1 M KCl buffered to pH 6.2 with 10 mM phosphate. Sinusoidal forcing was applied at 0.1 Hz and 200 mV amplitude. (This Figure was adapted from [6])

- Increasing overall conductivity (both polarities): Neutralizing charges in general in aerolysin, flattens the energy landscape in the pore lumen, and leads to a higher overall conduction and less rectification. This has the adverse effect that the overall structure of the pore loses stability with each amino acid that is substituted, leading to an increase in gating.
- Overall, the least amount of mutations possible is preferred since the pore stability is least compromised. More rational substitutions that take the pore stability into account and contribute to it through optimized hydrogen bonding patterns or disulfite bridges are likely possible as well.
- Non-lumen charges: Also, charges that do not point directly into the lumen, such as E237 in aerolysin, can influence the ion transport dramatically. E237A exhibits an increased

overall conductance (Fig. 3a) and gating rate [6]; the reason why is not yet understood.

- Crosslinking to suppress gating of MspA: Chemical cross-linking of the octamer eliminated gating while preserving rectification and conductance [6]. At the same time, cross-linking establishes structural stiffening as a control that isolates the role of lumen-charge distribution from barrel mechanics. Crosslinking of pores that undergo major conformational changes upon insertion, such as aerolysin, will not insert when crosslinked in solution.

3.1.2 Expression and Purification of β-Barrel Nanopores

A. **Transformation into chemically competent BL21 (DE3) or BL21 (DE3) pLys (all constructs)**

1. Thaw an aliquot (50 µL) of chemically competent cells on ice for 10–15 min. Keep tubes cold at all times.
2. Add 1–5 µL of plasmid DNA solution (concentration of DNA in water should be around 100ng/µL; wild-type or containing the desired mutations) to the cells. Flick gently to mix. Incubate on ice for 20–30 min to allow DNA–membrane association.
3. Heat-shock the cells by placing the tube at 42°C for 45 s, then immediately return to ice for 2 min. This thermal pulse facilitates DNA uptake.
4. Add 450–500 µL of pre-warmed LB to the tube. Incubate the cells at 37°C with shaking (200–250 rpm) for 45–60 min to allow multiplication.
5. Plate 50–200 µL of the recovered culture onto LB agar containing the appropriate antibiotic(s). If needed, spin down the remaining cells briefly, remove some supernatant, and plate the concentrated cell suspension to increase colony yield.
6. Incubate plates inverted at 37°C overnight.
7. Start an overnight preculture by inoculating 100 mL LB with the corresponding antibiotic(s) from a single colony that grew on the plate overnight. Incubate the pre-culture at 37°C with vigorous shaking overnight.

B. **Expression (all constructs)**

1. For production, transfer 30 mL of the preculture into a large flask (e.g., 2 L LB in a 5 L flask) and grow at 37°C until the culture reaches mid-log phase (OD600 = 0.6–1.0), which optimizes responsiveness to induction.
2. Induce protein expression by adding IPTG to a final concentration matched to the construct: 0.25 mM for

aerolysin and 0.1 mM for α-HL and MspA. Gentle mixing after addition ensures even induction.

3. Immediately shift the incubator to 18°C and continue shaking overnight. The lower temperature favors proper folding and solubility across these β-barrel systems and typically improves downstream purity (*see* **Note 2**).
4. Harvest cells by centrifugation at 4000×*g* for 15 min at 4°C. Decant the supernatant and either proceed directly to lysis or store the pellets at −20°C to pause the workflow without loss of yield.

C. **Lysis and clarification (construct-specific)**
Aerolysin

1. Gently resuspend the cold cell pellet in Aerolysin Buffer A (about 35 mL per 2 L culture) supplemented with an EDTA-free protease inhibitor tablet to protect the His–tag interaction in later steps.
2. Lyse cells by pulsed sonication on ice (for example, 10 s on/10 s off to a total of 2.5 min at ~70% amplitude), keeping the suspension cold to prevent proteolysis.
3. Reduce viscosity with a brief TurboNuclease treatment (10 min on ice), which improves column performance.
4. Clarify the lysate by centrifugation at 20,000×*g* for 30–60 min at 4°C and filter the supernatant through a 0.45 μm filter to protect the chromatography system.

α-HL

1. Resuspend the pellet thoroughly in α-HL lysis buffer, which includes 1% Triton X-100 to aid solubilization and add protease inhibitor.
2. Lyse by sonication on ice and follow with a 10 min TurboNuclease step to lower viscosity, ensuring smooth Ni-NTA loading.
3. Clarify by centrifugation at 20,000×*g* for 60 min at 4°C and filter the supernatant (0.45 μm) before loading.

MspA

1. Resuspend the pellet in MspA lysis buffer containing Genapol X-80 and protease inhibitor, as maintaining non-ionic detergent throughout is key for octamer stability (*see* **Note 3**).
2. Lyse by sonication on ice, then apply a controlled 60°C (*see* **Note 4**), 10 min heat step to precipitate heat-labile contaminants without compromising the octamer. Cool promptly.

3. Add TurboNuclease and incubate 10 min on ice to reduce viscosity. Clarify at 20,000 × *g* for 60 min at 4°C and filter the supernatant (0.45 μm).

D. **HisTrap purification (ÄKTA; construct-specific buffers)**

1. Equilibrate a 5 mL HisTrap column with the corresponding lysis/equilibration buffer for each construct to match the sample matrix and ensure stable binding.
2. Load the clarified lysate at approximately 1 mL/min, collecting the flow-through for record. A steady baseline indicates consistent binding.
3. Wash the column first with three to ten column volumes (CV) of equilibration buffer to remove unbound material. For aerolysin and α-HL, follow with a low-imidazole wash (for example, 10% of the elution buffer, ~50 mM imidazole, 10 CV) to strip weak binders; for MspA, use the dedicated wash buffer containing Genapol X-80. (*see* **Note 5**)
4. Elute the pore with His–tag using a linear gradient to 100% elution buffer (up to 500 mM imidazole), collecting fractions across the peak.
5. Evaluate fractions by SDS-PAGE and pool those containing the target assembly to carry forward.

E. **Desalting/buffer exchange and storage**

1. For aerolysin, perform buffer exchange into Buffer F (150 mM NaCl, 20 mM Tris–HCl pH 7.4) using a HiPrep 26/10 desalting column. Pool peak fractions, concentrate if needed to a final concentration of 0.5mg/ml and store at 4°C for short term use or aliquot and freeze at −20°C.
2. For α-HL, retain the heptamer in the elution/wash buffer mixture in which it elutes; this matrix is compatible with downstream bilayer experiments and avoids an extra exchange step.
3. For MspA, exchange pooled fractions into the final buffer containing OTOE using PD-10 columns according to the manufacturer's instructions. Store at 4°C or aliquot and freeze, avoiding repeated freeze–thaw.

F. **Verification of Purified Pores in SDS–PAGE**

Evaluate the final product by SDS–PAGE (Fig. 4). For aerolysin to trigger the conformational changes necessary for heptamerization incubate with trypsin in agarose at a ratio of 1:4 (trypsin/pore, v/v) and run both cleaved and not cleaved samples.

G. **Optional MspA chemical cross-linking**

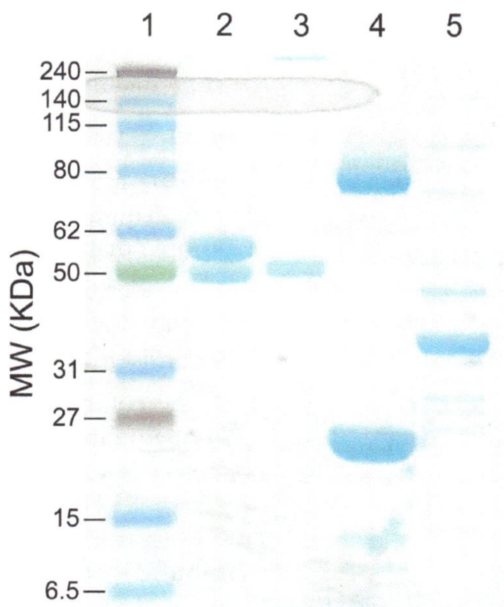

Fig. 4 SDS–PAGE analysis of purified pore-forming toxins. Bis-Tris 4%–12% gel showing **(1)** molecular weight ladder, **(2)** aerolysin wild type monomers before proteolytic cleavage of the C-terminal peptide with trypsin, **(3)** aerolysin wild type after trypsin cleavage showing monomers lacking the C-terminal peptide and higher-molecular-weight bands corresponding to pore assemblies, **(4)** purified MspA wild type showing both monomeric and oligomeric species, and **(5)** purified α-HL monomers

1. To generate a mechanically stiffened octamer, mix 100 μL of the MspA preparation with 10 μL of 8% (v/v) glutaral (final 0.8% v/v), incubate for 10 min at room temperature, and proceed directly to functional assays.
2. Alternatively, mix 100 μL of MspA with 5 μL of 20% (v/v) methanal (final 1% v/v), incubate 10 min at room temperature, and proceed. Both treatments reliably abolish gating while leaving rectification and conductance within the expected range.

3.2 Measuring and Quantifying Rectification

To measure the classical non-gating IV response of the pore, use a triangular signal of 1 or 2 hertz (depending on gating dynamics) at a relevant amplitude (Fig. 1c). At this speed, amplitudes of around 300 mV can be reached with a good membrane. This limit can be pushed to 500 mV or higher. Apply the triangular signal and start recording before adding the pores. Do not aim for a single pore; it is faster to add more pores and count the insertions in the recorded trace. Identify a cycle that does not contain insertions or gating events. Divide this cycle by the number of pores. Around three to ten pores are ideal for averaging while avoiding gating events.

Ensure that all pores are inserted in the same direction, by plotting different cycles and confirming that the shape of the curve does not change. Quantify rectification by the ratio of the conductance at both polarities at a given voltage. To distinguish concave and convex rectification, it can be expressed as a unitless value between −1 and 1:

$$\beta = (I_+ - |I_-|)/(I_+ + |I_-|) \quad (1)$$

3.3 Measuring and Quantifying Gating

3.3.1 Measuring Gating Data

To measure gating, frequencies of 1 to 0.01 hertz are relevant. About 0.1 Hz with a 200 mV amplitude will capture the gating of aerolysin (Fig. 1d) and alpha hemolysin. MspA pores tend to gate more and thus can be measured at 100 mV amplitude with ease. A sinusoidal signal has the advantage that a larger fraction of the cycle is spent at higher voltages, which makes the capture of the effect faster. While a triangular signal has the advantage that dV/dt is linear and thus can sometimes make the interpretation of data easier.

For an ensemble measurement, apply a signal at 50 mV lower than the amplitude of the final measurement to protect the membrane while monitoring the insertion of pores. Once the current has reached equilibrium switch to the amplitude of choice and start recording for around 20 cycles depending on frequency and number of pores. The more pores in the membrane the less cycles are needed. To prove ergodicity, measure a single pore gating for around 50 cycles and average over them. If the response is the same as the ensemble response, safely assume that the pores do not influence each other in their behavior.

3.3.2 Optimizing Membrane Quality for Gating Measurements

Store lipids in aliquots of around 0.2 mg under argon or nitrogen atmosphere. Do this by using 150 μL conical glass inserts that can be put in 1.5 mL glass vials that close with a screwcap lined with a PTFE coated rubber. Fill aliquots, dry under vacuum, and close them in a glovebox with inert gas before freezing at −80°C. Open the vial when it has reached room temperature to avoid water condensation when dissolving in solvent. Octane, hexadecane, squalene, decane, etc. likely lead to different membrane properties and thus potential differences in gating. The lipids themselves also influence gating. Choose a membrane support that is on the bigger end, around 100 to 150 μm. This makes it easier to have a larger "oil free" membrane.

3.3.3 Processing Data

Each recorded cycle needs to be selected by hand for further processing to exclude cycles where there is an excess of pores inserting as this biases the signal. Cycles where the membrane breaks or where there is visible membrane electroporation are obviously excluded.

Once the cycles are selected, they can be added and normalized by the number of pores in the membrane, which can be done by fitting the non-gating part of the non-gating IV curve to the data. This is also the confirmation of the homogeneous directionality of the pores in the membrane. If the non-gating IV curves shape is different in the overlaid gating data, then the experiment needs to be discarded as pores were inserted in both directions in the membrane. The experiments need to be repeated to see whether there is a spread in results caused by the membrane itself. Since each experiment is already an average of several pores the only change in between repeat experiments is the formed membrane (see above).

3.3.4 Quantifying Gating

Once the data of one experiment has been averaged and normalized, the resulting hysteresis loop can be integrated or a rate can be extracted. For integrating, it is good to integrate against the open pore current that is provided by the non-gating *IV* curve from the rectification quantification. This way the resulting gating values are more comparable between pores with different degrees of rectification. The hysteresis loop can be divided into four quadrants that are all separately integrated against the non-gating *IV* curve. This yields two gating values per polarity. To extract a gating rate the *IV* curve needs to be transformed into a gating probability (p) vs. V curve.

$$p\left(t\right) = \frac{1}{1 \quad - \in}\left(1 - \frac{I_{ens}(t)}{I_O(t)}\right)$$

I_{ens} is the average measured current of an ensemble of pores in the membrane, I_O is the non-gating pore current and ε the fraction of current that the closed state conducts. This fraction can be extracted by repeatedly measuring a single pore at a voltage where the pore frequently gates (Fig. 1b), and extracting an average residual current of the gated state divided by the open pore current. The maximum slope at each polarity of the pV curve is the gating rate in Hz (Fig. 1d).

3.3.5 On Changing the Gating Properties of a Nanopore

The most influential parameter on gating in beta-barrel forming nanopores is the charge of the pores' lumen as well as other charges on the pore. First, identify the charges in the pore and then start mutating them by switching to neutral amino acids. It does not matter for the most part whether the size or hydrophobicity of the amino acid changes. Although there might be differences in pore formation efficiency. The sites that have the most influence are located at constrictions and a few nanometers away from the entry and exit of the pore for aerolysin and presumably other cylindrical pores. In other words, locations in the pore that influence the flux of ions the most are the most relevant ones. The more homogeneously charged any of these influential sites becomes, the more will the pore gate and the gating polarity will depend on the polarity of the charge of that site.

The least number of mutations that leads to a neutral lumen should lead to the least amount of gating. This is probably because there is a tradeoff between pore stability when mutating and the reduction of gating effect. In other words, if the pore is less likely to gate but also less stable the observed change in gating itself might be negligible.

4 Notes

1. The actual charge distribution within nanopore lumens is more complex than it appears. Clustering multiple residues of the same charge in proximity is energetically unfavorable, which necessitates the formation of salt bridges, shielding by counterions, or shifts in protonation states to maintain local electroneutrality. Furthermore, parameters such as temperature, pH, and ionic strength may vary significantly at the nanoscale, especially within chemically heterogeneous environments.
2. Express at 18°C overnight after IPTG induction to improve solubility and reduce proteolysis for all three pores.
3. Detergent requirements: Use Triton X-100 for α-HL lysis; use Genapol X-80 throughout MspA lysis/wash/elution and OTOE in the final buffer to preserve the octamer.
4. A brief 60°C of MspA, 10 min treatment after sonication helps clear contaminants without disrupting the octamer. Do not exceed 60°C.
5. Imidazole concentration during purification wash steps: Include a low imidazole wash (~50 mM) before elution for aerolysin/α-HL. For MspA specifically, keep 5 mM imidazole with Genapol X-80 to balance specificity and solubility.

References

1. Ouldali H, Sarthak K, Ensslen T et al (2020) Electrical recognition of the twenty proteinogenic amino acids using an aerolysin nanopore. Nat Biotechnol 38:176–181
2. Huang G, Willems K, Soskine M et al (2017) Electro-osmotic capture and ionic discrimination of peptide and protein biomarkers with FraC nanopores. Nano Lett 8:935
3. Galenkamp NS, Soskine M, Hermans J et al (2018) Direct electrical quantification of glucose and asparagine from bodily fluids using nanopores. Nat Commun 9:4085
4. Zhang S, Huang G, Versloot RCA et al (2021) Bottom-up fabrication of a proteasome–nanopore that unravels and processes single proteins. Nat Chem 13 (12):1192–1199
5. Brinkerhoff H, Kang ASW, Liu J et al (2021) Multiple rereads of single proteins at single–amino acid resolution using nanopores. Science 374:1509–1513
6. Mayer SF, Mitsioni MF, Robin P et al (2025) Lumen charge governs gated ion transport in β-barrel nanopores. Nat Nanotechnol 21:116–124
7. Cao C, Cirauqui N, Marcaida MJ et al (2019) Single-molecule sensing of peptides and nucleic acids by engineered aerolysin nanopores. Nat Commun 10:4918
8. Manrao EA, Derrington IM, Laszlo AH et al (2012) Reading DNA at single-nucleotide resolution with a mutant MspA nanopore and phi29 DNA polymerase. 30:349–353
9. Sauciuc A, Morozzo della Rocca B, Tadema MJ et al (2024) Translocation of linearized full-length proteins through an engineered nanopore under opposing electrophoretic force. Nat Biotechnol 42:1275–1281

Chapter 4

Protein Analyte Sensing with an Outer Membrane Protein G (OmpG) Nanopore

Monifa A. V. Fahie, Bib Yang, Christina M. Chisholm and Min Chen

Abstract

Nanopore sensing is a powerful lab-on-a-chip technique that allows for the analysis of biomarkers present in small sample sizes. In general, nanopore clogging and low detection accuracy arise when the sample becomes more complex such as in blood or lysate. To address this, we developed an OmpG nanopore that distinguishes not only different proteins in a mixture but also protein homologues. Here, we describe this OmpG-based nanopore system that specifically analyzes target biomarkers in complex mixtures.

Keywords Outer membrane protein G, Sulfhydryl-maleimide chemistry, Protein–protein interactions, Nanopore gating, Extracellular loops, Loop dynamics, Sulfonamide, Carbonic anhydrase, Biotin, Antibody, Streptavidin, Avidin, , Electrophysiology

1 Introduction

Nanopore sensing is an attractive analytical tool in the field of diagnostics. It has been used to identify small molecules [1–4] and biomolecules [5–11]. Nanopores have been engineered from either pore-forming proteins, synthetic materials, or even 3D-DNA structures [12–26]· One of the most successful applications of nanopore sensing has been in nucleic acid sequencing [27]; however, nanopore studies for peptide or protein sensing have recently gained considerable interest and development [27–36]. Proteins can be detected by (i) translocation or interaction within the walls or constriction site of the nanopore (ii) binding to external adaptors where a change in the adaptor is transduced to the nanopore, and (iii) binding directly to the outer surface of the nanopore.

For a translocating peptide or protein, the ionic blockade that is generated in the nanopore is sensitive to a myriad of features on the protein such as its folding state, monomeric/multimeric state, post-translational modifications, or distribution of surface charge [10, 37–43]. Thus, the same protein could generate different ionic

Fanjun Li (ed.), *Nanopore Technology*, Methods in Molecular Biology, vol. 3072,
https://doi.org/10.1007/978-1-0716-5527-6_4,

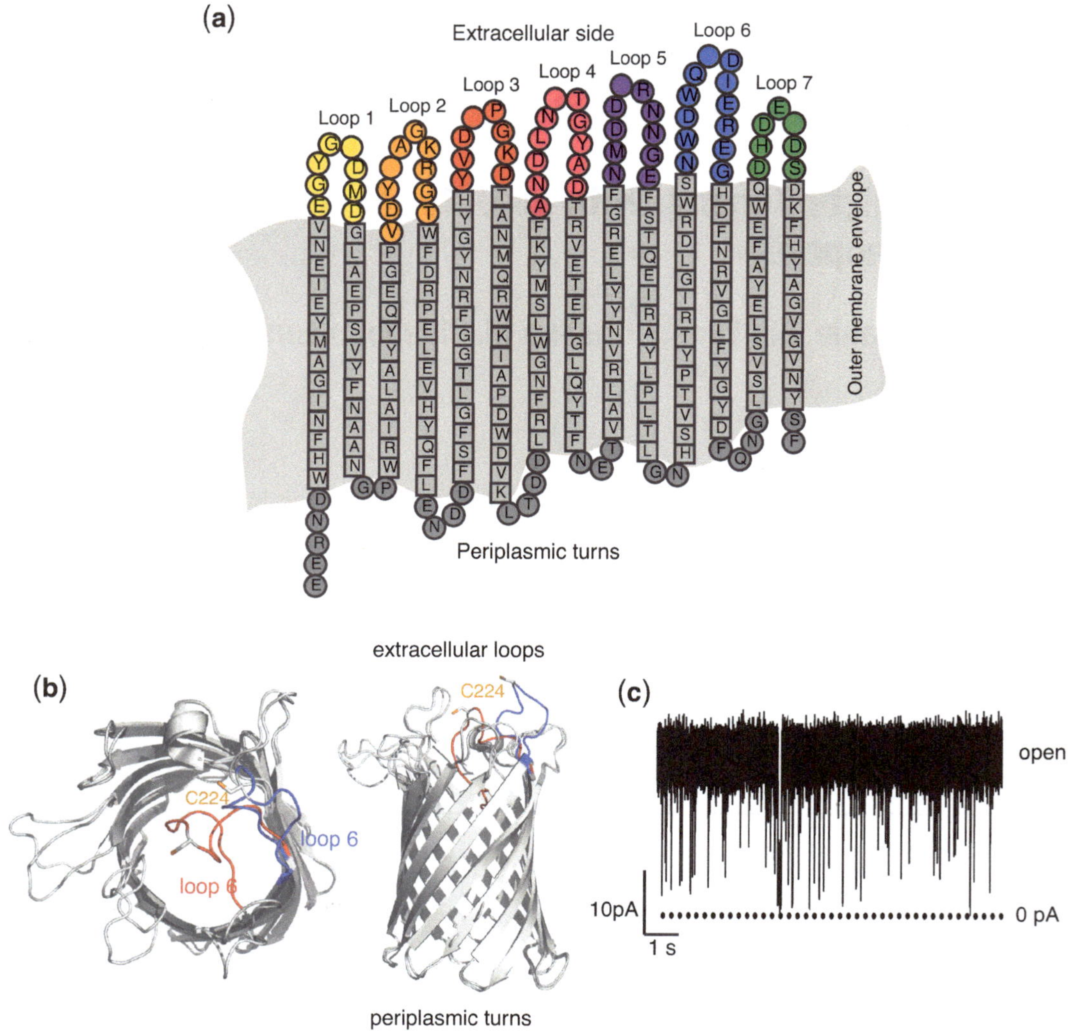

Fig. 1 Outer membrane protein G structure and characteristic gating pattern. (**a**) Flat representation of the OmpG protein amino acid sequence. The peptide sequence of the seven loops are highlighted in color, and the residues used to mutate into cysteine are highlighted in white bold lettering. (**b**) Top and side view of the OmpG crystal structure. (**c**) Characteristic OmpG gating signal in the current recording

blockades as it passes through a nanopore depending on its state. This can make detection specifically challenging.

Specialized nanopores are those decorated with high affinity molecules that make them specific to only a few analytes. These types of nanopores have been mainly used to detect the native/active state of protein analytes. Once the protein is bound to the high affinity molecule this can induce a change in the nanopore via an adaptor [7] or via direct binding [44–51]. The current method described in this chapter is an example of a specialized nanopore that discriminates among protein variants through direct interaction with analyte.

We engineered a nanopore based on outer membrane protein G (OmpG) from *Escherichia coli (E. coli)*. This nanopore has a unique feature where it generates a fluctuating current unlike other common nanopores. We exploited the dynamic nature of the seven flexible extracellular loops of OmpG to capture specific analytes (Fig. 1). Analytes are distinguished through changes in loop dynamics and protein–protein interaction with the OmpG loop surface [40, 46, 52, 53]. We observed that a single OmpG construct could distinguish among several proteins even structural homologs (Fig. 2). Because of the sensing power of OmpG, it is an exciting platform on which to build nanopore sensors for sensitive and specific analysis.

2 Materials

2.1 Cloning of Cysteine Mutant OmpG

1. pT7-OmpGwt plasmid.
2. Oligonucleotides for mutagenesis.
3. High fidelity polymerase.
4. Restriction enzymes such as DpnI.
5. Chemically Competent DH5α cells.
6. 100 mg/ml Ampicillin or carbenicillin.

2.2 Protein Expression of OmpG-Cysteine Mutant

1. Chemically competent BL21 DE3 cells.
2. LB media (Miller version): 10 g/l Tryptone, 5 g/l Yeast Extract, 10 g/l sodium chloride.
3. 0.5 M Isopropyl β-D-1-thiogalactopyranoside (IPTG).

2.3 Preparation of OmpG-Biotin or OmpG-Sulfonamide Nanopore

1. Storage buffer: 50 mM Tris-HCl, pH 8, 150 mM NaCl, 1 mM EDTA.
2. Lysis buffer: 50 mM Tris-HCl, pH 8, 150 mM NaCl, 200 μ g/ml lysozyme, 1 mM EDTA, 3 mM TCEP.
3. Inclusion Body (IB) wash buffer: 50 mM Tris-HCl, pH 8, 1.5 M Urea, 1 mM TCEP.
4. Binding buffer: 50 mM Tris-HCl, pH 8, 3 mM TCEP, 8 M urea.
5. Probe Sonication system.
6. 0.45 μm membrane filters.
7. Anion exchange columns (Q sepharose).
8. Desalting columns (5 ml, Cytiva).
9. Chromatography gravity/centrifuge columns (15 ml).
10. Falcon tubes (50 ml).
11. Wash buffer: 50 mM Tris-HCl pH 8, 3 mM TCEP, 8 M Urea, 75 mM NaCl.

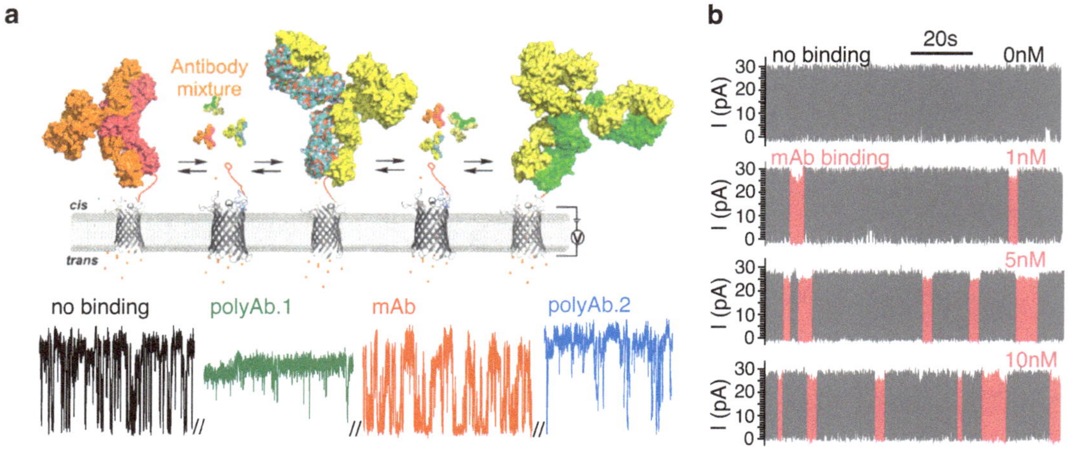

Fig. 2 Detection of different antibodies binding with OmpG-biotin construct (**a**) Schematic of OmpG and antibody reversible interactions and corresponding current recording traces of no binding and binding signals. (**b**) Concentration dependent binding of monoclonal antibody (mAb) with OmpG-biotin construct

12. Elution buffer: 50 mM Tris-HCl pH 8, 3 mM TCEP, 8 M Urea, 200 mM NaCl.
13. High salt buffer: 50 mM Tris-HCl pH 8, 3 mM TCEP, 8 M Urea, 500 mM NaCl.
14. Labelling buffer: 50 mM HEPES, pH 7.0, 150 mM NaCl, 8 M Urea.
15. Labelling reagents: Maleimide-PEG2-Biotin for preparing OmpG-biotin, Maleimide-PEG4-sulfonamide for preparing OmpG-sulfonamide.
16. Refolding buffer: 50 mM Tris-HCl, pH 9.0, 3.25% octyl glucoside (OG).
17. Trypsin enzyme. Trypsin activity buffer: 50 mM Tris-HCl pH 8, 1 mM $CaCl_2$.
18. Iodoacetamide.
19. Chymotrypsin enzyme. Chymotrypsin activity buffer: 100 mM Tris-HCl pH 8, 10 mM $CaCl_2$.
20. Bradford assay reagents.
21. SDS-PAGE materials and electrophoresis equipment.

2.4 Current Recording of OmpG-Biotin or OmpG-Sulfonamide Nanopore

1. Phospholipids: 10 mg/ml Diphytanoylphosphatidylcholine (DPhPC) dissolved in pentane.
2. Oil: 10% v/v hexadecane dissolved in pentane.
3. Polytetrafluoroethylene film (25 μm thickness) for the aperture.
4. Biotin binding analytes—(i) streptavidin, (ii) avidin, (iii) mouse anti-biotin monoclonal antibody clone BTN.4, and (iv) mouse anti-biotin monoclonal antibody clone SB58C.

Table 1
List of the oligonucleotides used to generate the OmpG mutants

OmpG mutations	Mutagenic Oligonucleotides
L1 (E24C)	Forward: 5'-TATGGCTGTGATATGGATGGGCTG-3'
	Reverse: 5'-CATATCACAGCCATAACCCTCGAC-3'
L2 (S58C)	Forward: 5'-GATTATTGCGCGGGTAAACGTGGAACG-3'
	Reverse: 5'-ACCCGCGCAATAATCTACCGGCCC-3'
L3 (E101C)	Forward: 5'-GTTGATTGTCCGGGTAAAGACACG-3'
	Reverse: 5'-ACCCGGACAATCAACGTAGTGATAAC-3'
L4 (T143C)	Forward: 5'-CTGAACTGTACCGGTTACGCTGATAC-3'
	Reverse: 5'-ACCGGTACAGTTCAGATCGTTGGC-3'
L5 (S182C)	Forward: 5'-GACGACTGCCGCAATAACGGTGAG-3'
	Reverse: 5'-ATTGCGGCAGTCGTCCATATTGAAG-3'
L6 (D224C)	Forward: 5'-GGGACTGGCAGTGTGATATTGAACGTGAAG-3'
	Reverse: 5'-GTTCAATATCACACTGCCAGTCCCAGTTAC-3'
L7 (G264C)	Forward: 5'-CGACGAATGCGACAGTGATAAATTC-3'
	Reverse: 5'-ACTGTCGCATTCGTCGTGATCCTG-3'

5. Sulfonamide binding analytes—(i) human carbonic anhydrase VII and (ii) human carbonic anhydrase II.
6. Human blood (10%).
7. Equipment: two chamber chip, Ag/AgCl electrodes, Digidata 1320A/D board (Axon Instruments), Axopatch 200B integrating patch clamp amplifier (Molecular Devices).
8. Software: latest versions of pClamp and Clampfit.

3 Methods

3.1 Substitution of a Single Cysteine Mutation on OmpG Protein

1. Introduce single cysteine substitutions on either of OmpG's seven loops by site directed mutagenesis. The plasmid template is pT7-OmpG (*see* **Note 1**). The mutagenesis primers to mutate residues on OmpG loops are described in Table 1 (*see* **Note 2**).
2. Use a standard polymerase chain reaction protocol from a commercially available high-fidelity enzyme such as Phusion to generate the PCR product with the desired mutation (*see* **Note 3**).

3. Digest the template DNA in the PCR reaction mix with DpnI enzyme (*see* **Note 4**).
4. Transform the DpnI digested PCR mix into competent *E. coli* DH5α cells (or an equivalent cell strain) onto LB-Agar plates with 100 μg/ml ampicillin.
5. Isolate single clones and identify the desired mutant construct with restriction enzyme digestion and DNA sequencing.

3.2 Expression of OmpG Proteins

1. Transform the validated pT7-OmpG cysteine mutant plasmid into BL21 DE3 *E. coli*competent cells and plate onto LB-Agar plates supplemented with 100 μg/ml ampicillin (LB-Agar-Amp). Incubate LB-Agar-Amp plate(s) for 14–18 h at 37°C (*see* **Note 5**).
2. Isolate one or more colonies from the BL21 transformation and inoculate a starter culture of LB media supplemented with 100 μg/ml ampicillin. Allow the starter culture to grow 37°C with agitation at 200 rpm for 12–16 h.
3. Inoculate a larger volume (at least ten times volume of starter culture) of LB+Amp media with the starter culture. Allow cells to grow at 37°C, 200 rpm shaking until the OD_{600} reaches between 0.5–0.8 (*see* **Note 6**).
4. Add inducer Isopropyl β-D-1-thiogalactopyranoside (IPTG) to culture at a final concentration of 0.5 mM to induce the protein expression and continue to shake culture at 37°C (*see* **Note 7**).
5. Harvest cells after 3–4 h of IPTG induction by centrifugation at 3000 × *g* for 20–30 min. Remove LB supernatant and resuspend cell pellet in either storage buffer or lysis buffer (*see* **Note 8**).
6. Centrifuge cell pellet to remove buffer and store in −20°C freezer for approximately 2 months or until ready to purify (Sect. 3.3, Step 1). Once cell pellet is resuspended in lysis buffer prepare the inclusion body (Sect. 3.3, Steps 2–5).

3.3 Isolation of OmpG Inclusion Body

1. Resuspend OmpG cell pellet in lysis/storage buffer at room temperature (*see* **Note 9**). Add lysozyme to a final concentration of 200 μg/ml and reducing agent (DTT or TCEP) to a final concentration of 3 mM. Incubate the OmpG containing lysate at room temperature for 20 min or 37°C for 10 min to allow the lysozyme to break down the cell wall.
2. Sonicate the cell mixture on ice to further break the bacterial membranes and shear the nucleic acid to reduce the lysate's viscosity. Pulse 5 s on/5 s off for a total of 7 min (*see* **Note 10**).
3. Centrifuge the lysate at 20,000 × *g* for 20 min. Discard the supernatant.

4. Resuspend the pellet in inclusion body (IB) wash buffer. Incubate at room temperature for 10 min.
5. Centrifuge pellet again at 20,000 × *g* for 15 min. Discard IB wash supernatant. The resulting inclusion body pellet can be stored at −20°C for up to 4 months.

3.4 Anion Exchange Purification of Denatured OmpG Cysteine Proteins

1. Solubilize the OmpG-containing inclusion body in binding buffer (*see* **Note 11**) for at least 30 min at room temperature or until pellet is ~90% solubilized (*see* **Note 12**)
2. Centrifuge the solubilized inclusion body at 20,000 × *g* for 15 min. Discard the pellet and pass the supernatant through a 0.45 μm filter before purification.
3. Equilibrate anion exchange Q beads with at least 5× column volume of binding buffer.
4. Apply the filtered inclusion body solution onto the equilibrated Q beads to allow the OmpG to bind. Collect the flow-through.
5. Wash the beads with 5× column volume of binding buffer. Collect wash.
6. Wash the beads with 5× column volume of wash buffer. Collect second wash.
7. Elute protein from beads with 3× column volume of elution buffer into clean centrifuge tubes.
8. Wash beads with 5× column volume of high salt buffer to remove any bound proteins or nucleic acids. Collect third wash.
9. Analyze protein content for each wash or elution collection with SDS-PAGE. A 12 or 15% gel is suitable for OmpG analysis.
10. For short-term storage (48 h or less), place eluted protein in 4°C (*see* **Note 13**).
11. Clean the Q beads: Wash beads with 10× column volume of distilled water. Check the manufacturer's instructions on appropriate cleaning solvents and protocols for the beads. Store the beads in 20% ethanol at 4°C for future use.

3.5 Conjugation of Maleimide Ligand to OmpG Cysteine Mutant

1. Determine the protein concentration of the eluted OmpG fraction(s) with a Bradford assay (*see* **Note 14**).
2. Use about 0.5–1.0 mg of protein for the labeling reaction. Buffer exchange the OmpG using size exclusion chromatography (SEC) column equilibrated with non-reducing labeling buffer. Refer to the manufacturer's recommendations for appropriate volume of protein to purify with the column. For e.g., if using a 5 ml HiTrap desalting column (GE), no more

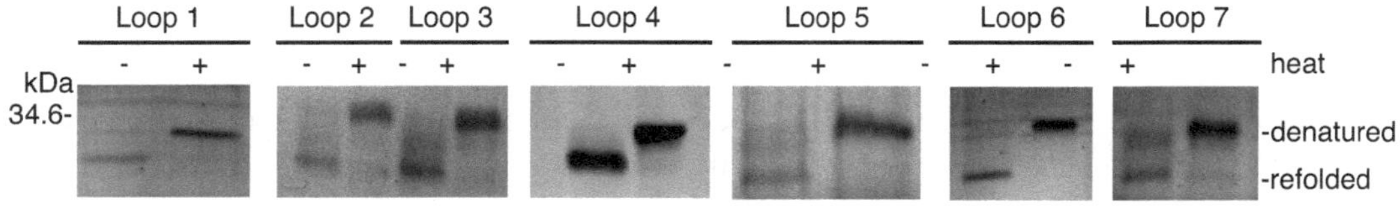

Fig. 3 Refolding efficiency gel analysis of OmpG-ligand constructs. Constructs were denatured at 95°C for 10–15 min and compared with refolded construct. Proteins were loaded onto a 12% or 15% SDS-PAGE gel and protein mobility in gel was analyzed as an indicator for their folding status

than 0.75 ml protein may be injected onto column in order to efficiently exchange the buffer (*see* **Note 15**).

3. To label the OmpG cysteine mutant with small ligands like maleimide-PEG$_2$-biotin or maleimide-PEG$_4$-sulfonamide, incubate the protein with excess ligand in a molar ratio of at least 1:5 (protein to ligand) at room temperature (~23°C) for 1–2 h with moderate shaking (~100 rpm) (*see* **Note 16**).
4. Add DTT to a final concentration of 10 mM to quench the excess maleimide ligand.
5. Buffer exchange in elution buffer without reducing agent by SEC to remove excess free ligand.

3.6 Refolding of Labeled OmpG Protein

1. Dilute the labeled OmpG mixture with refolding buffer until the final concentration of urea is approximately 3 M (*see* **Note 17**).
2. Incubate samples at 37°C for 48–72 h.
3. Test the refolding efficiency of the labeled OmpG proteins by heat denatured gel shift assay. Take one replicate of the refolded OmpG protein (at least 2 μg) and prepare with SDS-sample buffer. Take a second replicate of the OmpG protein, mix with SDS-sample buffer and boil for 10–15 min at 95°C.
4. Run both boiled and unboiled proteins on an SDS-PAGE gel and analyze the shift in apparent molecular weight of the boiled vs unboiled sample (Fig. 3).
5. Store refolded OmpG protein in small aliquots (10–20 μl) at −80°C for up to 1 month (*see* **Note 18**).

3.7 Test Labeling Efficiency of Biotin Labeled OmpG with a Gel Shift Assay

1. After diluting denatured OmpG into refolding buffer, mix OmpG and streptavidin in a 1:1 molar ratio and incubate at 23°C for 2–5 min (*see* **Note 19**).
2. Add SDS-sample buffer and run samples on a 12% or 15% SDS-PAGE gel (Fig. 4). As a negative control, run the labeled OmpG minus the streptavidin to compare.

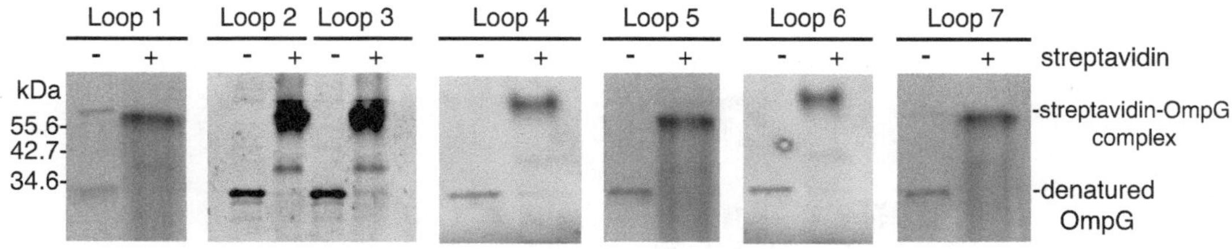

Fig. 4 Labeling efficiency of OmpG-biotin loop constructs. OmpG-biotin constructs were heat denatured in SDS-sample buffer at 95 C for 10–15 min. Thereafter, one sample was incubated with 1:1 mole ratio of streptavidin protein for 5 min at ambient temperature, while no streptavidin was added to the other sample. The constructs, ±streptavidin, were run on SDS-PAGE gels and gel mobility shifts in the presence of streptavidin were analyzed as a marker for labeling efficiency

3. Using an imaging software, such as Image J, perform a densitometric analysis on the OmpG only sample vs the labeled OmpG + streptavidin mixture to determine the labeling efficiency estimate.

3.8 Proteolytic Digestion of Labeled OmpG-Sulfonamide for Mass Spectrometric Analysis

1. Digest the labeled denatured OmpG-sulfonamide, i.e., OmpG in 8 M urea with trypsin or chymotrypsin based on the manufacturer's protocol for the specific protease. Trypsin digestion is suitable for analysis of peptides from loops 2, 3, 4, and 5. To analyze loops 1, 6, and 7 digest the OmpG with chymotrypsin instead.
2. Dilute denatured OmpG in trypsin activity buffer so that the urea concentration is 1 M or lower. Reduce any existing disulfide bonds with DTT at a final concentration of 10 mM (*see* **Note 20**). Incubate at 23°C for 30 min, then add iodoacetamide to a final concentration of 50 mM. Incubate at 23°C in the dark for 1 h with gentle agitation.
3. Use a trypsin:OmpG mole ratio range of 1:20–1:75 and incubate digestion mixture at 23°C for 12–16 h.
4. Alternatively, for cleavage with chymotrypsin, add DTT to a final concentration of 5 mM to reduce any unlabeled OmpG dimers in ~8 M urea. Incubate for 5 min.
5. Add 15 mM iodoacetamide and incubate for 15 min at room temperature in the dark. Dilute the sample with chymotrypsin activity buffer to obtain <1 M urea and add chymotrypsin to a final protease:protein mole ratio of 1:20. Allow the digestion mixture to incubate for 12–16 h at 23°C.
6. Perform mass spectrometric analysis on the digested OmpG peptides to identify the existence labeled peptides compared to unlabeled ones (*see* **Note 21**).

3.9 Current Recording of OmpG Proteins

1. Generate an artificial bilayer in a two-chamber chip separated by a 25 μm thick Polytetrafluoroethylene film with an aperture of 100 μm diameter in the center [40, 46].

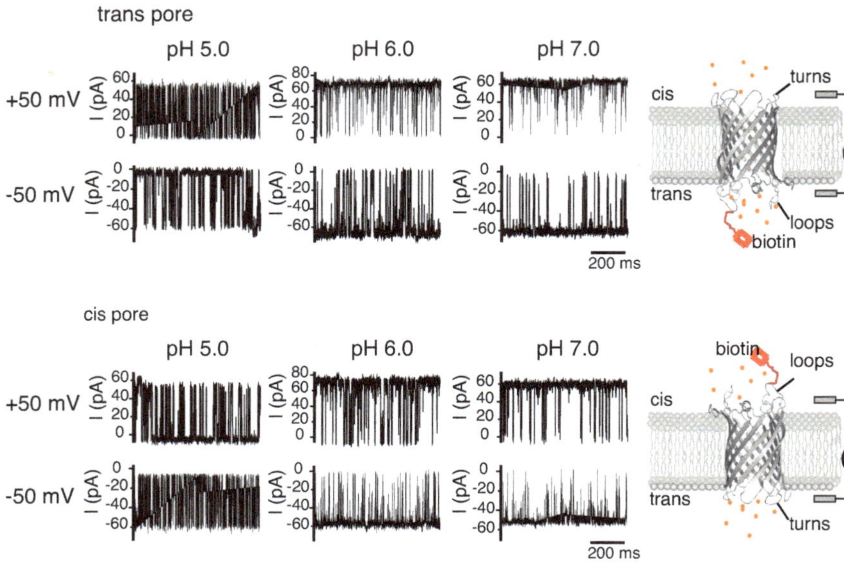

Fig. 5 Orientation of OmpG constructs inserted into lipid bilayer. OmpG has asymmetrical gating behavior termed quiet and noisy. A *trans* oriented pore is quiet (lower gating frequency) at negative voltage and noisy (higher gating frequency) at positive voltage

2. Gently apply 1–2 μl hexadecane/pentane (10% v/v) solution to the Polytetrafluoroethylene film to cover the aperture with a layer of hexadecane. Incubate at room temperature for 2–5 min (*see* **Note 22**).
3. Add appropriate volume of the buffer of your choice to each chamber (*see* **Note 23**).
4. Carefully drop 10–20 μl of 1,2-Diphytanoyl-sn-glycerol-3-phosphocholine (DPhPC) dissolved in pentane at 10 mg/ml onto the surface of the buffer to generate a monolayer of phospholipids. Do not pipet the DPhPC lipids directly into the buffer as the pentane needs to evaporate, therefore, keep pipet tip above the surface of the buffer. Allow the pentane to evaporate for 2–5 min.
5. Immerse an Ag/AgCl electrode into the buffer of each chamber, where one chamber is grounded. Slowly pipet up and down the buffer mixture in each chamber to assist in the formation of a monolayer in each chamber (*see* **Note 24**).
6. Once the bilayer is formed, test the stability at 200–300 mV. Add OmpG proteins to a final concentration of 0.1–1 nM to the chamber, which is grounded (*see* **Note 25**). Apply a voltage of 200–300 mV to assist pore insertion into the bilayer at the aperture.

7. Lower the voltage to 20–70 mV for characterization once a single OmpG protein has inserted into the bilayer. Filter the current using a Bessel filter at 2 kHz and acquire the current at a sampling rate of 50 μs.
8. Determine the orientation of the inserted OmpG pore [46, 54] by analyzing the asymmetrical gating pattern at positive and negative potentials (Fig. 5 and Table 2).
9. After determining the chamber in which the OmpG loops reside, add protein analytes to the specific chamber Table 3. Mix the solution containing the protein analyte by carefully pipetting up and down about 1/3rd of the volume of buffer in the chamber (*see* **Note 26**).
10. As a negative control use the unlabeled OmpG to prescreen for nonspecific interactions of desired protein analytes with the OmpG nanopore (*see* **Note 27**).
11. For the OmpG-sulfonamide sensor, prepare a human blood lysate sample. First resuspend blood in PBS buffer and centrifuge at 400 × *g* for 3 min. Discard the supernatant. Repeat wash one more time to remove any hemolyzed material. After washing blood, flash freeze blood cells in liquid nitrogen. Allow cells the thaw on ice. Centrifuge 10,000 × *g* to pellet any cell debris and keep supernatant containing blood lysate.
12. Add an appropriate volume of blood lysate to the loop facing chamber and slowly mix with a stir bar or by gently pipetting up and down. In general, up to 30% maximum of blood lysate is appropriate to maintain a stable bilayer. Record the OmpG nanopore before and after addition of blood lysate and take note of any changes in ionic signal.

3.10 Analysis of OmpG Current Recording Traces Post Recording

1. To determine the open conductance of OmpG, analyze at least 10 s of OmpG only trace and generate a histogram in Clampfit software analysis program. Apply a gaussian fit to the histogram to obtain the value for the open conductance.
2. To analyze specific features of OmpG gating activity, use the event detection tool in Clampfit to retrace the open and closed behavior of the nanopore. Choose a specific region of the trace to analyze using a cursor pair to mark the beginning and the end of the analysis region. Optional: filter the analysis region of the trace before applying the event detection program. Go to the filter option in the Analyze drop down and choose lowpass Gaussian at 1000 Hz−3 dB cutoff. Using single channel search, assign the current levels for the open (0) state. For example, if the maximum current of the pore is 60 pA, assign the open state at ~55 pA. A horizontal line labeled 0 marks the assignment.

Table 2
Asymmetrical characteristics of a *trans* OmpG nanopore at various pH in 1 M KCl

Gating Characteristic	pH 5		pH 6		pH 7	
	−50 mV	+50 mV	−50 mV	+50 mV	−50 mV	+50 mV
Open Probability	0.17±0.04	0.50±0.06	0.73±0.10	0.88±0.04	0.94±0.02	0.94±0.02
Gating frequency (s^{-1})	50.1±10.6	121.2±19.9	53.2±14.1	92.8±29.9	32.9±7.7	52.7±11.5
T_{on} (ms)	3.4±0.7	4.1±0.9	15.6±3.3	11.3±1.8	29.3±4.7	15.6±3.0
T_{off} (ms)	15.8±2.6	3.9±0.7	4.5±0.8	1.3±0.2	1.9±0.5	1.1±0.2

Assign the current level for the closed (1) state relative to the open level. For example, if the open (0) state is 55 pA, choose a desired closed state below this level, for example, −10 pA below or at 45 pA. A horizontal line labeled 1 marks the assignment. Allow the event detection to start by selecting Nonstop retracing. Once the event detection is complete, determine by eye whether the retracing satisfactorily matches the actual current trace. Adjust the current levels 0 and 1 to improve the event retracing if needed. In the Results window, the open (0) and closed (1) states are listed with the information about the amplitude of the state, when the state starts and ends in the trace, and therefore, the dwell time of the state. From this list of all open and closed states of the chosen analysis region, the gating frequency, the closed event duration and the open duration can be calculated.

3. To determine the gating frequency, calculate the total number of closed events and divide it by the total time of the trace analyzed.
4. To determine the event duration or the inter-event duration take only the closed (1) or open (0) dwell times from the Results window, respectively. Using the Extract Data subset option in the Analyze drop down, copy only the desired values from the total results into a new sheet. Create a histogram of either the closed (1) or open (0) dwell times. Fit the histogram to a standard exponential. Obtain the tau value from the fitting results, which is the average dwell time of either the closed (1) or open (0) state population.
5. To determine the open probability, take the sum of all the individual open dwell times and divide it by the total time of the analyzed trace region (which includes both open and closed states of the OmpG pore.

Table 3
Characteristic current traces of analyte binding to OmpG nanopore sensor

OmpG construct	Loops for sensing	Analyte	Characteristic signal
OmpG-biotin	L6 only	Streptavidin	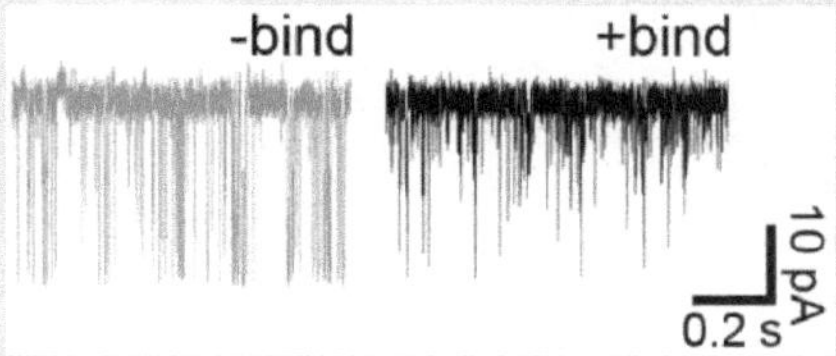
OmpG-biotin	All loops except L5	Avidin	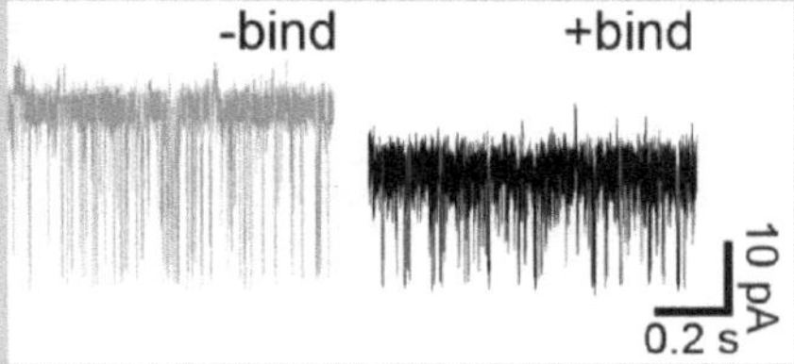
OmpG-biotin	All loops except L5	SB58C antibody	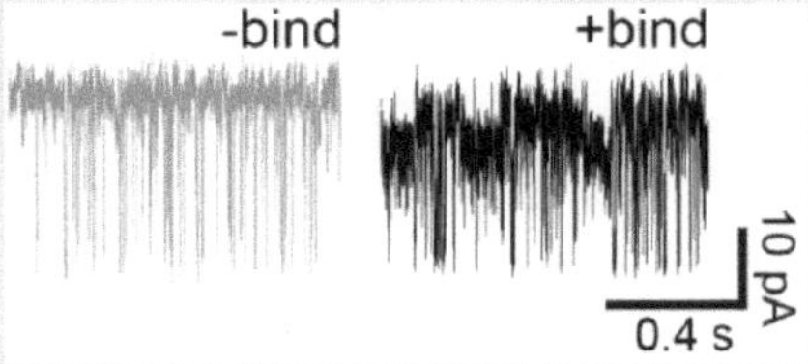
OmpG-biotin	L2, L4, L6, L7	BTN.4 antibody	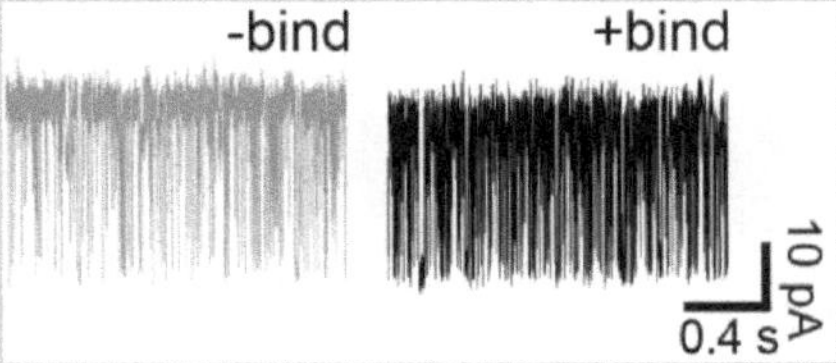
OmpG-biotin	L3	BTN.4 antibody	

Continued

Table 3
Continued

OmpG construct	Loops for sensing	Analyte	Characteristic signal
OmpG-sulfonamide	L3	hCAII	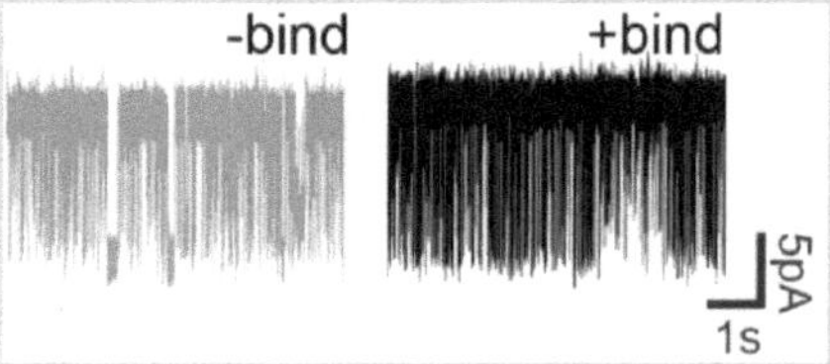
OmpG-sulfonamide	All loops	hCAVII	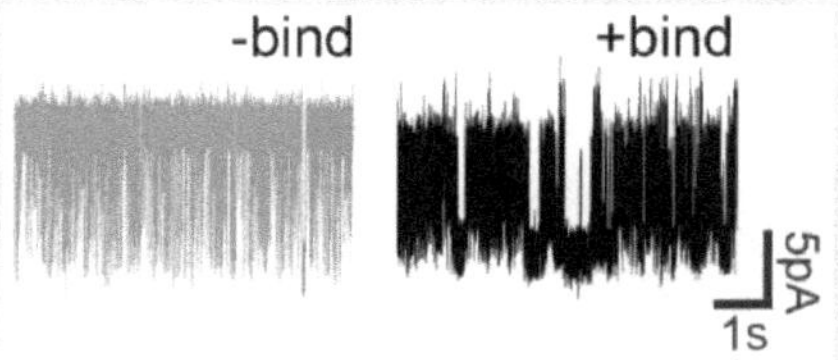

4 Notes

1. If OmpG expression in inclusion body is low, ensure that the signal peptide is removed from the open reading frame of the gene so that it expresses in the inclusion body and not secreted to the outer membrane. The signal peptide of OmpG consists of amino acids 1–21 MKKLLPCTALVMCAGMACAQA therefore maintain the start methionine and delete amino acids 2–21.
2. The mutagenic primers contain at least 15 bp sequence on its 3′ end that anneals the template. The 5′ end contains the mutation (either deletion, insertion or substitution). Also, at least 15 bp of sequence at the 5′ end of each primer anneals to the 5′ end of its primer pair. See Table 1 for examples.
3. An alternative PCR method uses primer SC47: 5′-CAGAAGTGGTCCTGCAACTTTATC (reverse) with mutagenic forward and SC46: 5′-ATAAAGTTGCAGGACCACTTCTG (forward) with mutagenic reverse to generate two complementary fragments, which when ligated generates the entire plasmid. Primers SC46 and SC47 anneal to the AmpR cassette in the plasmid pT7-OmpG. Mix the two PCR products in a 1:1 molar ratio and subject to DpnI digestion for 3 h to degrade the parental

plasmid. Optional steps are PCR or gel purification of the PCR products and in vitro ligation of PCR products with T4 DNA ligase.

4. PCR purify product before DpnI digestion if PCR buffer is not compatible with enzyme activity.
5. It is recommended to purify OmpG protein from a fresh transformation plate no more than 2 weeks old if stored at 4°C, not from BL21 (DE3) glycerol stocks.
6. To ensure a high yield of OmpG expression, shake culture at high speed (200–250 rpm) and use only up to 30% the volume of shaker flask, to keep cells well ventilated and growing aerobically.
7. OmpG yield lowers when induced at low temperatures, e.g., 16°C. An OD_{600} reading up to 1.0 is acceptable before IPTG induction in LB at 37°C.
8. Resuspend pellet in 10 ml buffer for every 250 ml culture.
9. When thawing an OmpG cell pellet, it is also acceptable to quickly warm at 37°C until thawed.
10. If a sonicator/microfluidizer is not available after lysis with lysozyme, add a DNAse I treatment step to break down the genomic DNA and reduce the lysate's viscosity. Add 50 kU DNase I and 3 mM $MgCl_2$ to the mixture and incubate for 30 min at room temperature.
11. OmpG can be purified in denatured form in 6.5–8 M urea. When preparing high urea concentrations, use caution not to heat the urea solution above 30°C where the urea starts to degrade.
12. For efficient solubilization, resuspend inclusion body in binding buffer and place in beaker with stir bar and stir at constant speed for at least 30 min. If the room temperature is higher than 30°C, solubilize the inclusion body on ice with stirring.
13. After purification, avoid freezing protein at −20°C or −80°C before labeling. The protein refolding efficiency is unchanged, but the labeling efficiency of the cysteine residue on OmpG may be reduced. Therefore, it is best to purify OmpG-Cys and then label within 2–3 h of purifying.
14. The Bradford assay is suitable because it is not affected by buffers containing urea and/or reducing agents.
15. HEPES buffer is used because the -NH_2 group on the Tris–HCl molecule may interfere with maleimide cysteine chemistry, competing with the cysteine. Also, the optimum pH range for maleimide cysteine conjugation is 6.5–7.5.
16. Another option is to incubate OmpG-Cys with maleimide ligand for 12–18 h at 4°C with gentle agitation. If the ligand

is in powder form, first dissolve in an appropriate solvent such as DMSO at an appropriate concentration according to manufacturer's instructions.

17. Refolding is most efficient when the protein concentration is between 0.35 mg/ml (10 μM) and 3.5 mg/ml (100 μM). Also, a quick refolding procedure is to dilute the denatured OmpG in a 2:3 v/v ratio, i.e., 2 volumes of denatured OmpG with 3 volumes of refolding buffer where the final octyl glucoside concentration is ~1.95%.
18. Supplement glycerol into the refolded OmpG protein sample to a final concentration of 20% v/v and mix well for homogeneity. Store in small aliquots at −80°C. The glycerol helps to extend the shelf life of the refolded OmpG protein for 3–6 months as opposed to 1–2 months without glycerol. Aim for a protein concentration of 0.3–0.8 mg/ml upon storage.
19. The streptavidin/biotin complex is resistant to SDS denaturation. It is recommended to lower the urea concentration in your denatured OmpG sample to below 6 M prior to mixing with streptavidin to ensure that the streptavidin-OmpG complex forms and remains intact during electrophoresis. Alternatively, you can test labeling efficiency with your refolded OmpG sample.
20. This step is to reduce any unlabeled OmpG dimer molecules.
21. The cysteine residues of unlabeled peptides would be modified with iodoacetamide, thus the mass is increased by ~57 Da due to the carbamidomethylation.
22. The incubation period is to allow the pentane to evaporate. The incubation time will depend on the ambient room temperature where at higher temperature incubate for shorter time.
23. Using a salt concentration lower than 100 mM is not advisable as the current produced from OmpG pores will be below 20 pA at 50 mV and thus decreases the signal to noise ratio. Typical buffers contain either NaCl or KCl ions in a range from 0.15–3.0 M.
24. The appropriate working OmpG concentration to allow single pore insertion should be determined for each batch of protein purified. Typically diluting OmpG samples to 0.01 mg/ml and using about 0.5 μl in a 900 μl chamber yields single pore insertions.
25. If too many pores continue to insert into the bilayer, lower the OmpG concentration in the chamber by removing some of the buffer and replacing with fresh buffer. Add more lipids if bilayer becomes unstable. Repeat the buffer renewal if too many OmpG pores continue to insert into the bilayer.

Another option is to discard buffer in both chambers, wash with water and 70%–85% ethanol alternately and prepare a new bilayer. Dilute the OmpG sample even further before adding to the new bilayer.

26. Ensure that your pipet tip is as far away from the aperture as possible so as not to disturb the bilayer or introduce more OmpG insertions.
27. If there is no signal indicating the detection of protein analyte, the inserted OmpG may be unlabeled, break and remake bilayer to induce insertion of another OmpG pore. Other options, use the appropriate loop(s) to sense proteins, refer to Table 3; use ligands with short linkers such as PEG2–PEG4 or no linker if possible.

References

1. Kasianowicz JJ, Bezrukov SM (1995) Protonation dynamics of the alpha-toxin ion channel from spectral analysis of pH-dependent current fluctuations. Biophys J 69:94–105. https://doi.org/10.1016/S0006-3495(95)79879-4
2. Braha O, Gu LQ, Zhou L, Lu X, Cheley S, Bayley H (2000) Simultaneous stochastic sensing of divalent metal ions. Nat Biotechnol 18:1005–1007. https://doi.org/10.1038/79275
3. Wu Y, Gooding JJ (2022) The application of single molecule nanopore sensing for quantitative analysis. Chem Soc Rev 51:3862–3885. https://doi.org/10.1039/D1CS00988E
4. Ouyang Y, Wang K, Jia W, Zhang P, Huang S (2024) Simultaneous identification of vitamins B1, B3, B5, and B6 by an engineered nanopore. Nano Lett 24(38):11944–11953. https://doi.org/10.1021/acs.nanolett.4c03329
5. Bezrukov SM, Vodyanoy I, Parsegian VA (1994) Counting polymers moving through a single ion channel. Nature 370:279–281
6. Kasianowicz JJ, Henrickson SE, Weetall HH, Robertson B (2001) Simultaneous multianalyte detection with a nanometer-scale pore. Anal Chem 73:2268–2272. https://doi.org/10.1021/ac000958c
7. Movileanu L, Howorka S, Braha O, Bayley H (2000) Detecting protein analytes that modulate transmembrane movement of a polymer chain within a single protein pore. Nat Biotechnol 18:1091–1095
8. Halverson KM, Panchal RG, Nguyen TL, Gussio R, Little SF, Misakian M, Bavari S, Kasianowicz JJ (2005) Anthrax biosensor, protective antigen ion channel asymmetric blockade. J Biol Chem 280:34056–34062. https://doi.org/10.1074/jbc.M507928200
9. Hornblower B, Coombs A, Whitaker RD, Kolomeisky A, Picone SJ, Meller A, Akeson M (2007) Single-molecule analysis of DNA-protein complexes using nanopores. Nat Methods 4:315–317. https://doi.org/10.1038/nmeth1021
10. Talaga DS, Li J (2009) Single-molecule protein unfolding in solid state nanopores. J Am Chem Soc 131:9287–9297. https://doi.org/10.1021/ja901088b
11. Bhatti H, Jawed R, Ali I, Iqbal K, Han Y, Lu Z, Liu Q (2021) Recent advances in biological nanopores for nanopore sequencing, sensing and comparison of functional variations in MspA mutants. RSC Adv 11:28996–29014. https://doi.org/10.1039/d1ra02364k
12. Braha O, Walker B, Cheley S, Kasianowicz JJ, Song L, Gouaux JE, Bayley H (1997) Designed protein pores as components for biosensors. Chem Biol 4:497–505. https://doi.org/10.1016/S1074-5521(97)90321-5
13. Wendell D, Jing P, Geng J, Subramaniam V, Lee TJ, Montemagno C, Guo P (2009) Translocation of double stranded DNA through membrane adapted phi29 motor protein nanopore. Nat Nanotechnol 4:765–772. https://doi.org/10.1177/0963721412473755. Surging
14. Butler TZ, Pavlenok M, Derrington IM, Niederweis M, Gundlach JH (2008) Single-molecule DNA detection with an engineered MspA protein nanopore. Proc Natl Acad Sci 105:20647–20652. https://doi.org/10.1073/pnas.0807514106

15. Stefureac R, Long Y-T, Kraatz H-B, Howard P, Lee JS (2006) Transport of alpha-helical peptides through alpha-hemolysin and aerolysin pores. Biochemistry 45:9172–9179. https://doi.org/10.1021/bi0604835
16. Cornell BA, Braach-Maksvytis VL, King LG, Osman PD, Raguse B, Wieczorek L, Pace RJ (1997) A biosensor that uses ion-channel switches. Nature 387:580–583. https://doi.org/10.1038/42432
17. Guo BY, Zeng T, Wu HC (2015) Recent advances of DNA sequencing via nanopore-based technologies. Sci Bull (Beijing) 60:287–295. https://doi.org/10.1007/s11434-014-0707-6
18. Han A, Schürmann G, Mondin G, Bitterli RA, Hegelbach NG, De Rooij NF, Staufer U (2006) Sensing protein molecules using nanofabricated pores. Appl Phys Lett 88:1–4. https://doi.org/10.1063/1.2180868
19. Firnkes M, Pedone D, Knezevic J, Döblinger M, Rant U, Do M, Rant U (2010) Electrically facilitated translocations of proteins through silicon nitride nanopores: conjoint and competitive action of diffusion. Nano Lett 10:2162–2167. https://doi.org/10.1021/nl100861c
20. Chen P, Gu J, Brandin E, Kim YR, Wang Q, Branton D (2004) Probing single DNA molecule transport using fabricated nanopores. Nano Lett 4:2293–2298. https://doi.org/10.1021/nl048654j
21. Burns JR, Stulz E, Howorka S (2013) Self-assembled DNA nanopores that span lipid bilayers. Nano Lett 13:2351–2356. https://doi.org/10.1021/nl304147f
22. Bell NAW, Keyser UF (2014) Nanopores formed by DNA origami: a review. FEBS Lett 588:3564–3570. https://doi.org/10.1016/j.febslet.2014.06.013
23. Dekker C (2007) Solid-state nanopores. Nat Nanotechnol 2:209–215. https://doi.org/10.1038/nnano.2007.27
24. Li J, Stein D, McMullan C, Branton D, Aziz MJ, Golovchenko JA (2001) Ion-beam sculpting at nanometre length scales. Nature 412:166–169. https://doi.org/10.1038/35084037
25. Bell NAW, Engst CR, Ablay M, Divitini G, Ducati C, Liedl T, Keyser UF (2012) DNA origami nanopores. Nano Lett 12:512–517. https://doi.org/10.1021/nl204098n
26. Messager L, Burns JR, Kim J, Cecchin D, Hindley J, Pyne ALB, Gaitzsch J, Battaglia G, Howorka S (2016) Biomimetic hybrid nanocontainers with selective permeability. Angew Chem Int Ed 55:11106–11109. https://doi.org/10.1002/anie.201604677
27. Kasianowicz JJ, Balijepalli AK, Ettedgui J, Forstater JH, Wang H, Zhang H, Robertson JWF (2016) Analytical applications for pore-forming proteins. Biochim Biophys Acta, Biomembr 1858:593–606. https://doi.org/10.1016/j.bbamem.2015.09.023
28. Movileanu L (2009) Interrogating single proteins through nanopores: challenges and opportunities. Trends Biotechnol 27:333–341. https://doi.org/10.1016/j.tibtech.2009.02.008
29. Majd S, Yusko EC, Billeh YN, Macrae MX, Yang J, Mayer M (2010) Applications of biological pores in nanomedicine, sensing, and nanoelectronics. Curr Opin Biotechnol 21:439–476. https://doi.org/10.1016/j.copbio.2010.05.002
30. Oukhaled A, Bacri L, Pastoriza-Gallego M, Betton JM, Pelta J (2012) Sensing proteins through nanopores: fundamental to applications. ACS Chem Biol 7:1935–1949. https://doi.org/10.1021/cb300449t
31. Shi W, Friedman AK, Baker LA (2017) Nanopore sensing. Anal Chem 89:157–188. https://doi.org/10.1021/acs.analchem.6b04260
32. Willems K, Meervelt VV, Wloka C, Maglia G (2017) Single-molecule nanopore enzymology. Philos Trans R Soc B 372:20160230. https://doi.org/10.1098/rstb.2016.0230
33. Zhang L, Gardner ML, Jayasinghe L, Jordan M, Aldana J, Burns N, Freitas MA, Guo P (2021) Detection of single peptide with only one amino acid modification via electronic fingerprinting using reengineered durable channel of Phi29 DNA packaging motor. Biomaterials 276:121022. https://doi.org/10.1016/J.BIOMATERIALS.2021.121022
34. Motone K, Kontogiorgos-Heintz D, Wee J, Kurihara K, Yang S, Roote G, Fox OE, Fang Y, Queen M, Tolhurst M, Cardozo N, Jain M, Nivala J, Allen PG (2024) Multi-pass, single-molecule nanopore reading of long protein strands. Nature 633:662–669. https://doi.org/10.1038/s41586-024-07935-7
35. Lu C, Bonini A, Viel JH, Maglia G (2025) Toward single-molecule protein sequencing using nanopores. Nat Biotechnol 43:312–322. https://doi.org/10.1038/s41587-025-02587-y
36. Galenkamp NS, van den Noort M, Maglia G (2025) Dynamics of single enzymes confined inside a nanopore. RSC Chem Biol 6(11):1686–1704. https://doi.org/10.1039/D5CB00149H
37. Nivala J, Mulroney L, Li G, Schreiber J, Akeson M (2014) Discrimination among protein variants using an unfoldase-coupled nanopore. ACS Nano 8:12365–12375. https://doi.org/10.1021/nn5049987
38. Nir I, Huttner D, Meller A (2015) Direct sensing and discrimination among ubiquitin and ubiquitin chains using solid-state

nanopores. Biophys J 108:2340–2349. https://doi.org/10.1016/j.bpj.2015.03.025
39. Rosen CB, Rodriguez-Larrea D, Bayley H (2014) Single-molecule site-specific detection of protein phosphorylation with a nanopore. Nat Biotechnol 32:179–181. https://doi.org/10.1038/nbt.2799
40. Fahie MA, Chen M (2015) Electrostatic interactions between OmpG Nanopore and analyte protein surface can distinguish between glycosylated isoforms. J Phys Chem B 119:10198–10206. https://doi.org/10.1021/acs.jpcb.5b06435
41. Versloot RCA, Lucas FLR, Yakovlieva L, Tadema MJ, Zhang Y, Wood TM, Martin NI, Marrink SJ, Walvoort MTC, Maglia G (2022) Quantification of protein glycosylation using nanopores. Nano Lett 22:5357–5364. https://doi.org/10.1021/acs.nanolett.2c01338
42. Cao C, Magalha P, Krapp LF, Bada Juarez JF, Finn Mayer S, Rukes V, Chiki A, Lashuel HA, Dal Peraro M (2024) Deep learning-assisted single-molecule detection of protein post-translational modifications with a biological nanopore. ACS Nano 18:34. https://doi.org/10.1021/acsnano.3c08623
43. Li M, Xiong Y, Cao Y, Zhang C, Li Y, Ning H, Liu F, Zhou H, Li X, Ye X, Pang Y, Zhang J, Liang X, Qing G (2023) Identification of tagged glycans with a protein nanopore. Nat Commun 14:1–12. https://doi.org/10.1038/s41467-023-37348-5
44. Rotem D, Jayasinghe L, Salichou M, Bayley H (2012) Protein detection by nanopores equipped with aptamers. J Am Chem Soc 134:2781–2787. https://doi.org/10.1021/ja2105653
45. Harrington L, Alexander LT, Knapp S, Bayley H (2015) Pim kinase inhibitors evaluated with a single-molecule engineered nanopore sensor. Angew Chem Int Ed 54:8154–8159. https://doi.org/10.1002/anie.201503141
46. Fahie M, Chisholm C, Chen M (2015) Resolved single-molecule detection of individual species within a mixture of anti-biotin antibodies using an engineered monomeric nanopore. ACS Nano 9:1089–1098. https://doi.org/10.1021/nn506606e
47. Pham B, Eron SJ, Hill ME, Li X, Fahie MA, Hardy JA, Chen M (2019) A nanopore approach for analysis of caspase-7 activity in cell lysates. Biophys J 117:844–855. https://doi.org/10.1016/j.bpj.2019.07.045
48. Kim M, Foster JC, Moore MD, Chen M (2023) Improving single-molecule antibody detection selectivity through optimization of peptide epitope presentation in OmpG nanopore. ACS Sens 8:2673–2680. https://doi.org/10.1021/acssensors.3c00528
49. Foster JC, Pham B, Pham R, Kim M, Moore MD, Chen M (2023) An engineered ompg nanopore with displayed peptide motifs for single-molecule multiplex protein detection. Angew Chem Int Ed Engl 62:e202214566. https://doi.org/10.1002/anie.202214566
50. Foster JC, Pham B, Pham R, Ryan P, Tong N, Sharp J, Inoue S, Liang J, Beis K, Chen M (2025) Barrel expansion of outer membrane protein G nanopore through β-hairpin duplication. Protein Sci 34:e70203. https://doi.org/10.1002/pro.70203
51. Hwang HJ, Kim JS, Lee J, Min JS, Jeong KB, Kim E, Lee MK, Chi SW (2022) Single-molecule sensing of an anticancer therapeutic protein-protein interaction using the chemically modified OmpG nanopore. Anal Chem 94:7449–7454. https://doi.org/10.1021/acs.analchem.1c04840
52. Fahie MA, Yang B, Mullis M, Holden MA, Chen M (2015) Selective detection of protein homologues in serum using an OmpG nanopore. Anal Chem 87:11143–11149. https://doi.org/10.1021/acs.analchem.5b03350
53. Fahie MA, Yang B, Pham B, Chen M (2016) Tuning the selectivity and sensitivity of an OmpG nanopore sensor by adjusting ligand tether length. ACS Sens 1:614–622. https://doi.org/10.1021/acssensors.6b00014
54. Chen M, Li Q-H, Bayley H (2008) Orientation of the monomeric porin OmpG in planar lipid bilayers. ChemBioChem 9:3029–3036. https://doi.org/10.1002/cbic.200800444

Check for updates

Chapter 5

A Selective Activity-Based Approach for Analysis of Enzymes with an OmpG Nanopore

Monifa A. V. Fahie, Bach Pham, Fanjun Li and Min Chen

Abstract

Many enzymatic activity assays are based on either (i) identifying and quantifying the enzyme with methods such as western blot or enzyme linked substrate assay (ELISA) or (ii) quantifying the enzymatic reaction by monitoring the changing levels of either product or substrate with techniques such as mass spectrometry, absorption or fluorescence spectroscopy, thin layer chromatography, luminescence, and more. We have generated outer membrane protein G (OmpG)-based nanopore approaches to distinguish enzyme identity and analyze the enzyme's catalytic activity. Here, we engineered an OmpG nanopore with a peptide cut site inserted into one of its loops to detect proteolytic behavior. In addition, we generated an OmpG nanopore with a single-stranded DNA attached to a loop for analyzing nucleolytic cleavage. These OmpG nanopore approaches may be highly useful in analyzing specific enzymes in complex biological samples, or in directly determining kinetics of enzyme-substrate complex association and dissociation.

Keywords Nanopore engineering/design, Enzyme activity sensor, OmpG, Nuclease, Caspase, Protease, DNA ligand, Peptide substrate, ENzyme-substrate complex, Single channel recording

1 Introduction

Enzymes with catalytic cleaving activities (-ases) are ubiquitous in biology. They are essential for catabolic and anabolic processes and therefore can help to (i) maintain the cell's survival such as with lysosomal enzymes during autophagic flux [1] or (ii) carry out a sequence of events leading to cell death such as with caspases during apoptosis [2]. Studies in enzyme activity are popular in biochemistry, where bulk activity assays that either monitor the loss of substrate or the gain of product are most common. For enzymes with catalytic cleaving activity a reduction in the substrate's molecular weight is a feasible readout of enzymatic activity. For peptide substrates with molecular weights too small to distinguish a reduction in size, assays that take advantage of chromogenic of fluorogenic substrates have been utilized.

Fanjun Li (ed.), *Nanopore Technology*, Methods in Molecular Biology, vol. 3072,
https://doi.org/10.1007/978-1-0716-5527-6_5,

Nanopores have been previously used to detect enzymes either directly [3–9] or indirectly through their enzymatic products, which can be made more specific with the use of nanopore coupled enzyme systems [10–22]. Although, nanopores have successfully detected either the enzymes or their products, there are still only few examples that can distinguish enzyme or protein homologs or differentiate between an enzyme's conformation states [8, 9, 23–27]. In addition, there have been little example of nanopores that can distinguish the different states of the enzyme-substrate interaction as they progress through a complete reaction cycle [8, 24, 28]. In this chapter, we describe the analysis of two different cleaving enzymes, an endonuclease and a protease, by an engineered outer membrane protein G (OmpG) nanopore.

Endonucleases are a class of enzymes that cleave the phosphodiester bond within a nucleic acid moiety. They can either cleave DNA or RNA whether single stranded or double stranded. These class of enzymes are involved in processes such as DNA replication [29], DNA repair [30], and programmed cell death [31]. Another class of enzymes, proteases, cleave peptide bonds within a protein molecule and their functions are found in diverse processes such as in viral infection and autophagy [2, 32].

Here, we describe an OmpG based nanopore platform for not only directly detecting enzymes but also capturing specific steps in their catalytic process, which can ultimately be used to selectively and specifically analyze enzyme behavior in a mixture containing related enzymes or homologs.

2 Materials

2.1 Cloning of Mutant OmpG Proteins

1. pT7-OmpGwt plasmid (amino acids 22–301).
2. Primers for mutagenesis (see Table 1).
3. High fidelity polymerase such as Phusion, Q5, or KOD.
4. Restriction enzymes such as DpnI.
5. Chemically competent *E. coli* cells such as DH5α.
6. 100 mg/ml Ampicillin or carbenicillin dissolved in sterile ultrapure water stock.
7. LB media.
8. SOC media.
9. LB Agar plates with 100 μg/ml ampicillin (or carbenicillin).

2.2 Protein Expression of OmpG Mutants

1. LB media (Miller high salt version): 10 g/l Tryptone, 5 g/l yeast extract, 10 g/l sodium chloride, 100 μg/ml ampicillin.
2. LB Agar plates with 100 μg/ml ampicillin (or carbenicillin).

Table 1
Oligonucleotides used to generate OmpG nanopores for enzyme sensing

OmpG construct	OmpG mutations	Sequence
OmpGL6neu-DNA	D221S, D224C, D225S, E227T, R228G, E229T	F1: CAGTGTAGTATTACCGGCACGGGCCATGATTTTAAC
		R1: GGTAATACTACACTGCCAGCTCCAGTTACTCCAGCG
OmpGcasp	H231A	F2: AGCGATTTTAACCGTGTAGGTTTATTTTAC
		R2: ACGGTTAAAATCGCTGCCCGTGCCGGTAATAC
	H261A	F3: GATGCCGACGAAGGCGACAGTGATAAATTC
		R3: ACTGTCGCCTTCGTCGGCATCCTGCCACTC
	Deletion of D215	F4: GGGCTGCGCTGGAGTAACTGGGACTGGCAG
		R4: ACTCCAGCGCAGCCCAATGCGCGTATACGG
	Insertion of GDEVDG, caspase cleavage site	F5: GATGAAGTCGACGGTGATATTGAACGTGAAGGCC
		R5: ATCACCGTCGACTTCATCTCCCTGCCAGTCCCAG

3. Chemically competent *E. coli* cells such as BL21 DE3.
4. 0.5 mM Isopropyl β-D-1-thiogalactopyranoside (IPTG) stock solution.

2.3 Preparation of OmpGcasp or OmpGL6neu-DNA Nanopore

1. Sonicator, French press, or microfluidizer.
2. Centrifuge capable of pelleting volumes of 250 ml or more at 3000 × *g*.
3. Centrifuge capable of pelleting volumes of 50 ml or less at 12,000 × *g*.
4. 0.45 μm filter for filtering lysate supernatant.
5. Anion exchange columns (GE, 5 ml).
6. HiTrap Desalting columns (GE, 5 ml).
7. Fast protein liquid chromatography (FPLC) such as AKTA (GE).
8. Lysis buffer: 50 mM Tris–HCl, pH 8, 150 mM NaCl, 1 mM EDTA.
9. Inclusion Body wash buffer: 50 mM Tris–HCl, pH 8, 1.5 M Urea.
10. Anion exchange low salt buffer: 50 mM Tris–HCl, pH 8, 50 mM NaCl, 8 M urea.
11. Anion exchange high salt buffer: 50 mM Tris–HCl pH 8, 500 mM NaCl, 8 M Urea.
12. Tris(2-carboxyethyl)phosphine (TCEP) or dithiothreitol (DTT) for purification of cysteine containing OmpG protein.
13. OmpG conjugation buffer: 50 mM HEPES pH 7, 150 mM NaCl, 8 M Urea.
14. Thiolated DNA oligonucleotide (23 mer).
15. Centricon with 3 kDa molecular weight cutoff.
16. TE buffer: 10 mM Tris·HCl pH 8, 0.1 mM EDTA.
17. Non-reducing elution buffer: 50 mM Tris–HCl pH 8, 8 M Urea, 200 mM NaCl.
18. 2-dipyridyl disulfide (2-PDS) dissolved in acetonitrile.
19. Diethyl ether.
20. Refolding buffer: 50 mM Tris–HCl, pH 9.0, 3.25% octyl glucoside (OG).
21. Glycerol (for long-term storage of refolded OmpG proteins).

2.4 Bulk Activity Assays for Caspase or Nuclease S1 Enzymes

1. Chromogenic or fluorogenic caspase 3/7 substrate such as Ac-DEVD-AFC.
2. Caspase enzyme.
3. Caspase activity buffer: 50 mM HEPES, 150 mM NaCl, pH 7.5.

4. Single-stranded complementary DNA oligonucleotide (23 mer).
5. Nuclease S1 enzyme.
6. Nuclease S1 activity buffer: 20 mM sodium acetate pH 5, 1 M NaCl, 5 mM $CaCl_2$.
7. 1 M Tris–HCl pH 8 (to quench the nuclease S1 activity).
8. SDS Polyacrylamide gel electrophoresis equipment and reagents.
9. UV-Vis spectrophotometer or plate reader.

2.5 Single-Channel Recording of OmpG Nanopores

1. 1, 2-diphytanoyl-*sn*-glycero-3-phosphocholine (DPhPC) lipids.
2. Hexadecane.
3. Pentane to dissolve DPhPC or hexadecane.
4. Teflon film.
5. Silicone Glue.
6. Piezoelectric spark generator.
7. Sandpaper.
8. Air table (to minimize vibrations that lowers signal to noise ratio).
9. A two-chamber chip made from a polyoxymethylene material such as Delrin.
10. Two Ag/AgCl electrodes.
11. Digidata 1320 A/D board (Axon Instruments).
12. Axopatch 200B integrating patch clamp amplifier (Molecular Devices).

3 Methods

3.1 Cloning of OmpG Mutants for Enzyme Sensing

1. For the nuclease S1 sensor, mutate six amino acid residues on loop 6, one of which includes a cysteine at position 224 (OmpGL6neu-C224) using primers F1 and R1 in Table 1.
2. For the caspase sensor, make three mutations: H231A, H261A, and a deletion of D215 on the OmpGwt backbone. For the fourth mutation, replace the amino acid residue D224 with a caspase cleavage site GDEVDG into loop 6. The mutagenic oligonucleotides to generate these four mutations are found in Table 1.

3. Perform mutagenesis PCR with the chosen primers and the pT7-OmpGwt plasmid as template. Remove template DNA from successful PCR product with DpnI digestion. Clean-up PCR reaction with silica based spin columns.
4. Use the appropriate competent cells such as DH5α for bacterial transformations of the desired PCR-generated DNA construct.
5. Pick colonies from transformation plates, grow overnight cultures, miniprep DNA, sequence DNA to verify successful implementation of the mutations.

3.2 Expression of the Desired Plasmids in BL21 (DE3) Competent Cells

1. Transform plasmids into BL21 (DE3) competent cells. Isolate single colonies to inoculate an overnight LB culture containing 100 μg/ml ampicillin (or carbenicillin) of about 5 ml. Use this overnight culture to start a larger liquid culture of 100–500 ml on the next day.
2. Grow culture at 37°C with 200–250 rpm shaking to an OD600 of 0.5–0.8. Depending on how much starter culture used, this may take 2–4 h shaking incubation.
3. Induce with a final concentration of 0.5 mM IPTG and harvest the cell pellet after 3–4 h of shaking at 37°C, 200–250 rpm. Remove the supernatant from the cell pellet. Freeze the cell pellet at −20°C for up to 6 months.

3.3 Lysis and Inclusion Body Preparation

N.B. If you are preparing the nuclease S1 sensor go directly to Sect. 3.6 for specific details on the preparation of this protein.

1. Thaw the harvested cell pellet and resuspend in lysis buffer.
2. Lyse the cell suspension using either a sonicator, French press or microfluidizer. Clear the lysate by centrifugation at 20,000 × *g*, 4°C for 20 min. Discard the supernatant.
3. Resuspend the pellet with inclusion body wash buffer and centrifuge again at previous speed, temperature, and time. Discard the supernatant. The inclusion body pellet can be stored in −20°C freezer for up to 6 months.

3.4 Purification from Inclusion Body

1. Dissolve the OmpG-containing inclusion body in anion exchange low salt buffer for 30 min using a stir bar and a stir plate to help dissolve the inclusion body. Optionally, you can also help dissolve the inclusion body by sonicating at low intensity.
2. Centrifuge at 20,000 × *g* for 20 min to pellet any undissolved inclusion body. Optionally, pass the supernatant through a vacuum filtration device using a 0.45 μm membrane filter. Purify the OmpG proteins with a Q-anionic exchange column. Wash and elute with anion exchange high salt buffer by applying

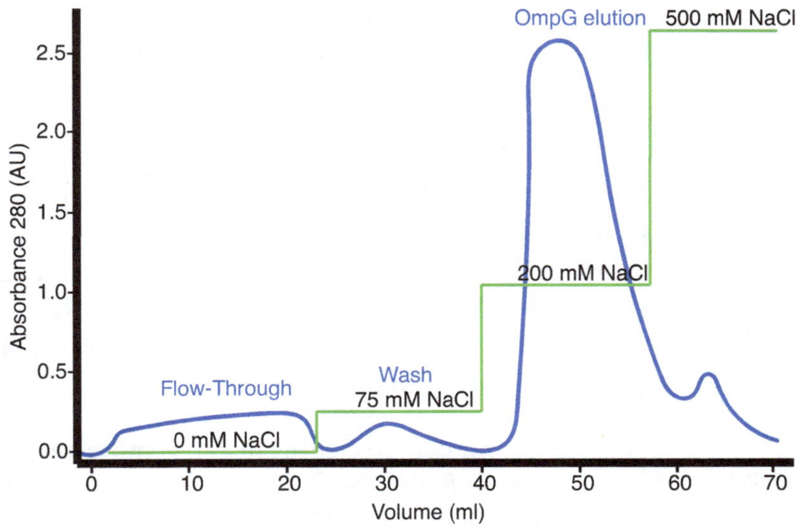

Fig. 1 Chromatogram of an OmpG anion exchange purification from inclusion bodies. A gradient of sodium chloride concentration in 8 M urea denaturing buffer is used to sequentially elute the OmpG protein from a Q sepharose column

a gradient on FPLC system (Fig. 1). OmpG protein elutes around 200 mM NaCl, thus around 30% gradient.

3. Run the elution fractions on SDS-PAGE to analyze protein quality. Choose the purest and highly concentrated fractions to refold (see Sect. 3.5). Denatured proteins may be stored in −80°C for up to 6 months (no cysteine mutation).

3.5 Refolding of Denatured OmpG Proteins

1. After collecting the purest OmpG proteins, mix in the refolding buffer dropwise until the final concentration of urea reaches 3.0 M or below. For example, with highly concentrated OmpG samples (50–100 μM) mix 1 ml of purified OmpG with about 2.5 ml of refolding buffer, this typically yields >90% refolded protein.
2. Incubate the refolding OmpG sample at 25°C or 37°C for at least 36 h before using the protein for single channel recording. Test refolding efficiency by a gel shift assay [33].
3. For long-term storage, mix refolded proteins with glycerol to a final concentration of 20% v/v. Dispense the protein in several small aliquots. Flash freeze in liquid nitrogen and store refolded proteins at −80°C for up to 6 months.

3.6 Preparation of Nuclease Sensor OmpG Nanopore

It is recommended that the DNA ligand is activated with 2-dipyridyl disulfide (2-PDS) (*see* **Note 1**) and stored at −20°C before purifying cysteine containing OmpG. After removing the reducing agent from purified OmpGL6neu-C224, keep protein on ice while determining protein concentration and subsequently label OmpG with the 2-PDS DNA ligand within 2 h to obtain at least 50% conjugated protein.

3.6.1 Activation of Thiolated DNA Ligand (Deprotection)

1. Quick centrifuge the dried 23 mer DNA at 10,000×*g* to reduce any loss upon opening the vial.
2. Add 100 µl of TE buffer and vortex to solubilize the DNA.
3. Add 100 µl of 0.5 M TCEP pH 7 to the DNA and incubate at room temperature for ~2 h to deprotect the thiol group.
4. Store the reduced DNA mixture in the −20°C freezer until further use or continue with the 2-dipyridyl disulfide (2-PDS) conjugation in Sect. 3.6.2 below.

3.6.2 Conjugation of 2-PDS to Thiolated DNA

1. Desalt the deprotected ligand to remove the excess reducing agent. Apply the DNA mixture to a 5 ml centricon with an appropriate kDa cutoff according to the DNA molecular weight, e.g., a 3 kDa cutoff.
2. Add 5 ml TE buffer to dilute the DNA. Concentrate the DNA to ~100 µl on the centricon by centrifuging at an appropriate speed according to manufacturer's instructions.
3. Add another 5 ml TE buffer and concentrate the DNA. Repeat the TE dilution and subsequent concentration two more times to bring the reducing agent concentration to a sub micromolar level.
4. During the final centricon-based desalting step, prepare a working solution of 50 mM 2-dipyridyl disulfide (2-PDS) solution in acetonitrile. For example, dissolve 23.3 mg 2-PDS in 106 µl acetonitrile for a concentration of 1 M. Then take 5 µl of the 1 M solution and dilute in 95 µl acetonitrile to make a 50 mM solution.
5. Resuspend the resulting ~100 µl desalted DNA in the centricon and transfer from the centricon into a 1.5 ml tube. Mix the desalted DNA with an equal volume of 50 mM 2-PDS. Incubate for 2 h at room temperature and mix by inverting every 20–30 mins.
6. To remove the excess 2-PDS, mix at least 200 µl of diethyl ether into the DNA mixture in order to distinguish the two layers.
7. Centrifuge at top speed for 30 s. Discard ~150–180 µl of the top layer (the diethyl ether layer) into an organic waste bottle.
8. Repeat the diethyl ether extraction another nine times.
9. Allow the residual diethyl ether in the DNA solution to evaporate for 4–16 h in a chemical hood.
10. Determine the DNA molar concentration and store the activated 2-PDS DNA in a −20°C freezer until ready to use for labeling (*see* **Note 2**).

3.6.3 Lysis and Inclusion Body Preparation of OmpGL6neu-C224

1. Resuspend the OmpG cell pellet in lysis buffer supplemented with 3 mM TCEP (or DTT). Lyse cells on ice using any mechanical/enzymatic lysis method of your choice.
2. Isolate the inclusion body pellet by centrifuging at 20,000 × *g* for 20 min. Resuspend the pellet with inclusion body wash buffer containing 3 mM TCEP and centrifuge again at previous speed and time. Discard supernatant. The resulting inclusion body pellet can be stored in −20°C freezer for up to 6 months.

3.6.4 Purification from Inclusion Body

1. As in Sect. 3.3, dissolve OmpG-containing inclusion body in anion exchange low salt buffer containing 3 mM TCEP for 30 min with stir bar and stir plate to help break up the inclusion body for dissolution.
2. Centrifuge to pellet any undissolved inclusion body. Pass supernatant through a 0.45 μm filter. Purify OmpG proteins with a Q-anionic exchange column under reducing conditions. Wash and elute with anion exchange high salt buffer containing 3 mM TCEP by applying a gradient (Fig. 1).
3. Analyze the protein fractions by SDS-PAGE. Determine the concentration of the purest OmpG containing fractions with an appropriate protein concentration assay such as the Bradford assay. Desalt the protein with a 5 ml desalting column into OmpG conjugation buffer (*see* **Note 3**).

3.6.5 Labeling of OmpG Cysteine Mutant with Activated 2-PDS DNA

1. Incubate the OmpGL6neu-Cys with activated 2-PDS DNA in a 1:1 molar ratio (protein:DNA).
2. Shake gently at room temperature for 2 h or alternatively for 12–18 h at 4°C.
3. Separate conjugated OmpG-Cys-DNA from unlabeled OmpG-Cys on a non-reducing SDS-PAGE gel with a percentage suitable for significant separation between conjugated vs the unlabeled OmpG-Cys (*see* **Note 4**).
4. Excise the band of the OmpGL6neu-DNA that runs slower than unconjugated, at about ~50 kDa (Fig. 2) and extract it from the gel by crushing gel with a pestle in 3× the gel volume of non-reducing elution buffer (*see* **Note 5**).
5. Incubate with gentle agitation at 25°C for 2–6 h. Continue to elute protein from crushed gel by storing at 4°C for 14–18 h.
6. Next day, centrifuge the crushed gel mixture at high speed (at least 10,000 × *g*) for 15 min at room temperature (*see* **Note 6**). Collect the eluted protein.
7. Eluted denatured OmpGL6neu-DNA can be flash frozen in liquid nitrogen and stored in −80°C for up to 6 months.

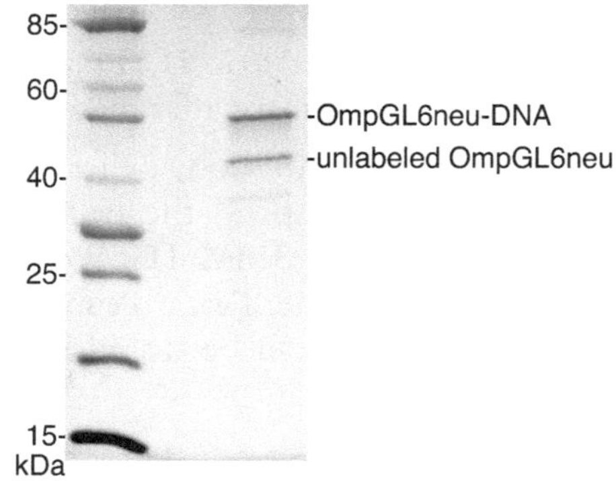

Fig. 2 SDS-PAGE analysis of labeled vs unlabeled OmpG construct. OmpG incubated with activated thiolated DNA was run on a 15% gel to separate labeled vs unlabeled OmpG proteins

8. Follow instructions in Sect. 3.5 above for refolding of the DNA labeled OmpG protein.

3.7 Cleavage Assay of OmpGcasp Protease Sensor

1. Mix caspase and refolded OmpGcasp together in caspase activity buffer in either 1:20 or up to a 1:1 caspase to OmpG mole ratio.
2. Incubate for 0.5–4 h at room temperature (22–25°C) (*see* **Note 7**). Heat denature all samples by boiling at 95°C for 15 min in SDS-PAGE Laemmli sample buffer and then run on appropriate percentage acrylamide gel (Fig. 3a).

3.8 Cleavage Assay of OmpG Nuclease Sensor

1. Incubate the refolded OmpGL6neu-DNA with nuclease S1 enzyme in nuclease activity buffer for 60 min at 25°C. Quench the reaction by increasing the pH to 8 by adding 1 M Tris–HCl pH 8.
2. Mix the protein reactions with SDS Laemmli sample buffer and boil at 90–95°C for 10 min to denature all proteins. Analyze nuclease activity by gel shift in SDS-PAGE.
3. As a positive control, treat OmpGL6neu-DNA in 10 mM DTT. Of note, when running this positive control on the gel ensure to leave empty lanes on both sides of this DTT-treated sample, as the DTT can bleed into other lanes as the voltage is applied (Fig 3b).

3.9 Single-Channel Recording of OmpG Proteins

1. Similar to previous studies, perform single-channel recording experiments in a two-chamber chip with a 25 μm thick Teflon film containing a 100 μm aperture separating the chambers [24] (Fig. 4).
2. To make an aperture for single channel recording go to Sect. 3.10. Add 1–3 μl solution of 10% v/v hexadecane dissolved in pentane to coat the Teflon sheet on both faces (*see* **Note 8**).

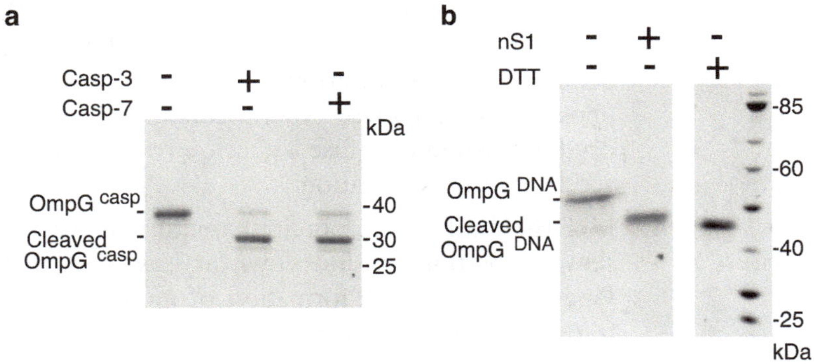

Fig. 3 Cleavage gels of OmpG substrates. (**a**) Caspases 3 and 7 were incubated with OmpGcasp construct in a 1:20 mole ratio. Proteins were boiled in SDS-sample buffer at 95°C for 10 min and run on gel electrophoresis to analyze cleavage efficiency. (**b**) OmpGL6neu-DNA construct was incubated with nuclease S1 (nS1) at pH 5. Proteins were heat denatured and analyzed by gel shift for cleavage efficiency

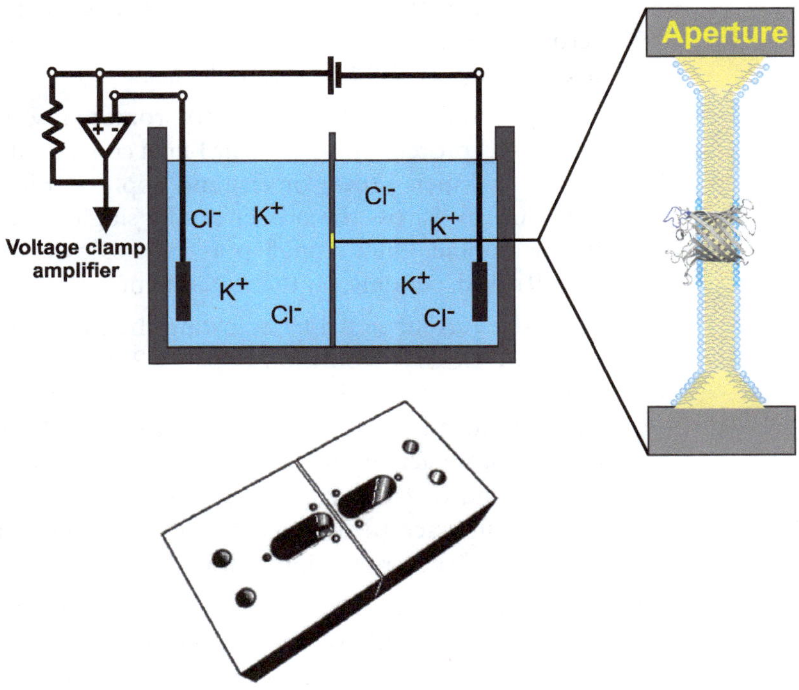

Fig. 4 Schematic of the two-chamber chip used to create DPHPC lipid bilayers. The two chambers are separated by a Teflon film (yellow) which has a 100 μm hole in the center on which lipid monolayers from each chamber fuse to create a lipid bilayer. The nanopore of interest nestles in the lipid bilayer creating a small current when voltage is applied

This will coat the aperture in a thin layer of oil. Allow the volatile pentane to evaporate in 2–5 min at ambient temperature.

3. Fill each chamber with buffer to completely submerge the aperture on both faces. Add approximately 15 μg of phospholipids

to each chamber to form a monolayer (*see* **Note 9**). 1, 2-diphytanoyl-*sn*-glycero-3-phosphocholine (DPhPC) dissolved in pentane at a concentration of 10 mg/ml is the recommended phospholipid for these experiments. Allow the pentane to air dry for 2–5 min or infuse a gentle stream of compressed air to accelerate the evaporation.

4. Mix the buffer/phospholipid mixture in each chamber by slowly pipetting up and down at least 75% of the mixture. Pipetting aids in the formation of monolayers covering each face of the aperture to create a bilayer.
5. Connect the chip to the Axopatch 200B integrating patch clamp amplifier by submerging the silver portion of a silver chloride electrode into the buffered mixture (not too close to the aperture) and the "gold" portion to the headstage.
6. When voltage is switched on, it is only applied to one chamber (*trans*), while the other chamber remains grounded (*cis*) thereby generating opposing positive and negative voltage across each face of the bilayer. To assess the stability of the bilayer, see Sect. 3.11 for more details.
7. After generating a stable bilayer, to promote insertion of a single OmpG protein, mix in 2–3 nM final concentration to either of the chambers (*trans* or *cis*) and apply a voltage of about 200–300 mV across the bilayer until a single pore inserts into the membrane. Once a single pore inserts, lower the voltage to 20–60 mV depending on the experiment.
8. Filter the current signal by adjusting the low pass Bessel filter (80 dB/DECADE) on the Axopatch 200B integrating patch clamp to 2 kHz. For signal acquisition, use a sampling at 100 μ s after digitization with the Digidata 1440 A/D board. Record the nanopore activity with pClamp software program at both positive and negative voltages for at least 5–10 min. Evaluate the current trace to determine if it is a single pore insertion (Fig. 5) and the orientation of the pore [33, 34].
9. After determining pore orientation and recording to characterize the apo condition of the nanopore, add the enzyme of interest into the loop facing chamber and gently mix so that the bilayer is not disturbed. Record the nanopore and enzyme mixture at both voltages and take note of any changes in current signal (Figs. 6 and 7).

3.10 Method for Making an Aperture for the Two-Chamber Chip

1. Cut an appropriate square of Teflon film and place it on the circumference of the chamber to determine the size of the hole. Draw the outline of the circumference of the opening that the Teflon film will sit between the two buffer chambers.
2. Using a sharp pin which can be made from a glass capillary, make an indent in the middle of the Teflon film. Take the

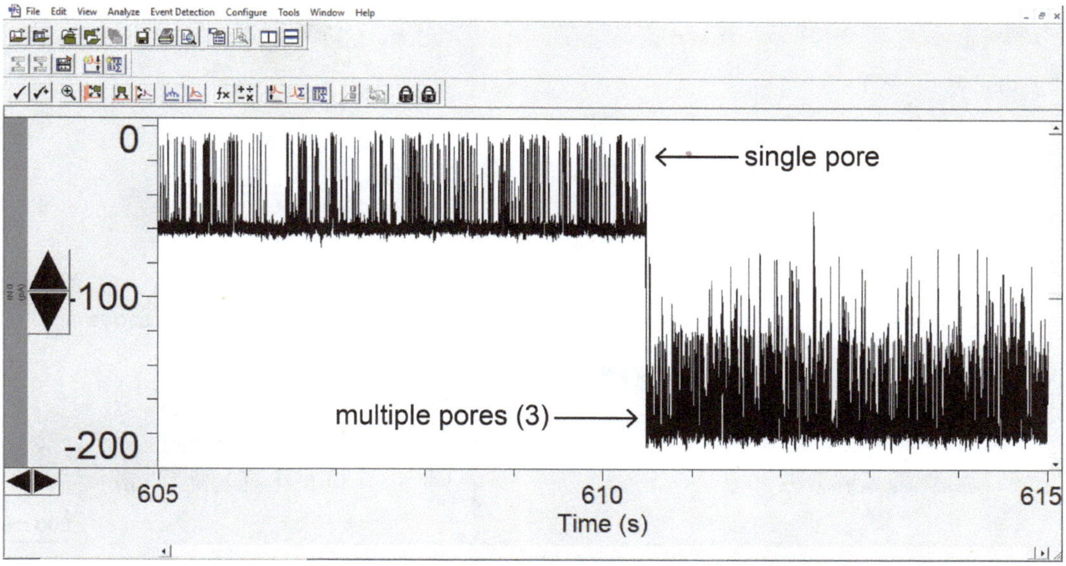

Fig. 5 Single-channel recording of a single OmpG pore vs multiple pore insertions. A raw trace showing the ionic behavior of a single pore at −50 mV until two other pores insert into the lipid bilayer. The ionic trace becomes noisy and difficult to analyze as opposed to a single OmpG pore

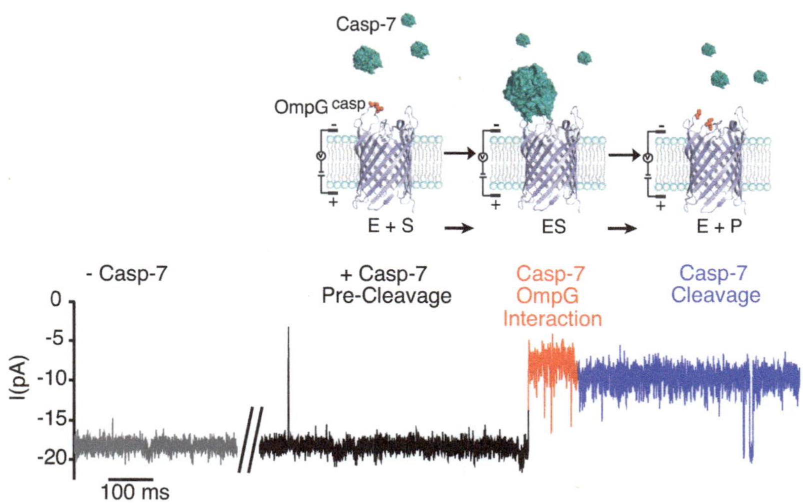

Fig. 6 A characteristic trace of OmpGcasp nanopore with and without caspase-7. The current signal of a single OmpGcasp nanopore showing a transient interaction with caspase-7 (in red) followed by the cleavage of the loop at the recognition site (in blue)

Teflon film and place on the piezoelectric spark generator so that the ident is positioned in the center of the electrical spark.

3. Generate 10–15 sparks with the clicker to create an aperture of approximately 100 μm in diameter. Fit the Teflon film onto

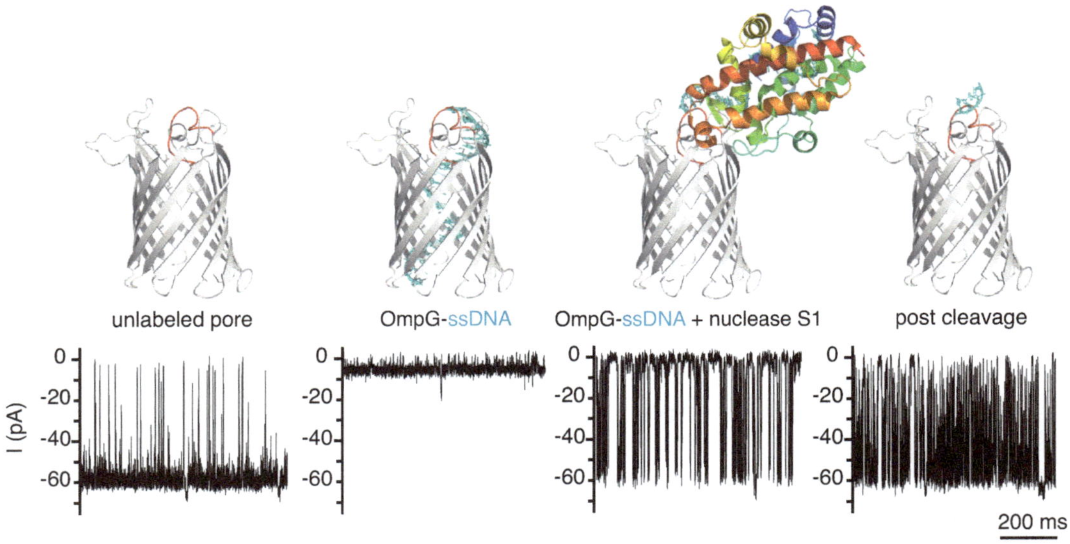

Fig. 7 A discontinuous trace of OmpGL6neu-DNA nanopore with and without nuclease S1. Conjugation of a single-stranded DNA oligonucleotide onto loop 6 of OmpG blocks the lumen at −50 mV, which can act as a positive marker for labeled pore. Binding of nuclease S1 to OmpGL6neu-DNA construct generates a unique signal that clearly distinguishes it from the final cleavage of the DNA molecule

the chamber and optionally use silicone glue to hold it in place. Assemble the second chamber.

4. Remove any excess Teflon film with razor blade. Allow the glue to dry for at least 8 h before using the chip. On the first use of newly sealed Teflon film, rinse the chip at least five times alternating between 70%–95% ethanol and distilled water, where your final rinse is 70%–95% ethanol. Allow chip to air dry or use a gentle stream of compressed air to let the ethanol evaporate.

3.11 Method for Testing the Bilayer Formation and Stability

1. Testing the bilayer formation: After slowly pipetting the buffer in each chamber to mix lipids, a consistent current of zero should be observed (Fig 8a). Evaluate the size and coverage of the bilayer with the bilayer test switch. An oscillating sin like curve is generated with a corresponding amplitude. An amplitude of 80–300 pA [when the scaled output/output gain (α) is set to 1] is indicative that a ≤100 μm bilayer has been formed (Fig 8b). The amplitude of the sigmoidal curve informs about the bilayer size. An amplitude no lower than 80 pA and no larger than 400 pA will usually generate a stable bilayer for several hours.

2. Test the bilayer stability: After evaluating bilayer formation with the bilayer seal test, turn off the seal test switch. Apply a voltage of ±200 to 300 mV across the bilayer for at least

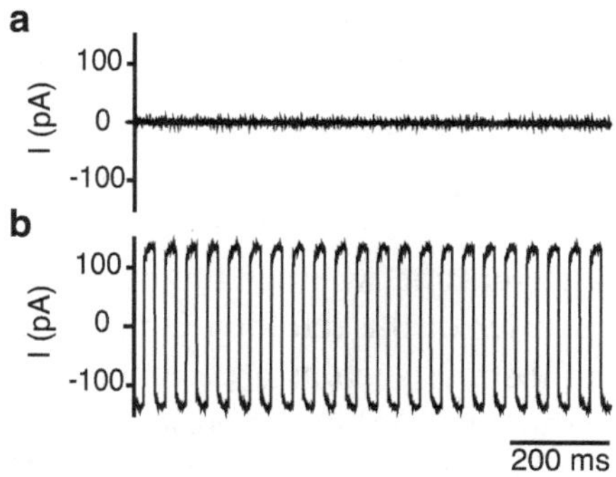

Fig. 8 Schematic trace of a stable bilayer. (**a**) The current with or without a voltage applied is stable at zero without any fluctuations in current. (**b**) Seal test signal of a formed bilayer in the chip aperture. Amplitude is above 100 pA

5 min and monitor the behavior of the current. A stable bilayer will remain at zero with little to no fluctuations in current. An unstable bilayer will either break/overload (i.e., a large current signal) or there may be several fluctuations in current seen. If the bilayer breaks, pipet the chamber solution to reform the bilayer. Apply a high voltage again. Once the bilayer stability is acceptable continue with experiment by acquiring and recording a single OmpG nanopore.

3. For any further issues with bilayer generation, refer to *see* **Notes 10–14** for troubleshooting tips.

4 Notes

1. The conjugation of 2-dipyridyl disulfide to the thiolated DNA is to make the DNA more reactive to the reduced cysteines on the OmpG sample. It, therefore, lowers the possibility of creating DNA-DNA disulfide dimers.
2. Activated DNA is stable in −20°C freezer for at least 6 months).
3. Reducing agents must be removed prior to conjugating OmpGL6neu-C224 with activated 2-PDS DNA.
4. Do not incubate the OmpGL6neu-DNA in SDS-sample buffer containing any reducing agent or run on a PAGE gel containing reducing agent. This will break the conjugated bond resulting in unlabeled OmpG.
5. An alternative to manually crushing the gel with a small pestle one can also shred the gel by placing gel in a 0.6 ml tube that has been punctured at the bottom with a 21 G needle to

create a small hole. Place the 0.6 ml tube into a 1.5 ml tube and centrifuge at high speed at room temperature.

6. Do not allow the eluted protein to warm up above 30°C, the temperature at which urea will start to degrade and lower the protein yield by reducing its solubility.

7. Determine whether the protease or nuclease is active in the buffer of the refolded OmpG sample. Use a positive control substrate such as a chromogenic substrate (or a ssDNA for the nuclease) in the presence of 2% octyl glucoside detergent, 3 M urea, or Tris–HCl pH 9. If the protein is not active in such a buffer, dilute the detergent and urea concentration to a level that retains the enzyme activity. Use this dilution ratio as a guide to dilute the refolded OmpG sample before incubating with enzyme. Diluting the detergent in OmpG sample after it has successfully refolded does not significantly denature the OmpG at room temperature.

8. Avoid directly touching the pipet tip to the Teflon film. This may puncture or bend the film as altering its shape integrity could make bilayer formation difficult.

9. Never apply dissolved phospholipid solution directly to the Teflon film. The phospholipids will adsorb onto the Teflon surface. An oil layer of hexadecane is needed to coat the Teflon film first. The phospholipids should always be added to the surface of the buffered solution. The oil coating the Teflon blocks the flow of aqueous solution from one chamber to the next and is used to correctly orient the phospholipids in the aperture, i.e., with the lipid tails facing the Teflon film and the polar phospho-heads facing the buffered solution.

10. A ~100 μm bilayer has a wave amplitude reading of ~100 pA when the seal test ext command on the Axopatch 200B is turned on.

    a) If the amplitude reading is smaller than 100 pA, the aperture may be smaller than 100 μm. Bilayers smaller than 100 μm may be too stable and unbreakable. Alternatively, there may be too much oil on the Teflon film. Pipet some of the buffered solution in the chamber towards the bilayer to wash some of the oil from the Teflon film. Pipet the solution (as far as possible from the aperture) to regenerate the bilayer. Measure the amplitude with the seal test again. Alternatively, add another 5–10 μl more phospholipids to each chamber, allow to evaporate and mix solution slowly. Retest with the seal test.

    b) If the amplitude is around 10 pA or lower, no bilayer has formed. There may be too little oil on the Teflon film. Wash the chip several times alternating between 70%–95%

ethanol and water. Dry the chip and remake the lipid bilayer.

c) If the amplitude consistently decreases (shrinking bilayer), break the bilayer and pipet to reform. There may not be enough lipids to sufficiently form a monolayer in each chamber, therefore add another 5–10 μl DPhPC lipids to each chamber.

d) Amplitudes consistently larger than 300 pA may be indicative of a deformed aperture possibly due to pipet puncturing. Discard the old Teflon film and remake a 100 μm aperture.

11. A stable bilayer has a consistent current of zero with or without voltage applied

a) If the current of the bilayer is non-zero (within 5 pA) without voltage applied, there may be a calibration error. Turn the pipet offset switch until the current signal returns to zero. If the pipet offset does not fix the calibration, then adjust the headstage offset whole cell $\beta = 1$, which is found in the back of the Axopatch 200B device.

b) If the trace fluctuates around zero causing a wobbly baseline, the silver electrode may be too long and therefore causing minute vibrations in the connection. Cut the wire shorter or wrap some of the wire length around itself so that it can be taut when connected to the headstage.

c) If the voltage is applied and a non-zero current is observed, the aperture may not have been properly cleaned from the previous experiment, where previous nanopore samples are contaminating the bilayer. Wash the chip extensively with 70%–95% ethanol and water. For a more intensive clean, sonicate the chip in either 0.1 M NaOH or 1%–5% detergent for 15–30 min. Rinse extensively with tap water, rinse again in distilled water, then wash with 70%–95% ethanol before drying. Remake bilayer.

12. A stable bilayer does not break when up to 200 mV is applied. If the bilayer breaks when high voltage (or low voltage) is applied, it is unstable. Check the expiration date of lipids and oil. Oxidized lipids or oil can cause unstable bilayers. Water contamination in the lipids or oil can also cause unstable bilayers. Alternatively, too much oil or lipids coating the Teflon film can cause instability in the bilayer. As the lipids or oil is used multiple times, the pentane evaporates resulting in more concentrated lipids or oil. It is recommended to seal vials containing lipids or oil with parafilm to reduce pentane evaporation and/or water contamination when stored in the freezer.

13. A stable bilayer does not break when a single or multiple nanopores insert. If the bilayer consistently breaks only when a nanopore inserts, there may be insufficient lipids. Add another 5–10 μl more phospholipids to each chamber. Re-test with the seal test. Alternatively, there may be too many nanopores in the chamber. Dilute the nanopores, by removing an appropriate volume of buffer from the chamber containing the nanopores. Replenish with new buffer and lipids proportional to what was removed. Retest bilayer stability and continue with recording a single nanopore.

14. A stable bilayer breaks when buffered solution is pipetted directly towards the aperture or when zapped.

    a) If the bilayer does not break when buffer solution is pipetted toward the aperture (do not place tip directly on Teflon film), or when the zap switch is used, there may be too much oil on the Teflon film. Wash the chip several times in alternating water and 70%–95% ethanol and remake bilayer with less oil.

    b) If the bilayer is unbreakable, the aperture may be smaller than 100 μm. With smaller bilayers, it is easy to overload the aperture with oil. Apply less oil to apertures smaller than 100 μm or use bilayers at least 100 μm in size.

    c) Another possible cause for the bilayer not breaking is a bad connection caused by the Ag/AgCl electrode. Over time the silver electrodes become turn white from multiple use in buffer. Regenerate Ag/AgCl by scraping off the white deposit with sandpaper to expose the silver underneath and then incubate the silver wire in 10% bleach solution for at least 30 min at room temperature, or several hours in 4°C.

References

1. Ravanan P, Srikumar IF, Talwar P (2017) Autophagy: the spotlight for cellular stress responses. Life Sci 188:53–67. https://doi.org/10.1016/j.lfs.2017.08.029
2. D'Arcy MS (2019) Cell death: a review of the major forms of apoptosis, necrosis and autophagy. Cell Biol Int 43:582–592. https://doi.org/10.1002/cbin.11137
3. Xie H, Braha O, Gu LQ, Cheley S, Bayley H (2005) Single-molecule observation of the catalytic subunit of cAMP-dependent protein kinase binding to an inhibitor peptide. Chem Biol 12:109–120. https://doi.org/10.1016/j.chembiol.2004.11.013
4. Cheley S, Xie H, Bayley H (2006) A genetically encoded pore for the stochastic detection of a protein kinase. ChemBioChem 7:1923–1927. https://doi.org/10.1002/cbic.200600274
5. Rotem D, Jayasinghe L, Salichou M, Bayley H (2012) Protein detection by nanopores equipped with aptamers. J Am Chem Soc 134:2781–2787. https://doi.org/10.1021/ja2105653
6. Harrington L, Cheley S, Alexander LT, Knapp S, Bayley H (2013) Stochastic detection of Pim protein kinases reveals electrostatically enhanced association of a peptide substrate. Proc Natl Acad Sci USA 110:E4417–E4426. https://doi.org/10.1073/pnas.1312739110

7. Van MeerveltV, Soskine M, Maglia G (2014) Detection of two isomeric binding configurations in a protein aptamer complex with a biological nanopore. ACS Nano 8:12826–12835
8. Li F, Fahie MA, Gilliam KM, Pham R, Chen M (2022) Mapping the conformational energy landscape of Abl kinase using ClyA nanopore tweezers. Nat Commun 13: 3541. https://doi.org/10.1038/s41467-022-31215-5
9. Li F, Foster JC, Nguyen L, Chen M (2025) Selective identification of allosteric inhibitors and co-drug combinations targeting kinases by using a nanopore tweezer approach. ACS Nano 19(39):34617–34627. https://doi.org/10.1021/ACSNANO.5C07468
10. Zhao Q, De Zoysa RSS, Wang D, Jayawardhana DA, Guan X (2009) Real-time monitoring of peptide cleavage using a nanopore probe. J Am Chem Soc 131:6324–6325. https://doi.org/10.1021/ja9004893
11. Mohammad MM, Iyer R, Howard KR, McPike MP, Borer PN, Movileanu L (2012) Engineering a rigid protein tunnel for biomolecular detection. J Am Chem Soc 134:9521–9531. https://doi.org/10.1021/ja3043646
12. Wang L, Han Y, Zhou S, Guan X (2014) Real-time label-free measurement of HIV-1 protease activity by nanopore analysis. Biosens Bioelectron 62:158–162. https://doi.org/10.1016/j.bios.2014.06.041
13. Kukwikila M, Howorka S (2015) Nanopore-based electrical and label-free sensing of enzyme activity in blood serum. Anal Chem 87:9149–9154. https://doi.org/10.1021/acs.analchem.5b01764
14. Nivala J, Marks DB, Akeson M (2013) Unfoldase-mediated protein translocation through an α-hemolysin nanopore. Nat Biotechnol 31:247–250. https://doi.org/10.1038/nbt.2503
15. Nivala J, Mulroney L, Li G, Schreiber J, Akeson M (2014) Discrimination among protein variants using an unfoldase-coupled nanopore. ACS Nano 8:12365–12375. https://doi.org/10.1021/nn5049987
16. Ho C-W, Van Meervelt V, Tsai K-C, De Temmerman P-J, Mast J, Maglia G (2015) Engineering a nanopore with co-chaperonin function. Sci Adv 1:e1500905. https://doi.org/10.1126/sciadv.1500905
17. Derrington IM, Craig JM, Stava E, Laszlo AH, Ross BC, Brinkerhoff H, Nova IC, Doering K, Tickman BI, Ronaghi M, Mandell JG, Gunderson KL, Gundlach JH (2015) Subangstrom single-molecule measurements of motor proteins using a nanopore. Nat Biotechnol 33:1073–1075. https://doi.org/10.1038/nbt.3357
18. Craig JM, Laszlo AH, Brinkerhoff H, Derrington IM, Noakes MT, Nova IC, Tickman BI, Doering K, de Leeuw NF, Gundlach JH (2017) Revealing dynamics of helicase translocation on single-stranded DNA using high-resolution nanopore tweezers. Proc Natl Acad Sci 114:11932–11937. https://doi.org/10.1073/pnas.1711282114
19. Kang X, Wu C, Alibakhshi MA, Liu X, Yu L, Walt DR, Wanunu M (2023) Nanopore-based fingerprint immunoassay based on rolling circle amplification and DNA fragmentation. ACS Nano 17:5412–5420. https://doi.org/10.1021/acsnano.2c09889
20. Li T, Song S, Zheng X, Lu W, Cao Y, xie H, Wang H (2025) DNA-polyarginine probe-enabled nanopore sensing for ultrasensitive detection of Dam methyltransferase activity and inhibition. Biosens Bioelectron 289:117859. https://doi.org/10.1016/j.bios.2025.117859
21. Guan Z, Sun Y, He Y, Cao J, Tian J, Ji Z (2025) Machine-learning-assisted nanopore sensing solution for the determination of matrix metalloproteinase. Biosens Bioelectron 288:117752. https://doi.org/10.1016/j.bios.2025.117752
22. Lu C, Bonini A, Viel JH, Maglia G (2025) Toward single-molecule protein sequencing using nanopores. Nat Biotechnol 43:312–322. https://doi.org/10.1038/s41587-025-02587-y
23. Soskine M, Biesemans A, Moeyaert B, Cheley S, Bayley H, Maglia G (2012) An engineered ClyA nanopore detects folded target proteins by selective external association and pore entry. Nano Lett 12:4895–4900. https://doi.org/10.1021/nl3024438
24. Pham B, Eron SJ, Hill ME, Li X, Fahie MA, Hardy JA, Chen M (2019) A nanopore approach for analysis of caspase-7 activity in cell lysates. Biophys J 117:844–855. https://doi.org/10.1016/j.bpj.2019.07.045
25. Kim M, Foster JC, Moore MD, Chen M (2023) Improving single-molecule antibody detection selectivity through optimization of peptide epitope presentation in OmpG nanopore. ACS Sens 8:2673–2680. https://doi.org/10.1021/acssensors.3c00528
26. Li X, Lee KH, Shorkey S, Chen J, Chen M (2020) Different anomeric sugar bound states of maltose binding protein resolved by a cytolysin A nanopore tweezer. ACS Nano 14:1727–1737. https://doi.org/10.1021/acsnano.9b07385
27. Galenkamp NS, van den Noort M, Maglia G (2025) Dynamics of single enzymes confined inside a nanopore. RSC Chem Biol 6(11):1686–1704. https://doi.org/10.1039/D5CB00149H

28. Fahie MAV, Li F, Palmer C, Yoon C, Chen M (2022) Modifying the pH sensitivity of OmpG nanopore for improved detection at acidic pH. Biophys J 121:731–741. https://doi.org/10.1016/j.bpj.2022.01.023
29. Okazaki T (2017) Days weaving the lagging strand synthesis of DNA—a personal recollection of the discovery of Okazaki fragments and studies on discontinuous replication mechanism. Proc Jpn Acad Ser B Phys Biol Sci 93:322–338. https://doi.org/10.2183/pjab.93.020
30. Wood RD (1996) DNA repair in Eukaryotes. Annu Rev Biochem 65:135–167
31. Enari M, Sakahira H, Yokoyama H, Okawa K, Iwamatsu A, Nagata S (1998) A caspase-activated DNase that degrades DNA during apoptosis, and its inhibitor ICAD. Nature 391:43–50
32. Garten W, Braden C, Arendt A, Peitsch C, Baron J, Lu Y, Pawletko K, Hardes K, Steinmetzer T, Böttcher-Friebertshäuser E (2015) Influenza virus activating host proteases: identification, localization and inhibitors as potential therapeutics. Eur J Cell Biol 94:375–383. https://doi.org/10.1016/j.ejcb.2015.05.013
33. Fahie M, Chisholm C, Chen M (2015) Resolved single-molecule detection of individual species within a mixture of anti-biotin antibodies using an engineered monomeric nanopore. ACS Nano 9:1089–1098. https://doi.org/10.1021/nn506606e
34. Chen M, Li Q-H, Bayley H (2008) Orientation of the monomeric porin OmpG in planar lipid bilayers. ChemBioChem 9:3029–3036. https://doi.org/10.1002/cbic.200800444

Check for updates

Chapter 6

Engineering a Monomeric Nanopore Scaffold for Multiplexed Protein Detection

Joshua C. Foster and Min Chen

Abstract

The complexity of biological samples often necessitates specialized processing and instrumentation to enable the assessment of disease-associated biomarkers, highlighting the need for simple and robust methods of multiplex detection. Nanopore biosensors represent a promising technology to address this challenge, with biological nanopores emerging as versatile scaffolds for sensor development. In particular, monomeric outer membrane proteins (OMPs) provide an accessible platform for site-specific engineering with reproducible, atomically precise fabrication. Outer membrane protein G (OmpG), a monomeric *Escherichia coli* porin possessing seven flexible extracellular loops that can be functionalized with affinity ligands, has been utilized as a highly selective sensor through a distinctive gating-pattern detection mode arising from loop-mediated sampling of the molecular surface of bound analytes. Here, we present a general protocol for the molecular grafting of two peptide affinity motifs onto OmpG, followed by purification and characterization of the engineered construct, to generate a single-channel multiplex sensor.

Keywords Nanopore, Multiplex, Protein, Sensor, Outer membrane protein G

1 Introduction

Nanopore sensing techniques record ionic current through a single nanometer-scale pore held under a transmembrane potential [1, 2]; when analytes interact with the sensor pore, they induce characteristic current perturbations that, depending on sensor and assay design may report such parameters as analyte identity [3–8], conformation [9–11], and kinetics [12–14], enabling real-time, single-molecule detection, and quantitation [15–22]. Engineering and preparation of a nanopore sensor is a complex multifield endeavor, relying on basic familiarity with many molecular biology and biochemistry techniques, in addition to the nanopore-specific knowledge. This chapter describes a general procedure to achieve multiplex nanopore sensing using Outer Membrane Protein G (OmpG) as a model monomeric nanopore scaffold [23]. Analyte detection using the OmpG system is based on changes to the sensor pore's gating pattern

Fanjun Li (ed.), *Nanopore Technology*, Methods in Molecular Biology, vol. 3072,
https://doi.org/10.1007/978-1-0716-5527-6_6,

induced by analyte interaction(s) with affinity reagent(s) displayed by one or more of the pore's seven flexible loops [24–26]. Here we discuss a multiplexdetection approach integrating two distinct binding motifs [27]; however, the OmpG scaffold can theoretically be further functionalized by leveraging additional of its seven loops for affinity-reagent display. Alternative amenable monomeric sensor pore scaffolds can theoretically be used in an analogous manner, although empirical determination of optimal presentation location and linker, etc., will likely still be required [26, 28]. This chapter will cover cloning, protein expression and purification, characterization, and assay set-up for multiplexprotein detection using model analytes streptavidin and the anti-FLAG monoclonal antibody FG4R.

2 Materials

Prepare reagent stocks at room temperature (unless otherwise indicated) using ultrapure water (Advantage A10 Milli-Q purified or equivalent) and analytical-grade reagents. Stock solutions should be stored under appropriate temperature-controlled conditions. Follow all applicable handling and disposal procedures, as indicated by the manufacturer's SDS. Mechanical pipettes & associated tips, electronic pipette controller(s) & associated serological tips, water ice, and tubes with appropriate sterilization are assumed to be available in the lab space.

2.1 Molecular Biology Reagents, Solutions, and Equipment

1. Phusion™ High-Fidelity PCR Kit or equivalent, store at −20°C.
2. Template plasmid DNA containing nanopore scaffold gene: pT7/OmpG^{WT} (50–100 ng/μL), store at −20°C (*see* **Note 1**).
3. Mutagenic primers (*see* **Note 2**).
4. Bio-Rad T100 ThermalCycler or equivalent.
5. Dimethyl sulfoxide (DMSO).
6. 50× TAE buffer: 2 M Tris base, 50 mM ethylenediaminetetraacetic acid (EDTA) disodium salt, 1 M glacial acetic acid. To prepare a 1 L solution, weigh 242 g Tris base and 18.61 g EDTA disodium salt dihydrate and transfer to a glass beaker. Add 250–700 mL water to the beaker and mix gently. Measure a 57.1 mL volume of 17.5 M glacial acetic acid using a glass graduated cylinder or serological pipette and transfer to the buffer solution. Mix and add water to a final 1 L volume. Store at room temperature.
7. Agarose Low Electroendosmosis (LE).
8. 100–250 mL Erlenmeyer flask.
9. 500 mL Erlenmeyer flask.

10. Bio-Rad 7 × 10 cm Mini-Gel Tray or equivalent.
11. Bio-Rad Sub-Cell GT Gel Caster or equivalent.
12. Bio-Rad 8- and 15-Well Combs or equivalent.
13. Mini ReadySub-Cell GT Horizontal Electrophoresis System or equivalent.
14. PowerPac™ Basic Power Supply or equivalent.
15. SYBR™ Safe DNA Gel Stain.
16. Gel Loading Dye, Purple (6X).
17. 1 kb DNA Ladder.
18. Gel Doc EZ System or equivalent.
19. UV/Stain-Free Sample Tray or equivalent.
20. Eppendorf 5417 R Centrifuge or equivalent.
21. Eppendorf F-45-30-11 fixed-angle rotor or equivalent.
22. Zymo DNA Clean & Concentrator-25 or equivalent.
23. DpnI restriction enzyme, store at −20°C.
24. DH10β Competent Cells or equivalent, store at −80°C.
25. LB Agar (Lennox L Agar): To prepare 250 mL solution, weigh 8.75 g of powdered media and transfer to a glass 500 mL Erlenmeyer flask. Add 250 mL water and cover with aluminum foil (35 g/L concentration). Autoclave for ≥ 30 min using liquid cycle. After sterilization, either allow the media to cool slightly for immediate use or store at room temperature and re-melt using a hot plate or microwave. Following antibiotic addition, the LB + antibiotic plates can be stored at 4°C for several weeks, with exact lifetimes depending on the stability of the specific antibiotic(s) used.
26. Fisherbrand™ Petri Dishes with Clear Lid, 100 mm × 15 mm, sterile or equivalent.
27. LB Broth (Lennox): To prepare 900 mL solution, weigh 18 g powdered media and transfer to a 1 L screw-top bottle. Add 900 mL water and lightly, but not completely, secure the cap to the bottle (20 g/L concentration). Autoclave for ≥ 30 min using liquid cycle. Allow to cool before use; store at room temperature.
28. 100 mg/mL Ampicillin stock solution: To prepare 20 mL, 100 mg/mL Ampicillin stock solution, weigh 2 g ampicillin sodium and transfer to a glass beaker. Add 20 mL water to the beaker and mix until fully dissolved (clear to pale-yellow solution), filter for sterility using a 0.22 μm PES filter, and aliquot to 1.5- or 2-mL Eppendorf tubes. Store at −20°C.
29. Monarch® Spin Plasmid Miniprep Kit or equivalent.

2.2 Protein Biochemistry Reagents, Solutions, and Equipment

1. BL21 (DE3) Competent Cells or equivalent, store at −80°C.
2. LB Agar (Lennox L Agar), follow Sect. 2.1 preparation specifications.
3. Fisherbrand™ Petri Dishes with Clear Lid, 100 mm × 15 mm, sterile polystyrene or equivalent.
4. LB Broth (Lennox), follow Sect. 2.1 preparation specifications.
5. Ampicillin: 100 mg/mL, follow Sect. 2.1 preparation specifications.
6. 2×YT Broth: To prepare 900 mL 2×YT solution weigh 27.9 g powdered media and transfer to a 1 L screw-top bottle. Add 900 mL water and lightly, but not completely, secure the cap to the bottle (31 g/L concentration). Autoclave for ≥ 30 min using liquid cycle. Allow to cool before use; store at room temperature.
7. 100 mL Erlenmeyer flask (sterile).
8. 1 L Fernbach or Erlenmeyer flask (sterile).
9. Isopropyl β-D-1-thiogalactopyranoside (IPTG): 500 mM. To prepare 20 mL, 500 mM IPTG solution, weigh 1.19 g IPTG and transfer to a glass beaker. Add water to 20 mL final volume, filter for sterility using 0.22 μm PES filter, and aliquot to 1.5- or 2-mL Eppendorf tubes. Store at −20°C.
10. Eppendorf 750 mL Wide-Neck Bottle or equivalent.
11. Eppendorf 5810R Centrifuge or equivalent.
12. Eppendorf swing-bucket rotor S-4-104 or equivalent.
13. Q Sepharose™ Fast Flow Resin or HiTrap™ Q HP anion exchange chromatography column or equivalent, store at 4°C.
14. Econo-Pac® Chromatography Columns 14 cm high, 1.5 × 12 cm polypropylene or equivalent.
15. Nalgene™ Oak Ridge High-Speed PPCO Centrifuge Tubes, or equivalent.
16. Beckman Avanti JXN-26 IVD Refrigerated Centrifuge or equivalent.
17. Beckman Coulter JA25.50 fixed-angle rotor, 8 × 50 mL or equivalent.
18. Lysis buffer: 50 mM Tris-HCl pH 8.0 1 mM EDTA disodium. To prepare a 1 L solution, weigh 6.06 g Tris base and 372 mg EDTA disodium salt dihydrate and transfer to a glass beaker. Add ~700 mL water and adjust pH to 8.0 using concentrated HCl. Bring final volume to 1 L via addition of water and perform final pH adjustment to 8.0. Store at room temperature.

19. Pellet Wash buffer: 1.5 M urea, 50 mM Tris-HCl pH 8.0. To prepare a 1 L solution, weigh 90.1 g urea and 6.06 g Tris base and transfer to a glass beaker. Add ~700 mL water and adjust pH to 8.0 using concentrated HCl. Bring final volume to 1 L via addition of water and perform final pH adjustment to 8.0. Store at 4°C.
20. Solubilization buffer: 8 M urea, 50 mM Tris-HCl pH 8.0. To prepare a 1 L solution, weigh 480.5 g urea and 6.06 g Tris base and transfer to a glass beaker. Add ~700 mL water and adjust pH to 8.0 using concentrated HCl. Bring final volume to 1 L via addition of water and perform final pH adjustment to 8.0. Store at 4°C.
21. Wash buffer: 8 M urea, 50 mM Tris-HCl pH 8.0, and 75 mM NaCl. To prepare a 1 L solution, weigh 480.5 g urea, 4.38 g NaCl, and 6.06 g Tris base and transfer to a glass beaker. Add ~700 mL water and adjust pH to 8.0 using concentrated HCl. Bring final volume to 1 L via addition of water and perform final pH adjustment to 8.0. Store at 4°C.
22. Elution1 buffer: 8 M urea, 50 mM Tris-HCl pH 8.0, and 200 mM NaCl. To prepare a 1 L solution, weigh 480.5 g urea, 11.69 g NaCl, and 6.06 g Tris base and transfer to a glass beaker. Add ~700 mL water and adjust pH to 8.0 using concentrated HCl. Bring final volume to 1 L via addition of water and perform final pH adjustment to 8.0. Store at 4°C.
23. Elution2 buffer: 8 M urea, 50 mM Tris-HCl pH 8.0, and 500 mM NaCl. To prepare a 1 L solution, weigh 480.5 g urea, 29.22 g NaCl, and 6.06 g Tris base and transfer to a glass beaker. Add ~700 mL water and adjust pH to 8.0 using concentrated HCl. Bring final volume to 1 L via addition of water and perform final pH adjustment to 8.0. Store at 4°C.
24. 40% (w/v) Acrylamide/bis-Acrylamide, 29:1 Ratio Solution, store at 4°C.
25. Resolving buffer: 1.5 M Tris-HCl pH 8.8. To prepare 1 L solution, weigh 181.7 g Tris base and transfer to a glass beaker. Add 700 mL water and pH adjust to 8.8 using concentrated HCl solution. Bring final volume to 1 L via addition of water and perform final pH adjustment to 8.8. Store at room temperature.
26. Stacking buffer: 500 mM Tris-HCl pH 6.8. To prepare 1 L solution, weigh 60.6 g Tris base and transfer to a glass beaker. Add 700 mL water and pH adjust to 6.8 using concentrated HCl solution. Bring final volume to 1 L via addition of water and perform final pH adjustment to 6.8. Store at room temperature.

27. Ammonium persulfate (APS): 10% (w/v). To prepare 10 mL solution, weigh 1 g APS and transfer to a glass beaker. Add 10 mL water and mix. Aliquot to 1 mL volumes and store at −20°C.
28. Sodium Dodecyl Sulfate (SDS): 10% (w/v). To prepare 10 mL solution, weigh 1 g SDS and transfer to a glass beaker. Add 10 mL water and mix. Aliquot to 1 mL volumes and store at −20°C.
29. Laemmli (4×) buffer: 250 mM Tris-HCl pH 6.8, 8% (w/v) SDS, 0.02% (w/v) bromophenol blue, and 40% (v/v) glycerol. To prepare 40 mL solution weigh 1.21 g Tris base, 4.2 g SDS, 8 mg bromophenol blue, and transfer to a glass beaker. Measure a 16 mL volume of glycerol using a graduated cylinder or serological pipette and transfer to the glass beaker. Add 10 mL MQ water and mix, then pH adjust to 6.8 using concentrated HCl. Bring the final volume to 40 mL using MQ water and perform final pH adjustment to 6.8. Store at room temperature.
30. Running buffer (10×): 250 mM Tris base, 1.92 M glycine, and 1% (w/v) SDS. To prepare a 2 L solution weigh 60.57 g Tris base, 269.1 g glycine, and 20 g SDS and transfer to a large beaker. Add 1.8 L water and mix using a stir bar until all ingredients are dissolved. Solution pH should be 8.3–8.6; bring volume to final 2 L. Store at room temperature.
31. N,N,N',N'-tetramethylethane-1,2-diamine (TEMED).
32. 2,2,2-Trichloroethanol (TCE).
33. Mini-PROTEAN Tetra Cell Casting Stand & Clamps or equivalent.
34. Mini-PROTEAN® Spacer Plates with 1.0 mm Integrated Spacers or equivalent.
35. Mini-PROTEAN® Short Plates or equivalent.
36. Mini-PROTEAN® Tetra Vertical Electrophoresis Cell or equivalent.
37. Mini-PROTEAN® Tetra Electrode Assembly or equivalent.
38. PowerPac™ Basic Power Supply or equivalent.
39. Unstained Protein Standard, Broad Range (10–200 kD) or equivalent, store at −20°C.
40. Gel Doc EZ System or equivalent.
41. UV/Stain-Free Sample Tray or equivalent.
42. n-Octyl-β-D-Glucopyranoside (OG), Anagrade: 10% (w/v). To prepare 10 mL solution, weigh 1 g of OG and transfer to a glass beaker. Add 10 mL of water. Aliquot to 1 mL volumes, store at −20°C.

43. Tris buffer: 50 mM Tris-HCl pH 9.0. For 50 mL solution, weigh 303 mg Tris base and transfer to a glass beaker. Add 30 mL water and pH adjust the solution to 9.0 using a concentrated HCl solution. Adjust the final volume to 50 mL using water. Store at room temperature.
44. Refolding buffer: 20 mM Tris-HCl pH 9.0, 3.33% OG (114 mM). For 1 mL of refolding buffer, add 400 μL "Tris buffer" to a fresh 1.5- or 2.0-mL Eppendorf tube. Add 333 μL of the 10% OG solution to the same tube and bring the final volume to 1 mL via addition of 267 μL water. Make fresh the day of refolding and only prepare enough for the anticipated refolding volume; do not store.

2.3 Single-Channel Current Recording Reagents, Solutions, and Equipment

1. Polytetrafluoroethylene (PTFE) film, thickness 0.025 mm, size 150 × 150 mm or equivalent.
2. 18-gauge disposable needle or borosilicate glass with filament or equivalents.
3. Daedalon Universal spark generator or equivalent.
4. DOWSIL™ 3145 RTV MIL-A-46146 Adhesive/Sealant or equivalent.
5. Custom non-insulating polymer chambers.
6. Silver wire 99.998%, 1.5 mm diameter or equivalent.
7. Red Machine Tool Wire PVC, 16 AWG, 26 Strand, or equivalent.
8. Conn pin 20-24 AWG gold crimp.
9. Harris Solder wire, 1/32 in diameter, 97% tin/3% copper, or equivalent.
10. Weller WLC200 Soldering Station.
11. HAKKO Benchtop Soldering Fume Absorber.
12. Concentrated HCl or Clorox germicidal bleach.
13. Axopatch 200B Microelectrode Amplifier.
14. Digidata® 1440A Low-noise Data Acquisition System.
15. Aperture Pretreatment solution: 10% (v/v) Hexadecane in pentane. Measure a 900 μL volume of pentane, 99% and transfer into a glass screw top vial. Measure a 100 μL volume of hexadecane, 99% and mix into the pentane. Solution can be stored at −20°C to 4°C.
16. Recording buffer: 300 mM KCl 50 mM sodium phosphate dibasic (*see* **Note 3**). Weigh 2.34 g potassium chloride (KCl) and 1.34 g sodium phosphate dibasic heptahydrate (Na_2HPO_4) and transfer to a glass beaker. Add 80 mL water and pH adjust the solution to 6.0. Bring volume to 100 mL using

water and filter for sterility using 0.22 μm PES filter. Store at room temperature.

17. Lipid solution: 10 mg/mL 1,2-diphytanoyl-sn-glycero-3-phosphocholine (4ME 16:0 PC, DPhPC) in pentane. Purchase powder and resuspend in pentane to indicated concentration. Solution can be stored at −20°C to 4°C for several days to weeks.
18. Streptavidin stock: 29 μM in 20% glycerol. Prepare a 20% (v/v) glycerol solution. Measure 650 μL of the 20% glycerol solution and add it to the 1 mg protein aliquot to achieve the desired stock concentration. Solution can be stored at −20°C long-term and 4°C temporarily.
19. DYKDDDDK Tag monoclonal antibody (FG4R) RRID AB_1957945. Solution can be stored at −20°C long-term and 4°C temporarily.

3 Methods

3.1 Scaffold Engineering—Motif Grafting to Sensor Pore

This section describes the molecular biology procedures used to generate a multiplex nanopore sensor construct by integrating affinity peptides directly into the coding sequence of the selected scaffold, in this case utilizing loops 3 and 6 of the OmpG scaffold (Fig. 1a). Integration is performed using overlap-extension PCR (Fig. 1b)

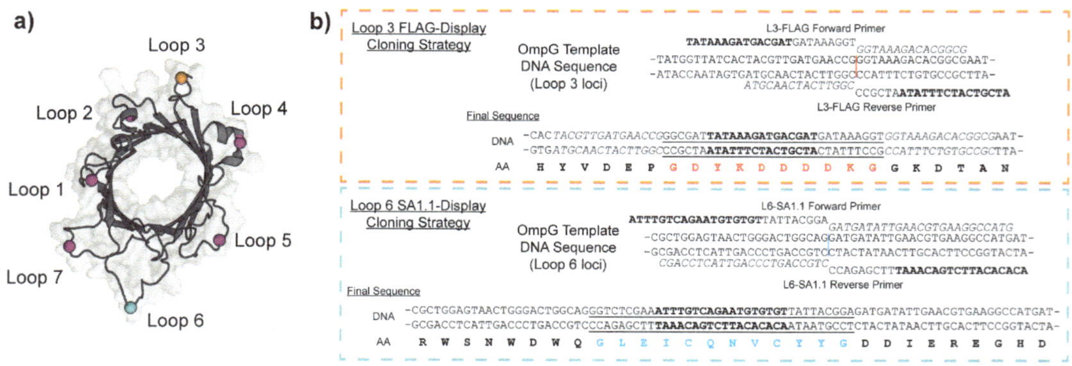

Fig. 1 OmpG Motif Insertion Sites and Cloning Strategy. (**a**) Top-down view of OmpG, with the "center" of each loop denoted by colored spheres to indicate possible motif insertion sites (PDB: 5MWV). The loops utilized in this detection strategy, loops 3 and 6, are highlighted in orange and blue, respectively. (**b**) Schematic of the overlap-PCR cloning strategy employed to introduce the indicated affinity motifs into the designated loop loci. Motif insertion sections are color-coded to match their respective loops, with the insertion position in the sequence indicated by a correspondingly colored vertical line. Italicized sequences indicate primer-template homology, bolded sequences indicate overlap homology generated in the inserted region, and underlined sequences indicate the final inserted motif

with Phusion™ polymerase as a representative high-fidelity cloning method (though any preferred technique may be substituted). Scaffold functionalization strategies vary widely in their yield, folding stability, and target accessibility; therefore, selecting appropriate affinity ligand(s) and precise insertion site(s) is critical in achieving optimal sensor performance (*see* **Notes 4 and 5**). Once affinity ligand(s) and insertion site(s) have been determined, proceed with the following steps to construct a multiplex-capable nanopore sensor:

1. Set up a 25 μL PCR reaction (scalable based on replicate number) using Phusion™ or another high-fidelity DNA polymerase (*see* **Note 6**): 5 μL 5X Phusion HF or GC buffer, 0.5 μL 10 mM dNTP mix, 1 μL Forward Primer(s) at 10 μM (*see* **Note 7**), 1 μL Reverse Primer(s) at 10 μM (*see* **Note 7**), an arbitrary volume of template DNA to a final concentration 0.3 ng/μL, 0.3 μL Phusion™ polymerase (add last), 0.75 μL DMSO (3%, optional) and nuclease-free water to 25 μL in a 0.2 mL PCR tube. Then put the tube in a ThermalCycler for PCR reactions with the following thermocycling parameters. Lid temperature ≥105°C for all steps. Step 1, Initial denaturation: 98°C for 30 s. Step 2, Denaturation: 98°C for 5–10 s. Step 3, Annealing: XX°C (set according to user's primer Tm) for 10–30 s. Step 4, Elongation: 72°C for 15–30 s/kb of product. Repeat Steps 2–4 for 25–35 cycles. Step 5, Final elongation: 72°C for 5 min. Step 6, Storage: 4–12°C indefinite hold.
2. Prepare a 1% TAE agarose gel by mixing 0.4 g LE agarose with 40 mL 1× TAE buffer in a 100–250 mL flask. Dissolve the agarose by heating, either using hot plate or microwave; ensure gentle mixing to mitigate rapid boiling. Cool slightly, add 1.7 μ L SybrSafe visualization reagent, mix gently by swirling and pour onto leveled Bio-Rad Sub-Cell GT Gel Caster holding a Bio-Rad 7×10 cm Mini-Gel Tray. Insert well-comb (10–15) carefully to avoid introducing air bubbles. Wait ~20 min for gel to cool and polymerize. Remove the well-comb and transfer the 1% gel into a Mini ReadySub-Cell GT Horizontal Electrophoresis System, ensure the gel is fully submerged in 1×TAE. Mix 3–10 μL of PCR reactions ratiometrically with Gel Loading Dye, Purple (6×) to achieve a 1× solution based on the user's desired input volume. Load sample into corresponding lane(s), leaving at least one lane available for nucleic acid ladder. Perform electrophoresis until the dye front has reached the bottom of the gel.
3. Following electrophoresis, carefully transfer the gel to an applicable imaging device such as a Gel Doc EZ System or similar, and visualize the PCR product and its purity using the UV/Stain-Free Sample Tray or equivalent method.

4. If the reaction is successful, as indicated by molecular weight and purity, proceed with product purification and subsequent DpnI digestion to remove template DNA. Purify PCR product(s) using Zymo Clean & Concentrate kit, following manufacturer's instructions, or an equivalent method. Measure DNA concentration using a NanoDrop. Following manufacturer's instructions, perform DpnI digestion on PCR product(s) to remove template DNA. Re-purify digested DNA using the same method and measure concentration.
5. Transform template-depleted DNA into an *E. coli* cloning strain (DH10β or other similar) and plate on LB agar supplemented with plasmid-appropriate antibiotic in Fisherbrand™ Petri Dishes with Clear Lid or similar vessel. Allow plates to incubate 12–16 h (overnight) at 30°C–37°C for colony growth. If desired, perform serial dilution plating to ensure single colony dispersion if high efficiency is anticipated.
6. Using a sterile toothpick or other implement, pick an independent colony to replica-streak onto an LB agar plate supplemented with the appropriate antibiotic, followed by using the same implement to subsequently inoculate a 3–5 mL volume of LB media supplemented with the same antibiotic contained in a sterile 50 mL Falcon tube or similar vessel. Repeat with as many colonies as desired, typically one to three is sufficient. Allow these cultures to incubate at 30°C–37°C/220–250 rpm for 6+ h (overnight is fine) to ensure adequate cellular density for plasmid harvesting. Allow plates to incubate 12–16 h (overnight) for colony growth. After colonies are observed plates can be stored at 4°C for days to weeks.
7. After cultures have reached sufficient OD_{600} cellular density, follow manufacturer's instructions on your miniprep kit of choice or other plasmid extraction protocol to perform plasmid extraction. Perform Sanger, or other sequencing, on extracted plasmid(s) to verify that sensor scaffold now carries all mutation(s)/affinity motifs of interest (*see* **Notes 8 and 9**).

3.2 Expression, Purification, and Refolding of Engineered Sensor Pore

The section describes the biochemical protocols for the expression and purification of an engineered monomeric nanopore sensor. OmpG, or other scaffolds, are expressed as an insoluble inclusion body (*see* **Note 10**), which is solubilized in 8 M urea, purified in a denatured form using a Q-Sepharose column (leveraging OmpG's high negative charge; Ni-NTA may be used if the construct carries a His-tag) and then refolded (*see* **Note 11**) by dilution of the urea and introduction of the detergent octyl glucoside (OG) above its critical micelle concentration (*see* **Note 12**). The refolded protein is then evaluated by SDS-PAGE gel-shift assay and/or single-channel current recording. This refolding strategy is broadly applicable to monomeric β-barrel sensor pores [29], although detergent choice and dilution kinetics may require scaffold-specific optimization. Oligomeric pores impose additional purification and assembly challenges that fall outside the scope of this chapter.

1. Transform your sequence-confirmed plasmid of interest containing the engineered pore into an applicable *E. coli* expression strain, such as BL21 (DE3) (*see* **Note 13**), and plate on LB agar supplemented with the plasmid-appropriate antibiotic selection in Fisherbrand™ Petri Dishes with Clear Lid or similar vessel. Allow to incubate 12–16 h (overnight) at 30°C–37°C for colony formation.
2. Once colonies can be observed, pick, using a sterile toothpick or pipette tip, a single isolated colony and utilize it to inoculate a 10–50 mL volume of 2xYT (or other rich media) contained in a sterile 50 mL Falcon tube or 100 mL Erlenmeyer flask and supplemented with the applicable antibiotic (*see* **Note 14**). Incubate this "Primary culture" at 30°C–37°C/220–250 rpm for 6+ h (overnight is fine) until significant density is observed, monitor cell density at OD_{600}.
3. Prepare 300–500 mL 2xYT or other rich media supplemented with the applicable antibiotic, contained in a sterile 1 L Fernbach or Erlenmeyer flask, to be referred to as the "Expression Culture." Measure the OD_{600} of the dense "Primary culture" and transfer a calculated volume to the 300–500 mL media "Expression Culture" such that an OD_{600} value of 0.01–0.075, or other low-density value is achieved.
4. Incubate the "Expression culture" at 30°C–37°C/220–250 rpm for 2–5 h or until an OD_{600} value of 0.4–0.7 is reached, with 0.6 considered ideal. Once the desired density is reached, induce protein expression using the appropriate induction compound, with most systems induced by IPTG to a final concentration of 100–500 μM. After induction continue incubating at 30°C–37°C/280 rpm for 3–4 h.
5. Once the induction timeframe is completed transfer the entirety of the culture to an Eppendorf 750 mL Wide-Neck Bottle, generate a counterbalance using an equivalently sized container filled with water. Load your culture and balance into an Eppendorf 5810 R centrifuge equipped with an Eppendorf S-4-104 swing-bucket rotor and harvest the culture via centrifugation at 4000 rcf/4°C/10–20 min. Decant the supernatant (*see* **Note 15**).
6. Prepare an affinity column: either pack a new Q-Sepharose anion-exchange column by adding the resuspended bead slurry into a prepared Econo-Pac column for gravity purification, or mount a prepacked Q-Column for FPLC run. A 2–3 mL bead volume is sufficient for purification(s) at the discussed scale; larger bead volumes or additional columns

can be used if larger/multiple expressions are utilized. Equilibrate the beads with three to five column volumes (CV) of "Solubilization buffer."

7. Resuspend the cell pellet from your expression in 30 mL "Lysis buffer" and transfer to glass beaker or 50 mL Falcon tube. Agitate using a stir bar at high speed or vortex to homogenize. Measure the OD_{600} density of the resuspended cells; a decrease in this value can be monitored during the processing as a metric of lysis progress/efficiency. Lyse the resuspended cells via sonication or other user-selected lytic method: ½ inch tip, 40% amplitude, 2 s on/4 s rest, 2 min process time repeated to OD_{600} decreases to $\leq$ 75% starting value (*see* **Note 16**).
8. Centrifugation to remove soluble protein. After the cells have been lysed transfer the solution to a respectively labeled 50 mL Nalgene tube for centrifugation to pellet the insoluble proteins for extraction. Centrifuge at 50000 rcf/4°C/20–30 min. For counterbalance preparation measure the weight of your sample in its container and use that value to tare your weight station. Obtain a fresh container of the same style as you are currently using, transfer it to the weight station along with the cap, and then fill using water until the weight station reads zero. Following centrifugation, remove all the supernatant. Retain some volume for analysis as "Soluble fraction."
9. Pellet wash. Resuspend the remaining cell pellet in 30 mL of "Pellet wash buffer" to remove any easily denatured proteins. Break up the pellet in solution using a glass stir rod then transfer to a 50 mL beaker with a stir bar and mix at 1000 rpm for ~20–30 min to ensure solution is homogenous. After the resuspension step is completed transfer the solution to a new 50 mL Nalgene tube and centrifuge again using the same parameters. Following centrifugation, remove the supernatant. Retain for analysis as "Pellet wash fraction."
10. Pellet solubilization. Resuspend the remaining cell pellet/inclusion body pellet in 30 mL "Solubilization buffer". Break up the pellet in solution using a glass stir bar then transfer to a 50 mL beaker with a stir bar and mix at 1000 rpm for ~20–30 min to ensure solution is homogenous. After the solubilization step is completed, transfer the solution to a new 50 mL Nalgene tube and centrifuge again using the same parameters.
11. Column binding. Following centrifugation, apply the supernatant to the equilibrated Q-Sepharose column, either allowing the solution to flow under gravity or using the system sample pump. Any residual protein/compounds in the Nalgene tube can be resuspended in ~30 mL of water via

vortexing and retained for analysis (labeled "Residual pellet fraction"). Retain the flow-through from this step for analysis (labeled as "FT fraction" assess column loading/binding).

12. Column-bound protein is now washed via the addition of 10 CVs "Wash buffer." Fraction collected as "Wash fraction" for analysis.
13. Column-bound protein is now further treated via the addition of 1.5 CVs "Elution1 buffer." Fraction collected as "Elution1 fraction" for analysis.
14. Any remaining column-bound protein is now eluted via the addition of 1.5 CVs of "Elution2 buffer." Fraction collected as "Elution2 fraction" for analysis.
15. SDS-PAGE preparation, 4/12% SDS-PAGE 1 mm: Prepare 12% resolving gel— recipe for one 5 mL gel but volume can scale with number of gels desired. Into a disposable container add 2.15 mL water, 1.25 mL "Resolving buffer", 1.5 mL 40% acrylamide, 25 μL TCE, 50 μL 10% APS, 50 μL10% SDS, and 5 μL TEMED and mix by inverting 5–10×, polymerization is initiated by TEMED, so add last and proceed rapidly from that point. Cast mixed solution into 1 mm Mini-PROTEAN® Spacer Plate and Mini-PROTEAN® Short Plate held within a Mini-PROTEAN Tetra Cell Casting Stand & Clamps (*see* **Note 17**). Prepare a 4% stacking layer — recipe for 2.5 mL gel but volume can scale with number of gels desired. Mix 1.59 mL water, 250 μL 40% acrylamide, 630 μL "Stacking buffer," 25 μL 10% APS, 25 μL 10% SDS, and 2.5 μL TEMED into a disposable container, such as a 50 mL Falcon tube. Pouring the solution over polymerized 12% layer, insert well comb carefully to avoid introducing any air bubbles.
16. SDS-PAGE electrophoresis. Transfer cast gel to a Mini-PROTEAN® Tetra Electrode Assembly and latch in using built in clamps. Load into Mini-PROTEAN® Tetra Vertical Electrophoresis Cell, remove well-comb, and submerge in 1× "Running buffer". Mix 5–25 μL of fractions of interest ratiometrically with Laemmli (4X) buffer to achieve a 1× solution based on the user's desired input volume. Load samples into corresponding lanes, leaving at least one lane available for protein ladder. Perform electrophoresis at 180 V for 60 min or until the dye front has reached the bottom of the gel.
17. Gel visualization & analysis. Following electrophoresis carefully transfer the gel to an applicable imaging device to visualize the protein product and its purity (*see* **Note 18**). If desired yield and purity are achieved, retain the fraction(s) of interest for refolding of the OmpG construct. Denatured protein can be stored at -80°C for months to years.

18. Protein refolding. Transfer 625 µL of "Refolding Buffer" into a fresh 1.5–2 mL Eppendorf tube. Add 375 µL of the OmpG fraction of interest and mix gently by pipetting. This step reduces the urea concentration in the sample and introduces OG to a final concentration of ~72 mM (*see* **Note 19**). Allow the refolding solution to incubate overnight at 37°C.
19. Refolding assessment. Prepare another 4% stacking/12% resolving SDS-gel as in Step 15 and load samples representing the OmpG construct in the denatured and detergent-refolded states (*see* **Note 20**). Perform electrophoresis until the dye front has reached the bottom of the gel and visualize the gel as before. Successfully refolded protein is SDS-resistant and will exhibit altered apparent electrophoretic mobility (Fig. 2, *see* **Note 11**). Successfully refolded protein can be utilized in single-channel current recordings for subsequent characterization and multiplex sensing; refolded protein samples can be stored at 4°C for several weeks to months.

3.3 Single-Channel Current Recording of Engineered OmpG Multiplex Sensor

This section describes the procedure for using engineered monomeric nanopore sensors in single-channel current recording multiplex protein detection. Following successful refolding, the engineered OmpG or alternative pore-scaffold multiplex sensor is assessed by single-channel current recording. For optimal control comparisons, constructs containing individual affinity motifs used in the multiplex sensor, as well as the wild-type pore scaffold, should also be expressed, purified, and characterized similarly.

1. Aperture preparation. An 18-gauge needle or thin glass filament is used to pierce a 25 µm thick PTFE film. The resulting hole is then treated with 3–7 applications of a high-voltage

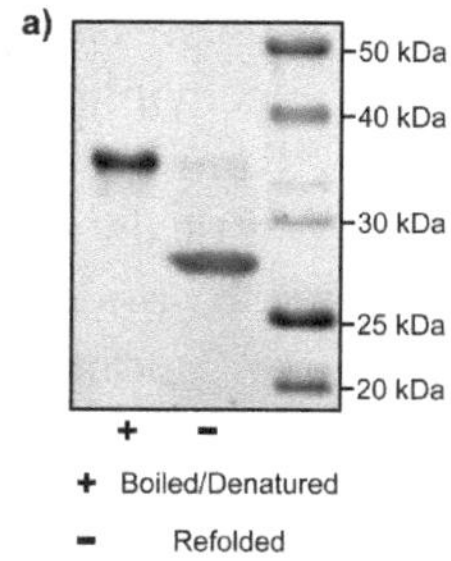

Fig. 2 Refolding Assessment of $OmpG^{WT}$. (**a**) SDS-PAGE (12%) with TCE-mediated protein visualization, showing the altered electrophoretic mobility between heat-denatured (+) and detergent-refolded (–) samples. The NEB broad range protein ladder was included for molecular weight reference. The gel was run at 180 V for 60 min, followed by TCE activation for 5 min, with imaging achieved using a Bio-Rad Gel Doc EZ system with UV/Stain-Free Sample Tray under auto-settings (~5 s exposure)

spark to smooth the geometry, producing an aperture of 100 μm diameter (*see* **Note 21**).

2. Chamber assembly. The PTFE film is mounted between two half-bowl chambers made of a nonconductive, waterproof polymer, sealing the interface between the chambers such that the aperture serves as the only shared access point. A recessed O-ring groove filled with Dowsil™ 3145 ensures a watertight seal at the PTFE interface, and the assembled "chip" is secured with screws.
3. Electrode preparation. Silver(Ag)/Silver-chloride(AgCl) electrodes are generated by soldering silver to one end of copper wire and a gold connector to the other. Following electrode assembly, the silver should be immersed in a concentrated HCl or bleach solution to build up the necessary chloride coating at the electrode surface (*see* **Note 22**). Caution: Ensure chloride solutions do not contact solder joints.
4. Data acquisition setup. Recordings are acquired using an Axopatch 200B amplifier and Digitdata 1440A digitizer, controlled via Clampex software on a Windows workstation. A custom Faraday cage was assembled into which the Axopatch headstage was mounted. Electrodes are inserted into the Faraday cage-mounted headstage: one electrode serves as ground ("*cis*") and the other applies voltage ("*trans*") (see **Note 23**). Axopatch will be operated in "Voltage-clamp" mode, sampling at 10 kHz, with a 2 kHz Bessel filter applied.
5. Aperture pretreatment. Add 0.5–1.5 μL of 10% (v/v) hexadecane in pentane to each side of the aperture to pretreat the surface. Place the chip into the Faraday cage immediately after treatment.
6. Buffer addition. Fill each chamber with "Recording buffer" until the aperture is fully submerged. The required volume depends on the chip design, but generally 300–600 μL will be sufficient.
7. Lipid bilayer formation. Apply 10–20 μL of "Lipid solution" to the surface of each chamber. Monolayer and bilayer formation is achieved by pipetting to modulate the chamber volume [30]: Keep "Lipid solution" on ice. Start with one chamber and modulate the chamber volume using a pipette of applicable volume capacity while monitoring the current readout on the connected computer. When the volume is brought below the aperture point, if properly treated, the lipids should adhere and form a mono/bilayer, which is visualized as the system going from saturating current to a zero-current state. Repeatedly modulate the chamber volume until a stable current-occluding layer can be formed multiple times. Verify that any current-occluding layers observed are "zappable"

by the high-current overload function of the Axopatch. Unzappable, oily layers can be removed by angled pipetting "washing" of the aperture one to three times. Once stable but "zappable," layers can be formed in one chamber repeat the process in the opposite chamber until the same result is achieved. Capacitance Check: A stable bilayer should exhibit a capacitance between 60–120 pF for a ~100 μm aperture.

8. Pore insertion. After successful planar lipid bilayer formation a volume of the refolded sensor pore solution is added to the *cis* chamber to achieve a final concentration of 100–500 pM. Apply +250 mV to promote insertion via membrane thinning. Alternative methods of membrane thinning, such as varying current waveforms, may be preferential for different sensor pores. The system current should be monitored in real time for fluctuations indicative of pore insertion, at which point the voltage is immediately reduced, ±50 mV for OmpG, to reduce bilayer destabilization disfavoring further insertion (*see* **Note 24**). If excessive or multi-pore insertions occur, the chamber should be re-prepared, a smaller amount of protein added, and the insertion protocol repeated.
9. Initiate recording. A short duration of the pore behavior without any analytes (Fig. 3a) should be recorded for quality control purposes (Fig. 3b); thirty seconds to several minutes of the recording voltage potential at both polarities is generally sufficient (*see* **Note 25**). Proceed with sample dosing to one of the chambers determined from the pore orientation and sensing method in the system (*see* **Note 26**). The sensor in this work is based on Foster et al. and was capable of mediating the detection of model analytes streptavidin and mAB FG4R (Fig. 3a) [27]. A representative experiment carried out using these analytes would add a volume of 29 μM stock to achieve a final concentration of ~100–1600 nM streptavidin and/or a volume of 6.6 μM stock FG4R to achieve a final concentration of 10–30 nM FG4R. Following the addition of these or other analytes, the chamber would then be gently but thoroughly mixed by pipetting 1/10th to 1/5th the chamber volume 10–20× times. Single-channel current recording can proceed for a user-defined duration (Fig. 3c–g), typically based on the sampling of multiple discrete interactions of the analyte(s) with the sensor pore sufficient to perform fitting assessments, or until bilayer breakage, additional pore insertion, or other detrimental occurrence renders the subsequent recording duration non-useful.
10. Analyze single-channel recording data. Common analytical parameters include: interevent duration, or "capture frequency",

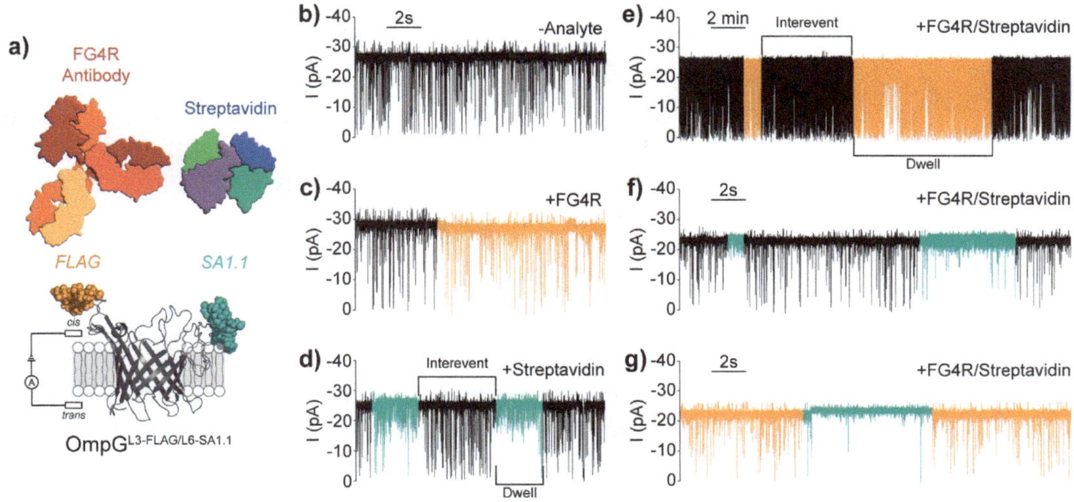

Fig. 3 Multiplex detection of two protein analytes by OmpG$^{L3\text{-}FLAG/L6\text{-}SA1.1}$ nanopore. (**a**) Schematic representation of OmpG$^{L3\text{-}FLAG/L6\text{-}SA1.1}$ sensor for single-channel multiplex detection. (**b–d**) Representative OmpG$^{L3\text{-}FLAG/L6\text{-}SA1.1}$single-channel current recordings: (**b**) basal behavior, (**c**) FG4R binding (20 nM), and (**d**) streptavidin binding (800 nM). (**e** and **f**) Representative single-channel current recording traces showing recapitulation of individual analyte signals when analytes are present in a simple mixture. (**g**) Concurrent observation of both binding signals. Boundary regions used for determination of common analytical parameters (interevent interval and dwell time) are indicated for each signal class. Experiments were performed in 50 mM Na_2HPO_4 pH 6.0 buffer containing 300 mM KCl at an applied potential of ±50 mV. Traces were filtered using a 500 Hz lowpass digital Gaussian filter.(Figure was adapted from [27])

defined as the duration between event interactions — i.e., from when binding interaction of a defined type stops until the next event of the same type begins (Fig. 3d, e). Dwell, or the duration that each discrete event lasts, defined from when the signal perturbation that defines an interaction type begins until it ends (Fig. 3d, e). I_{res}, or the current level occupied by the bound event or I_B divided by the average unitary current of the sensor pore I_0 ($I_B/I_0 = I_{res}$). The event I_{res} itself may have associated parameters such as RMSD or even multiple levels/states with their own RMSD and transition kinetics. It is likely analysis parameters and the optimal weight/consideration of each will be empirically determined by the specific functionalized sensor and analyte(s) assessed, although these general parameters can be used as guidelines. The specific analysis of these analytes will not be discussed here; for a more detailed breakdown of the experimental analysis, see Foster et al. [27]. The interpretation of the data will depend on the user's systems and potentially the biological relevance of detected analytes or any relationship of their presence/ratio.

4 Notes

1. The pT7 vector used here confers ampicillin resistance, with a typical selective concentration of 100–200 μg/mL, normally achieved via a 1:1000 or 1:500 dilution of a 100 mg/mL stock into the applicable media. When using alternative vectors, co-transformations, or specific expression strains, different or additional antibiotics may be required, each with its own effective concentration range.
2. Primers described here were designed using overlap extension PCR to introduce the desired amino acid sequences into the respective loops. Following this strategy, users should design primers containing both loop sequence homology and the intended motif sequence for their scaffold and affinity motif, though alternative cloning approaches may also be employed. The specific primers used for the OmpG mutagenesis are as follows: Loop 3 (FLAG motif insertion)—forward 5′-TATAAA GATGACGATGATAAAGGTGGTAAAGACACGGCG-3′ & reverse 5′-ATCGTCATCTTTATAATCGCCCGGTTCATCA ACGTAG-3′. Loop 6 (SA1.1 motif insertion); forward 5′-AT TTGTCAGAATGTGTGTTATTACGGAGATGATATTGAA CGTGAAGGCCATG-3′ & reverse 5′-ACACACATTCTGA CAAATTTCGAGACCCTGCCAGTCCCAGTTACTCCAG C-3′.
3. Alternative buffers or salts may be substituted depending on system or user requirements; however, such changes can significantly affect pore conductance and gating behavior [31, 32]. Therefore, when buffer conditions are altered, new baseline recordings must be established to ensure accurate comparisons. A more detailed discussion of recording conditions and considerations is provided in **Notes 22–24**.
4. Outer membrane protein G (OmpG) is a monomeric scaffold that utilizes its seven flexible loops as display sites for affinity motifs. The following are a few critical aspects that should be considered during sensor development. (1) Display loop selection: the local environment of each loop can favor or disfavor the presentation of specific affinity ligands or motifs based on their chemical properties (e.g., a highly positive motif may not be optimal on a negatively charged loop). Additionally, motif insertion can significantly affect the pore's gating behavior depending on the selected loop [26, 27], such that an otherwise effective display site may exhibit unfavorable gating. The optimal loop to display an affinity reagent generally requires empirical determination. (2) Display site within a loop: While the central region of a loop is commonly

used for motif integration, positional shifts within the loop can modulate resolution and sensor performance, with a central display site not always performing best [26], indicating that both the loop and motif position may require empirical optimization to achieve optimal sensor function. (3) Motif length or linker length/flexibility: The bound analyte must be close enough to the sensor pore to be sampled by the flexible loops and accordingly alter their gating behavior [26]. If an affinity ligand is connected via a linker that is too long analyte binding will not translate to altered pore behavior [28]. Relatedly, motif accessibility by flexible linkers was shown to improve binding with certain analytes [26]. (4) Multiplex arrangement: As a single-molecule technique, spatial accessibility can limit simultaneous analyte binding, making strategic motif placement critical to minimize perturbation of detection kinetics in mixtures. In OmpG, positioning motifs on maximally distant loops 3 and 6 allowed concurrent binding without affecting the kinetics of either analyte, likely influenced by analyte identity (streptavidin and mAb) and their binding modes. In contrast, placing motifs closer together (e.g., loops 3 and 4) resulted in one analyte dominating occupancy, determined by motif affinity and analyte concentration.

5. For oligomeric scaffolds, heterogeneous motif display can be achieved by co-expressing monomeric subunits engineered with different affinity reagents [33, 34]. However, this approach adds complexity, such as controlling subunit stoichiometry and assembly fidelity, and is beyond the scope of this chapter.
6. Reaction setup should be initiated by transferring the water volume, with primers and polymerase added last to minimize unwanted dimerization and nonspecific amplification; the order of addition for other components is less critical.
7. When incorporating two motifs, either a single PCR reaction using all primer sets (if melting temperatures are compatible) can be performed, or constructs carrying individual motifs can be generated sequentially, with the first serving as the template for a second round of cloning. Although slower, the sequential approach also provides single-motif controls for comparison.
8. Sequencing primers should flank any region(s) targeted for mutagenesis or insertion/deletion. For most expression plasmids utilizing the T7 promoter system, primers binding to the T7 promoter [5′-TAATACGACTCACTATAGGG-3′] and terminator [5′-CTAGTTATTGCTCAGCGGT-3′] sequences are convenient for this purpose.

9. If sequencing results show only the parental (template) sequence, additional DpnI digestion may be required. This may involve increasing the amount of enzyme, extending the digestion time, or reducing template input to avoid overloading the reaction. Alternatively, a second PCR with minimized template DNA can further reduce the likelihood of template carryover.
10. For nanopore sensing purposes, OmpG and other bacterial membrane proteins are often expressed without their native N-terminal signal sequence (OmpG, -KKLLPCTALVMCAGMACAQA-). Without this secretion motif, the proteins do not integrate into the outer membrane and instead accumulate in the cytoplasm as insoluble inclusion bodies. While this expression mode yields significantly more protein than membrane integration would allow, the resulting material is misfolded.
11. OmpG and other β-barrel proteins can be solubilized in denaturant and subsequently refolded into a stable barrel structure upon dilution of the denaturant and addition of an appropriate detergent. Refolded OmpG exhibits altered electrophoretic mobility, typically running at a lower apparent molecular weight than the denatured form. When working with novel scaffolds, empirical determination of this shift may be necessary [29].
12. If an alternative scaffold is used, OG may not be the optimal detergent for protein refolding. Commonly employed detergents include Lauryldimethylamine N-oxide (LDAO), n-dodecyl β-D-maltoside (DDM), and Genapol® X-080. Few empirical rules exist to guide detergent selection for novel membrane proteins, so a trial-and-error approach using standard detergents is typically employed.
13. Ensure compatibility between the expression vector's promoter and the expression strain used. Vectors containing a T7 promoter (e.g., pET series) require a strain with T7 RNA polymerase expression, such as BL21 (DE3). If expression toxicity is anticipated, alternative strains like C43 or use of suppressor plasmids (e.g., pLysS) may be beneficial.
14. Primary culture volume is generally arbitrary, with 10–50 mL sufficient for most expressions, although it can be adjusted based on user requirements.
15. The protocol can be paused after cell harvesting. The pellet may be stored directly at −20°C for weeks, or 4°C for several days. Alternatively, the pellet may be resuspended in lysis buffer and stored at 4°C overnight for next-day processing.
16. Selecting the proper probe size is important for optimal lysis sonication, as both sample volume and vessel geometry

determine the applicable probe. For example, a ½″ probe is suitable for volumes of approximately 20–250 mL, which should be placed in a narrow rather than wide vessel to maximize energy transfer. For the lysis described above, transfer the resuspended pellet to a 50 mL glass beaker and position it centrally below the sonication tip in a Styrofoam box filled with ice. Using a jack, raise the beaker all the way up until the bottom of the container is flush with the tip then—using the displacement of the tip as a metric—gently lower the jack until the tip of the sonicator probe is within the middle-to-upper third of the cell resuspension volume. Initiate lysis using the indicated pulse mode settings to minimize heat generation, alternative amplitude and pulse duration settings may be applied at the user's discretion. During and between cycles, realign the beaker as needed to account for shifting in the ice, ensuring that the probe never contacts the beaker walls.

17. To ensure a flat interface between the resolving and stacking layers, a small volume of isopropanol may be applied to the top of the resolving gel before polymerization to "self-level" the surface. Once polymerization is complete, pour off or otherwise remove this top isopropyl alcohol containing layer to prevent issues with polymerization of the 4% stacking gel.
18. This protocol uses a pre-stained gel containing TCE, which reacts with tryptophan residues for protein visualization [35]. If the protein lacks tryptophan or an alternative method is preferred, a traditional Coomassie stain/destain protocol can be used.
19. Volumes may be adjusted as needed; however, maintaining the final component concentrations and ratios is critical. The urea concentration should be reduced to ~3 M, while OG should be maintained at ~72 mM (~3× its critical micelle concentration).
20. To confirm that SDS-PAGE mobility shifts reflect conformation rather than proteolysis, a boiled control of the refolded protein can be included. Restoration of the original electrophoretic profile upon boiling verifies that the mobility shift is reversible and structure dependent.
21. Smaller (60–80 μm) and larger (120–150 μm) apertures may be used depending on experimental goals. Smaller apertures generally produce thicker bilayers, reducing insertion efficiency but enabling longer recordings, while larger apertures increase the likelihood of multiple pore insertions and are suited for population-level studies. Using multiple chips with different aperture sizes is recommended when working

with different sensor pores or pursuing varied experimental objectives.

22. Over time, electrode performance may degrade. Rinse with deionized water to remove debris and perform routine re-chlorination (minutes to hours) to restore function. If performance does not improve, sand down the silver surface to restore it and repeat chlorination. Persistent issues may require full electrode replacement by removing the old, soldered silver and following the original preparation procedure.

23. This *cis/trans* designation is critical, and many pores exhibit asymmetric current-voltage profiles useful for assigning pore orientation [36]. OmpG can insert bidirectionally, and accordingly the orientation must be determined from the observed asymmetric voltage and gating profile, indicating which chamber contains the affinity-reagent functionalized loops. The difference in gating probability between voltage polarities (+/−) is referred to as quiet (Q, less gating) and noisy (N, more gating) behavior. N-traces exhibit higher conductance when compared with the Q-trace at the opposite potential. For example, a pore exhibiting a Q-trace at positive potential and an N-trace at negative potential is designated Q^+/N^-; the opposite orientation is Q^-/N^+. In our system, a Q^+/N^- pore is considered *cis*, while Q^-/N^+ is considered *trans*.

24. The unitary conductance of your sensor pore must be known to determine if you are working with a single pore; it may be necessary to establish this value empirically prior to single-channel recordings with analyte(s). A standard approach involves adding higher concentrations of protein to the system to promote multiple pore insertions into the membrane, which appear as stepwise increases in current when monitored at a fixed voltage. The average size of these steps corresponds to the conductance of individual pore units.

25. If the sensor has not been previously characterized, basal recordings of its conductive and gating behavior should be performed, with recording duration guided by intrinsic gating and expected signal frequency. These recordings serve as controls to confirm that signal perturbations following analyte addition are specific and that the frequency of these observed perturbations should demonstrate dose dependency. Further negative controls, including analyte dosing to pores lacking motifs or containing the opposite motif to the dosed analyte verify that observed signals are specific and selective.

26. As OmpG sensing is mediated by interaction(s) between loop-displayed affinity reagents and the target analyte, the analyte or sample should be added to the chamber exposing these

loops; for a *cis*-oriented pore (loops facing the *cis* grounded electrode), the sample is added to the *cis* chamber. Alternative motif presentation strategies, scaffolds, or nanopore sensing modes (e.g., trapping, threading) will influence both the chamber for analyte addition and the optimal voltage magnitude and polarity to mediate interaction with the sensor pore.

Acknowledgments

This work was supported by NIH grants T32-GM008515 and T32-GM139789 (to J.C.F), and R01-GM115442 (to M.C), in addition to the US Department of Agriculture National Institute of Food and Agriculture Grant AFRI 2018-07590.

References

1. Varongchayakul N, Song J, Meller A, Grinstaff MW (2018) Single-molecule protein sensing in a nanopore: a tutorial. Chem Soc Rev 47:8512–8524
2. Wang Y, Zhang Y, Chen X, Guan X, Wang L (2021) Analysis with biological nanopore: on-pore, off-pore strategies and application in biological fluids. Talanta 223:121684
3. Liu L, Li T, Zhang S, Song P, Guo B, Zhao Y et al. (2018) Simultaneous quantification of multiple cancer biomarkers in blood samples through DNA-assisted nanopore sensing. Angew Chem Int Ed 57:11882–11887
4. Restrepo-Pérez L, Wong CH, Maglia G, Dekker C, Joo C (2019) Label-free detection of post-translational modifications with a nanopore. Nano Lett 19:7957–7964
5. Zernia S, van der Heide NJ, Galenkamp NS, Gouridis G, Maglia G (2020) Current blockades of proteins inside nanopores for real-time metabolome analysis. ACS Nano 14:2296–2307. https://doi.org/10.1021/acsnano.9b09434
6. Straathof S, Di Muccio G, Yelleswarapu M, Alzate Banguero M, Wloka C, van der Heide NJ et al. (2023) Protein sizing with 15 nm conical biological nanopore YaxAB. ACS Nano 17:13685–13699
7. Ahmad M, Ha J-H, Mayse LA, Presti MF, Wolfe AJ, Moody KJ et al. (2023) A generalizable nanopore sensor for highly specific protein detection at single-molecule precision. Nat Commun 14:: 1374
8. Wang K, Zhang S, Zhou X, Yang X, Li X, Wang Y et al. (2024) Unambiguous discrimination of all 20 proteinogenic amino acids and their modifications by nanopore. Nat Methods 21:92–101
9. Van Meervelt V, Soskine M, Singh S, Schuurman-Wolters GK, Wijma HJ, Poolman B et al. (2017) Real-time conformational changes and controlled orientation of native proteins inside a protein nanoreactor. J Am Chem Soc 139:18640–18646
10. Galenkamp NS, Biesemans A, Maglia G (2020) Directional conformer exchange in dihydrofolate reductase revealed by single-molecule nanopore recordings. Nat Chem 12:481–488
11. Li F, Gilliam K, Pham R, Chen M (2022) Mapping the conformational energy landscape of Abl kinase using ClyA nanopore tweezers. Biophys J 121:339a
12. Wloka C, Van Meervelt V, van Gelder D, Danda N, Jager N, Williams CP et al. (2017) Label-free and real-time detection of protein ubiquitination with a biological nanopore. ACS Nano 11:4387–4394
13. Shorkey SA, Du J, Pham R, Strieter ER, Chen M (2021) Real-time and label-free measurement of deubiquitinase activity with a Mspa nanopore. ChemBioChem 22:2688–2692
14. Galenkamp NS, Maglia G (2022) Single-molecule sampling of dihydrofolate reductase shows kinetic pauses and an endosteric effect linked to catalysis. ACS Catal 12:1228–1236
15. Clarke J, Wu H-C, Jayasinghe L, Patel A, Reid S, Bayley H (2009) Continuous base identification for single-molecule nanopore DNA sequencing. Nat Nanotechnol 4:265–270
16. Manrao EA, Derrington IM, Laszlo AH, Langford KW, Hopper MK, Gillgren N et al. (2012) Reading DNA at single-nucleotide

resolution with a mutant MspA nanopore and phi29 DNA polymerase. Nat Biotechnol 30:349–353
17. Pham B, Eron SJ, Hill ME, Li X, Fahie MA, Hardy JA et al. (2019) A nanopore approach for analysis of caspase-7 activity in cell lysates. Biophys J 117:844–855. https://doi.org/10.1016/j.bpj.2019.07.045
18. Yan S, Zhang J, Wang Y, Guo W, Zhang S, Liu Y et al. (2021) Single molecule ratcheting motion of peptides in a Mycobacterium smegmatis porin A (MspA) nanopore. Nano Lett 21:6703–6710
19. Cardozo N, Zhang K, Doroschak K, Nguyen A, Siddiqui Z, Bogard N et al. (2022) Multiplexed direct detection of barcoded protein reporters on a nanopore array. Nat Biotechnol 40:42–46
20. Jeong K-B, Ryu M, Kim J-S, Kim M, Yoo J, Chung M et al. (2023) Single-molecule fingerprinting of protein-drug interaction using a funneled biological nanopore. Nat Commun 14: 1461
21. Martin-Baniandres P, Lan W-H, Board S, Romero-Ruiz M, Garcia-Manyes S, Qing Y et al. (2023) Enzyme-less nanopore detection of post-translational modifications within long polypeptides. Nat Nanotechnol 18:1335–1340
22. Mayer SF, Cao C, Dal Peraro M (2022) Biological nanopores for single-molecule sensing. Iscience 25(4):104145
23. Conlan S, Zhang Y, Cheley S, Bayley H (2000) Biochemical and biophysical characterization of OmpG: a monomeric porin. Biochem 39:11845–11854. https://doi.org/10.1021/bi001065h
24. Fahie MA, Yang B, Mullis M, Holden MA, Chen M (2015) Selective detection of protein homologues in serum using an OmpG nanopore. Anal Chem 87:11143–11149. https://doi.org/10.1021/acs.analchem.5b03350
25. Fahie MA, Candido J, Andree G, Chen M (2021) Tuning protein discrimination through altering the sampling interface formed between the analyte and the OmpG nanopore. ACS Sens 6:1286–1294. https://doi.org/10.1021/acssensors.0c02580
26. Kim M, Foster JC, Moore MD, Chen M (2023) Improving single-molecule antibody detection selectivity through optimization of peptide epitope presentation in OmpG nanopore. ACS Sens 8:2673–2680
27. Foster JC, Pham B, Pham R, Kim M, Moore MD, Chen M (2023) An engineered OmpG nanopore with displayed peptide motifs for single-molecule multiplex protein detection Angew. Chem Int Ed 62:e202214566
28. Fahie MA, Yang B, Pham B, Chen M (2016) Tuning the selectivity and sensitivity of an OmpG nanopore sensor by adjusting ligand tether length. ACS Sens 1:614–622. https://doi.org/10.1021/acssensors.6b00014
29. Noinaj N, Kuszak AJ, Buchanan SK (2015) Heat modifiability of outer membrane proteins from gram-negative bacteria in the BAM complex. Springer, pp 51–56
30. Huemoeller WA, Tien HT (1970) A simple set-up for black lipid membrane experiments. J Chem Educ 47:469
31. Chen K, Bell NA, Kong J, Tian Y, Keyser UF (2017) Direction-and salt-dependent ionic current signatures for DNA sensing with asymmetric nanopores. Biophys J 112:674–682
32. Perez-Rathke A, Fahie MA, Chisholm C, Liang J, Chen M (2018) Mechanism of OmpG pH-dependent gating from loop ensemble and single channel studies. J Am Chem Soc 140:1105–1115
33. Wang Y, Zhang S, Jia W, Fan P, Wang L, Li X et al. (2022) Identification of nucleoside monophosphates and their epigenetic modifications using an engineered nanopore. Nat Nanotechnol 17:976–983
34. Satheesan R, Janeena A, Mahendran KR (2024) Hetero-oligomeric protein pores for single-molecule sensing. J Membr Biol 257:535–545
35. Ladner CL, Yang J, Turner RJ, Edwards RA (2004) Visible fluorescent detection of proteins in polyacrylamide gels without staining. Anal Biochem 326:13–20
36. Chen M, Li Q-H, Bayley H (2008) Orientation of the monomeric porin OmpG in planar lipid bilayers Chembiochem: a European. J Chem Biol 9:3029

Chapter 7

A Cyclodextrin-Mediated Host–Guest Interactions Within α-Hemolysin Nanopore Platform for PFAS Identification

Jaeon Cho, Lijun Peng and Chang Liu

Abstract

Per- and polyfluoroalkyl substances (PFAS) are a category of persistent and extensively spread environmental contaminants that have raised significant global concerns. Liquid chromatography–tandem mass spectrometry (LC–MS/MS), a conventional detection gold standard, can accurately identify them, but its drawbacks include high costs, complicated procedures, and limited portability. Therefore, we developed a nanopore-based single-molecule sensing platform that combines cyclodextrin (CD) host–guest interactions to achieve precise identification of PFAS. By employing an α-hemolysin (α-HL) biological nanopore, when each CD–PFAS complex interacts with the pore, it creates a unique ionic current blockade that discriminates homologous compounds for single-molecular sensing, such as perfluorooctanoic acid (PFOA) and perfluorooctane sulfonate (PFOS). Molecular dynamics simulations combined with experimental data reveal the structural and energetic foundations of CD–PFAS interactions. This work demonstrates a sensitive and selective platform for PFAS monitoring, paving the way toward environmental protection.

Keywords PFAS, Cyclodextrin, Nanopore sensing, Host–guest interaction, Single-molecule analysis, Environmental monitoring

1 Introduction

PFAS are a large class of synthetic fluorinated organics characterized by robust carbon–fluorine bonds that offer remarkable thermal stability, chemical resistance, and surfactant properties [1–3]. Since the 1940s, up to 6420 tons of these chemicals have been extensively used in water- and oil-repellent coatings, firefighting foams, stain-resistant textiles, personal care products, and various electronic applications [4–6]. However, their persistence and bioaccumulation potential have caused global environmental contamination by entering organisms through contaminated drinking water, food (e.g., aquatic products, dairy, and meat), air, and daily contact with consumer products [7, 8]. Then, they are propagated and biomagnified through the food chain, resulting in higher burdens in top trophic level species, such

Fanjun Li (ed.), *Nanopore Technology*, Methods in Molecular Biology, vol. 3072,
https://doi.org/10.1007/978-1-0716-5527-6_7,

as fish and mammals, posing a huge threat to the ecosystem and all kinds of organisms [9, 10]. Long-chain PFAS (e.g., PFOA and PFOS) exhibit high resistance to degradation and are now subject to restriction or ban on use in many countries [11, 12]. Even higher water soluble short-chain variants, introduced as alternatives, remains environmentally mobile and hard to get rid of, leading to potential health and environmental risks [13]. The US Environmental Protection Agency (EPA) implemented legally enforceable drinking water limits for PFOA and PFOS at four parts per trillion (ppt) and a hazard index threshold of 1 for covering four additional PFAS, including perfluorononanoic acid (PFNA), hexafluoropropylene oxide dimer acid (HFPO-DA, commonly known as GenX Chemicals), perfluorohexane sulfonic acid (PFHxS), and perfluorobutane sulfonic acid (PFBS) [2].

An accurate measurement of PFAS is essential for environmental surveillance and health threat assessment. LC–MS/MS, the prevailing gold standard for laboratory analysis, involving solid-phase extraction (SPE) for target enrichment and matrix elimination, liquid chromatography for PFAS compound separation, and tandem mass spectrometry for identification and quantification in monitoring multiple reactions mode, demonstrates excellent sensitivity and accuracy [14–18]. However, it requires expensive equipment, trained technicians, and long cycles for preparation and analysis [16, 19, 20]. These limitations hinder rapid field deployment and large-scale screening in severe environmental pollution solutions, motivating the development of simple, rapid, portable, and cost-effective sensing technologies [19, 21–23]. Novel electrochemical and optical sensors convert PFAS molecules into quantifiable electrical and optical signals for pursuing the development of portable, user-friendly, and cost-effective sensing technologies [24–29]. Although these techniques facilitate rapid and visual detection, many remain experiencing limitations with selectivity owing to the similarities in structure across PFAS homologs. The current recognition elements, molecularly imprinted polymers and other synthetic and biological receptors (e.g., optimized aptamers, MOFs, guanidiniocalix[5]arene) can enhance detection of PFAS [26, 27], but the capability to distinguish between closely related congeners, especially short-chain PFAS, is still insufficient.

Nanopore sensing proposes an innovative approach for the analysis of individual molecules of PFAS [30–34]. As analytes flow through a lipid membrane implanted with a nanometer-sized pore, this approach identifies single molecules by monitoring transient ionic current blockades. Each individual molecule generates a distinctive fingerprint determined by its size, charge, conformation, and interaction with the pore, which makes it possible for label-free and real-time monitoring. However, PFAS cannot be detected directly in nanopores due to their tiny size and rapid translocation without

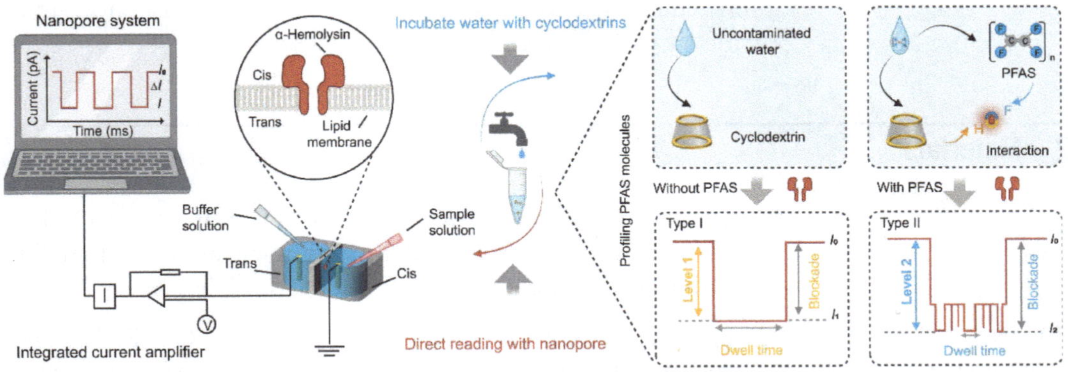

Fig. 1 Schematic illustration of the CD-mediated nanopore assay workflow for detecting PFAS molecules

sufficient dwell time. To address this problem, molecular recognition elements such as cyclodextrins (CDs) can be employed as host molecules. Through hydrophobic and van der Waals interactions, CDs and PFAS form inclusion complexes, which effectively extend dwell time and improve signal resolution [35–37].

Figure 1 illustrates the α-HL nanopore setup and the CD-mediated sensing mechanism for single-molecule PFAS detection [20]. The sensor consists of a self-assembled α-HL protein nanopore embedded in a lipid bilayer membrane, which separates the system into two chambers (*cis* and *trans*). Ag/AgCl electrodes placed in the two chambers are linked to a current signal amplifier, enabling simultaneous application of a transmembrane potential and detection of ionic current through the nanopore. During detection, CDs are incubated with the possibly contaminated water to generate CD-PFAS complexes first, subsequently injected into the *cis* chamber. Under a constant transmembrane potential without any analytes present, the unblocked nanopore generates a stable baseline current I_0. When CD molecules pass or block the nanopore, the current decreases to I_1, resulting in a Level 1 blockade ($\Delta I_1 = I_0 - I_1$, type I). In the presence of PFAS, the current further reduces to I_2, producing a Level 2 blockade ($\Delta I_2 = I_0 - I_2$, type II). This yields a characteristic two-step signal pattern (level 1 Δ I and level 2 Δ I), derived from host–guest interactionsbetween the PFAS and the CD molecules, operating as unique fingerprints to enable discrimination of different PFAS species at the molecular level.

According to these principles, this chapter systematically examines the interactions between representative PFAS (e.g., PFOA and PFOS) and various cyclodextrins (α-, β-, γ-, and hydroxypropyl-γ-CD) within α-HL nanopores. We clarify how cavity size, substituents, and PFAS chain length affect binding behavior and current response by combining experimental single-molecule detection with molecular dynamics simulations. The CD-mediated nanopore platform achieves a strong, sensitive, selective, and mechanistically interpretable way to detect PFAS that combines basic host–guest chemistry with advanced environmental sensing.

2 Materials

Prepare all work solutions with deionized (DI) water from a Milli-Q purification system (resistivity 18.2 megohm·cm at 25°C), then filter through 0.02 μm filters.

2.1 Per- and Polyfluoroalkyl Substances (PFAS) Stock Solution

1. Prepare Perfluorobutanoic acid (PFBA, $C_4HF_7O_2$), perfluoropentanoic acid (PFPA, $C_5HF_9O_2$), perfluorohexanoic acid (PFHxA, $C_6HF_{11}O_2$), perfluoroheptanoic acid (PFHpA, $C_7HF_{13}O_2$), perfluorooctanoic acid (PFOA, $C_8HF_{15}O_2$), perfluorononanoic acid (PFNA, $C_9HF_{17}O_2$), perfluorodecanoic acid (PFDA, $C_{10}HF_{19}O_2$), perfluorobutane sulfonic acid (PFBS, $C_4HF_9O_3S$), perfluorohexane sulfonic acid (PFHxS, $C_6HF_{13}O_3S$), and perfluorooctane sulfonate (PFOS, $C_8HF_{17}O_3S$) from SynQuest Laboratories.
2. Prepare an individual stock solution at 10 mg/ml by dissolving each PFAS sample in DI water. Store the stock solution at 4°C for future use.

2.2 Sample Solution

1. Prepare CDs from Sigma-Aldrich: α-CD, β-CD, γ-CD, and HP-γ-CD.
2. Incubate CD and PFAS molecules (1:1 molar ratio) at room temperature on a shaker. For example, in the case of the sample HP-γ-CD-PFOA:
   (a) Dissolve 0.0032 g (2.4×10^{-6} mol) of HP-γ-CD powder in 400 μl of DI water using an ultrasonic device in a 1.5-ml centrifugation tube.
   (b) Mix the dissolved solution with 100 μl of the PFOA stock solution and incubate the mixture on a shaker at room temperature for 12 h.
3. Store the remaining sample solution at 4°C for further testing.

2.3 Nanopore Work Solution

1. Prepare tris-hydrochloride (tris-HCl), potassium chloride (KCl), HCl, and sodium hydroxide (NaOH).
2. Dissolve KCl (22.37 g) and Tris-HCl (0.121 g) in 80 ml DI water.
3. Adjust the pH by using NaOH (2 mol/l) and HCl (2 mol/l) (*see* **Note 1**). The target pH is 8.0.
4. Add 20 ml of DI water to dilute the work solution. The target concentrations and pH are 3 mol/l KCl and 10 mmol/l Tris-HCl, with a pH of 8.0.

2.4 Nanopore Setup

1. Prepare α-HL from Staphylococcus aureus (lyophilized powder, protein ~60% by Lowry, ≥10,000 U/mg protein) from Sigma-Aldrich.
2. Prepare hexadecane (Reagent Plus, 99%) and n-pentane (anhydrous, ≥99%) from Sigma-Aldrich.
   (a) Prepare a 1 ml stock solution of 1:9 (100 μl:900 μl) hexadecane/pentane in a 1 ml glass vial (*see* **Note 2**).
   (b) Store the stock solution at −20°C for future use.
3. Prepare lipids containing 1,2-diphytanoyl-*sn*-glycero-3-phosphocholine (DPhPC, 4 ME 16:0 PC) from Avanti Polar Lipids.
   (a) Prepare a 10 mg/ml stock solution of DPhPC in pentane within a 1 ml glass vial (*see* **Note 3**).
   (b) Store the stock solution at −20°C for future use.
4. Prepare a planar bilayer chamber and a 25 μm-thick Delrin wall with a 200 μm aperture, separating the chamber into *cis* and *trans* chambers, as provided by Planar Membranes LLC.
5. Prepare Ag/AgCl electrodes for applying an electrical potential to the *trans* chamber (*see* **Note 4**).
6. Prepare the Planar Lipid Bilayer Workstation from Warner Instruments (It includes an FC series Faraday cage and a vibration isolation table) (*see* **Note 5**).
7. Prepare a Digidata converter, a bilayer clamp amplifier, and pClamp software (*see* **Note 6**).

3 Methods

3.1 pClamp Software Setup (One-Time Setup)

1. Turn on the power switches of a Digidata converter and a bilayer clamp amplifier.
2. Open pClamp software.
3. To set up the digitizer, go to Configure → Digitizer → Change → Digitizer Type → Scan (*see* **Note 7**).
4. To set up the configuration of the input signal, go to Configure → Lab Bench → Add → Current → pA units → Scale factor: 0.001 V/pA.
5. To set up the telegraphed instrument, Configure → Telegraphed Instrument → Input Channels: Analog IN #0 → Telegraphed Instrument: the bilayer clamp amplifier (*see* **Note 8**).
6. To set up protocol, Acquire → Edit Protocol → Acquisition Mode: Gap-free → Trial Length: Use available disk space → Sample Interval: 10 μs → Inputs tab → Analog IN Channels:

Channel #0: current → Units: pA → OK. If you want to save the protocol, which can be used as a customized.pro file in the next section, go to File → Save Protocol As → Enter a file name → Save as a .pro file

3.2 pClamp Software Run

1. Open pClamp software.
2. Acquire → Open Protocol → Double-click customized.pro file.
3. Click ▶ (View Only) to see real-time current traces.
4. To set the data file name of nanopore recordings, go to File → Set Data File Names.

3.3 Precoating and Lipid Formation

1. Precoat the aperture (200 µm in diameter) on a 25 µm-thick Delrin wall with 1 µl of 1:9 hexadecane/pentane (100 µl:900 µ l) (*see* **Note 9**).
2. Put 1 ml of nanopore work solution (3 mol/l KCl and 10 mmol/l Tris-HCl, with a pH of 8.0) in each chamber.
3. Place the chamber inside the FC series Faraday cage on the vibration isolation table.
4. Connect Ag/AgCl electrodes to a Planar Lipid Bilayer Workstation and submerge them in 1 ml of the nanopore work solution in each chamber. The *cis* chamber was grounded.
5. Add 20 µl of 10 mg/ml DPhPC. This bilayer is self-assembled (*see* **Note 10**).
6. Apply an electrical potential to the *trans* chamber via Ag/AgCl electrodes and test membrane stability by applying a voltage in the range of ±100 to ±200 mV.
7. Maintain the membrane capacitance at 160–180 pF across different voltage biases to ensure lipid membrane stability.

3.4 Single Pore Insertion and Recording

1. Add α-HL protein to the *cis* chamber to a final concentration of 0.01 to 0.05 ng/ml under +100 mV (*see* **Note 11**).
2. Once a single α-HL protein has been inserted on the *cis* side, add 20 µl of sample solution (*see* **Note 12**).
3. Record ionic current traces at +100 mV by clicking the Recording (red circle) button (*see* **Note 13**).
4. When the pore is broken, stop the recording and pipette up and down the working solution on the *cis* side to re-establish the membrane. Verify that the membrane's capacitance remains within the range of 160–180 pF. The protein will self-assemble again.
5. Capture and amplify electronic signals using a Digidata converter and pClamp software.

3.5 Data Analysis

1. Process all recordings (ABF files from pClamp) using an in-house MATLAB-based algorithm (Aperture) (*see* **Note 14**).
2. Extract relevant features (the blockade and the dwell time) of each detected signal (*see* **Note 15**).

4 Notes

1. Adjust pH using a pH meter and double-check with pH paper strips.
2. Mix 100 µl of hexadecane and 900 µl of pentane. If multiple nanopore recordings are conducted, prepare a few more vials of stock solutions for future use and store them at −20°C.
3. Likewise, if multiple nanopore recordings are required, prepare an additional vial of stock solution for future use and store it at −20°C. Especially, since 20 µl of the stock solution will be used for a nanopore recording, it is better to prepare multiple vials of stock solutions when making them. For example, dissolve 200 mg of DPhPC in 20 ml of pentane and distribute it into 1 ml glass vials.
4. Ag/AgCl wires and gold-plated jacks can be purchased directly from VWR. The silver wire and the gold-plated jack can be connected using soldering techniques. The process involves the use of rosin paste, soldering flux, and lead-free solder wire.
5. These are primarily intended to minimize unnecessary noise from current signals.
6. In this experiment, Axon Digidata 1440 A Low-Noise Data Acquisition System plus HumSilencer from Axon Instruments: Molecular Devices Electrophysiology, BC-535 Bilayer Clamp Amplifier from Warner Instruments, and Clampex 10.7 were used.
7. In this experiment, the Digidata 1440 A Series was selected.
8. In this experiment, Warner BC-535 was selected as the telegraphed instrument. The low-pass Bessel filter was set to 5 kHz.
9. Use a pipette to fully coat the dented area of the Delrin wall. Make sure all areas are wet by 1 µl of 1:9 hexadecane/pentane. Once pre-coated, let the wall dry for 1 to 2 min before adding the nanopore work solution (3 M KCl and 10 mM Tris-HCl at pH of 8.0) to each chamber.
10. Monitor the membrane capacitance. If it is too low (below 100 pF), α-HL protein is hardly inserted into the lipid bilayer.

Break the membrane (either by ramping up the voltage or using a 200 µl pipette and pipetting up and down near the aperture), and reform it in the aperture by pipetting up and down the working solution on the *cis* side.

11. Prepare α-HL in DI water at 1 mg/ml (5 µl per tube) and store at −80°C. To use in a nanopore setup, add 95 µl of the nanopore work solution to the tube, then withdraw 10 µl and transfer it to another tube. Next, add another 200 µl of the nanopore work solution. Insert 5 µl of this solution near the 200 µm aperture.
12. The α-HL protein is also self-assembled.
13. At +100 mV, a single-pore ionic current should be around 260 pA. If the current exceeds 300 pA, the pore should not be used; break and reform the membrane by pipetting up and down the entire working solution, then wait for another single-pore insertion.
14. The Aperture app first plots the recorded raw data (ABF file) in the system. Based on parameters (e.g., baseline current, threshold to consider a drop as an event, and event duration length), the app detects the signals, removing most baseline noise. Once the algorithm detects an event satisfying the parameters, it captures it and calculates the current blockade and the dwell time associated with it. PFAS signals have two levels of peaks. The algorithm was applied for each level.
15. The blockade represents the interaction between a single analyte and the nanopore. It is defined as $\Delta I/I_0$ ($\Delta I = I - I_0$; I: the average current blockade measured when the analyte is inside the pore; I_0: the average baseline current when the analyte is not near the pore). The dwell time represents the duration of interaction between the nanopore and a single analyte.

References

1. He Y, Chai Y, Jiao Z, Li H, Gu Z, Tian J, Wang W, Chi X (2025) Functional supramolecular hosts as platforms for the detection and removal of per- and polyfluoroalkyl substances. Matter 8:102246
2. United States Environmental Protection Agency (US EPA). Per-and Polyfluoroalkyl Substances (PFAS) Final PFAS National Primary Drinking Water Regulation. https://www.epa.gov/sdwa/and-polyfluoroalkyl-substances-pfas
3. Gao C, Drage DS, Abdallah MA-E, Quan F, Zhang K, Hu S, Zhao X, Zheng Y, Harrad S, Qiu W (2024) Factors influencing concentrations of PFAS in drinking water: implications for human exposure. ACS EST Water 4:4881–4892
4. Tansel B (2022) PFAS use in electronic products and exposure risks during handling and processing of e-waste: a review. J Environ Manage 316:115291
5. Wang Z, Cousins IT, Scheringer M, Buck RC, Hungerbühler K (2014) Global emission inventories for C4–C14 perfluoroalkyl carboxylic acid (PFCA) homologues from 1951 to 2030, Part I: production and emissions from quantifiable sources. Environ Int 70:62–75
6. Figuière R, Miaz LT, Savvidou E, Cousins IT (2025) An overview of potential alternatives

for the multiple uses of per- and polyfluoroalkyl substances. Environ Sci Technol 59:2031–2042
7. Wang Z, DeWitt JC, Higgins CP, Cousins IT (2017) A never-ending story of per- and polyfluoroalkyl substances (PFASs)?. Environ Sci Technol 51:2508–2518
8. Glüge J, Scheringer M, Cousins IT, DeWitt JC, Goldenman G, Herzke D, Lohmann R, Ng CA, Trier X, Wang Z (2020) An overview of the uses of per- and polyfluoroalkyl substances (PFAS). Environ Sci Process Impacts 22:2345–2373
9. Armitage JM, MacLeod M, Cousins IT (2009) Comparative assessment of the global fate and transport pathways of long-chain perfluorocarboxylic acids (PFCAs) and perfluorocarboxylates (PFCs) emitted from direct sources. Environ Sci Technol 43:5830–5836
10. Guo H, Xing M, Sun X, Liu T (2024) Bibliometric analysis and scientific trends on perfluorinated and polyfluoroalkyl substances (PFASs) as an emerging pollutant. Water-Energy Nexus 7:187–199
11. Frömel T, Knepper TP (2010) Biodegradation of fluorinated alkyl substances. In: De Voogt P (Ed.)Eds. Reviews of environmental contamination and toxicology, vol 208: perfluorinated alkylated substances. Springer, New York, NY, 161–177
12. Wang Z, Cousins IT, Scheringer M, Hungerbühler K (2013) Fluorinated alternatives to long-chain perfluoroalkyl carboxylic acids (PFCAs), perfluoroalkane sulfonic acids (PFSAs) and their potential precursors. Environ Int 60:242–248
13. Alazaiza MYD, Alzghoul TM, Ramu MB, Abu Amr SS, Abushammala MFM (2025) PFAS contamination and mitigation: a comprehensive analysis of research trends and global contributions. Case Stud Chem Environ Eng 11:101127
14. Al Amin Md, Sobhani Z, Liu Y, Dharmaraja R, Chadalavada S, Naidu R, Chalker JM, Fang C (2020) Recent advances in the analysis of per- and polyfluoroalkyl substances (PFAS)—a review. Environ Technol Innov 19:100879
15. Newton SR, Bowden JA, Charest N, Jackson SR (2025) Filling the gaps in PFAS detection: integrating GC-MS non-targeted analysis for comprehensive environmental monitoring and exposure assessment. Environ Sci Technol Lett 12:2
16. Joshi N, Garapati K, Ghose V, Kandasamy RK, Pandey A (2024) Recent progress in mass spectrometry-based urinary proteomics. Clin Proteom 21:14
17. Stecconi T, Tavoloni T, Stramenga A (2024) A LC-MS/MS procedure for the analysis of 19 perfluoroalkyl substances in food fulfilling recent EU regulations requests. Talanta 266:125054
18. Ryan CRM, Nazdrajić E, Campbell JL, Bell KY, Hopkins WS (2025) Two-dimensional LC × DMS analysis of 34 PFAS compounds. Anal Chem 97:21050–21059
19. Trojanowicz M, Koc M (2013) Recent developments in methods for analysis of perfluorinated persistent pollutants. Microchim Acta 180:957–971
20. Wei XJ, Choudhary A, Leon Y, Chang L (2024) Single-molecule profiling of per- and polyfluoroalkyl substances by cyclodextrin mediated host-guest interactions within a biological nanopore. Sci Adv 10:eadp8134
21. Lamichhane HB, Arrigan DWM (2023) Electroanalytical chemistry of per- and polyfluoroalkyl substances. Curr Opin Electrochem 40:101309
22. Clark RB, Dick JE (2020) Electrochemical sensing of perfluorooctanesulfonate (PFOS) using ambient oxygen in river water. ACS Sens 5:3591–3598
23. Gondhiya N, Rehman AU, Andreescu D, Andreescu S (2025) Portable electrochemical sensors for per- and polyfluoroalkyl substances: design, challenges, and opportunities for field deployment. Curr Opin Electrochem 53:101725
24. Fang C, Zhang X, Dong Z, Wang L, Megharaj M, Naidu R (2018) Smartphone app-based/portable sensor for the detection of fluoro-surfactant PFOA. Chemosphere 191:381–388
25. Menger RF, Beck JJ, Borch T, Henry CS (2022) Colorimetric paper-based analytical device for perfluorooctanesulfonate detection. ACS EST Water 2:565–572
26. Cennamo N, Zeni L, Tortora P, Regonesi ME, Giusti A, Staiano M, D'Auria S, Varriale A (2018) A high sensitivity biosensor to detect the presence of perfluorinated compounds in environment. Talanta 178:955–961
27. Gong J, Fang T, Peng D, Li A, Zhang L (2015) A highly sensitive photoelectrochemical detection of perfluorooctanic acid with molecularly imprinted polymer-functionalized nanoarchitectured hybrid of AgI–BiOI composite. Biosens Bioelectron 73:256–263
28. Hong Y, Chen X, Zhang Y, Zhu Y, Sun J, Swihart MT, Tan K, Dong L (2022) One-pot hydrothermal synthesis of high quantum yield orange-emitting carbon quantum dots for sensitive detection of perfluorinated compounds. New J Chem 46:19658–19666
29. Cui Y, Wang S, Han D, Yan H (2024) Advancements in detection techniques for per- and polyfluoroalkyl substances: a comprehensive review. TrAC Trends Anal Chem 176:117754
30. Lu X, Zhong D, An Q, Kang L, Fan N, Cao J, Li R, Cao Q, Zhou Y, Du X, Yan S, Li J, Qu X, Wang Y (2025) Multiplex nanopore detection

of structurally diverse per- and polyfluoroalkyl substances in surface water. ACS Nano 19:37281–37295
31. Shrikrishna NS, Gandhi S (2025) Nanopore-based sensing for biomarker detection: from fundamental principles to translational diagnostics. J Nanobiotechnol 23: 663
32. Parambil AM, Priyadarshini E, Paul S, Bakandritsos A, Sharma VK, Zbořil R (2025) Emerging nanomaterials for the detection of per- and poly-fluorinated substances. J Mater Chem A 13:8246–8281
33. Ren M, Liu D, Qin F, Chen X, Ma W, Tian R, Weng T, Wang D, Astruc D, Liang L (2025) Single-molecule resolution of macromolecules with nanopore devices. Adv Colloid Interface Sci 338:103417
34. Gu L-Q, Cheley S, Bayley H (2001) Capture of a single molecule in a nanocavity. Science 291:636–640
35. Chen J, Peng Q, Peng X, Zhang H, Zeng H (2022) Probing and manipulating noncovalent interactions in functional polymeric systems. Chem Rev 122:14594–14678
36. Meo PL, Lazzara G, Liotta L, Riela S, Noto R (2014) Cyclodextrin–calixarene co-polymers as a new class of nanosponges. Polym Chem 5:4499–4510
37. Wang W, Shao H, Zhou S, Zhu D, Jiang X, Yu G, Deng S (2021) Rapid removal of perfluoroalkanesulfonates from water by β-cyclodextrin covalent organic frameworks. ACS Appl Mater Interfaces 13:48700–48708

Chapter 8

Optical and Electrical Characterization of Nanopores in Freestanding Lipid Bilayers Formed on Microelectrode Cavity Arrays

Madison A. Handyside, Joseph S. Harvey, SeeEun Lee, Kathryn A. Michael, Kayla R. Sennese, Annika E. Evenson, Sarah E. Koeneman, Daniel L. Burden and Lisa M. Keranen-Burden

Abstract

Single-channel electrical recording remains a fundamental technique for investigating biological nanopores embedded in lipid bilayer membranes. However, this method cannot reveal information from nanopores that fail to induce transmembrane ionic currents. Here, we present procedures that integrate a microelectrode cavity array with a single-molecule fluorescence microscope to probe labeled nanopores on freestanding lipid bilayers with light and electricity at the same time. Because oligomeric prepores are frequently employed in nanopore sensing applications, we explore the conversion of prepores into fully conducting pores. The protocol includes modified instructions for preparing and labeling alpha-hemolysin (αHL) heptamers, as well as a description of the instrument configuration and associated data analysis tools. We employ an Excel template (available as a supplemental file) to calculate a membrane-bound, but uninserted, heptameric nanopore diffusion coefficient of $8.1 \pm 0.3\,\mu m^2\,s^{-1}$ and an open-to-closed nanopore ratio of 10^{-3} to 10^{-4}. The methods described here provide a general framework for extending orthogonal single-molecule optical and electrical measurements in microelectrode cavity arrays to a broad range of biological nanopores.

Keywords Nanopore, Single molecule fluorescence, Bilayer, Biomembrane, Ion-channel electrophysiology, Single particle tracking, Diffusion, Microelectrode cavity array, Alpha hemolysin

---

These authors Madison A. Handyside, Joseph S. Harvey, SeeEun Lee, Kathryn A. Michael, Kayla R. Sennese are contributed equally to this work.

Supplementary Information The online version contains supplementary material available at https://doi.org/10.1007/978-1-0716-5500-9_8

Fanjun Li (ed.), *Nanopore Technology*, Methods in Molecular Biology, vol. 3072,
https://doi.org/10.1007/978-1-0716-5527-6_8,

1 Introduction

Biological nanopores have emerged as powerful single-molecule sensors capable of detecting and characterizing a wide range of analytes by monitoring changes in ionic current as molecules translocate through the pore [1, 2]. This technology has achieved significant success in DNA and RNA sequencing [3], including direct detection of base modifications [4], and is increasingly being adapted for protein and peptide analysis [5–7], where challenges such as translocation control and signal complexity are being actively addressed. Ongoing advances in protein engineering, motor protein integration, and signal processing continue to expand the scope and resolution of these platforms [8–10].

While electrical recordings are central to nanopore sensing, their interpretive power is constrained by bandwidth limitations, instrumental noise, and signal ambiguity [11]. These challenges extend to studies of nanopore interactions with lipid membranes, where electrical signals cannot distinguish between nanopore insertion and structural rearrangement. Moreover, single-channel recordings probe only one conducting nanopore at a time and therefore fail to capture the broader binding and translational dynamics across the membrane surface. To overcome these challenges, orthogonal single-molecule fluorescence methods performed concurrently with single-channel electrical recordings can yield complementary insights that disentangle and clarify behavior [12–25].

A previous single-molecule fluorescence study of the alpha-hemolysin (αHL) nanopore, performed in a droplet on hydrogel bilayer, elucidated a rapid multi-step self-assembly process from monomers to heptamers [26]. However, in sensing applications, preassembled oligomers are commonly used [27–31]. This practice eliminates the planar-bilayer assembly steps from monomers, resulting in improved insertion control and better conductance reproducibility. The use of preassembled nanopores also facilitates analysis of translocation kinetics and ionic current signatures due to fewer complications from intermediate assembly states.

The protocol presented here employs preassembled αHL heptamers and compares the number of bilayer-associated prepores to the number of electrically conducting nanopores. Additionally, the translational motion of these structures within the bilayer is observed and quantified. Because measurements are performed using freestanding bilayers, lateral motion is not influenced by an underlying support. We describe modified methods for producing preassembled heptamers, as well as a technique for acquiring optical and electrical signals in a microelectrode cavity array. The chip-based array mitigates inherent limitations in data acquisition, temporal synchronization, and bilayer mechanical stability that are traditionally associated with freestanding bilayers. We find that the fraction of conducting nanopores relative to the number of bilayer-associated prepores can be low. Specific open-to-closed ratios vary from trial to trial and can be significantly affected by the choice of mutant and the location and type of fluorescent label. Given that the assembly and insertion mechanisms of αHL parallel those of other pore-forming proteins, the experimental framework, protocol descriptions, and analysis tools supplied here should be broadly applicable to a range of nanopores studies.

2 Materials

2.1 Solutions and Reagents

1. Electrophysiology recording buffer (used for MECAopto-inv chip): 1.0 M KCl, 10.0 mM Tris-HCl, and pH 7.5. Weigh 0.788 g Tris-HCl and transfer to a 1 L volumetric flask. Weigh 37.272 g KCl and transfer to the same flask. Add 900 mL of ultrapure water (18.2 MΩ cm). Adjust solution to pH 7.5 with concentrated NaOH. Fill up to 1 L with water. Sterile filter buffer through 0.22 μm cellulose acetate filter before use.
2. Lipid Solution (50 mg/mL in octane): Distribute a 4 mL ampule of 1,2-diphytanoyl-sn-glycero-3-phosphocholine (DPhPC, 25 mg/mL in chloroform, Avanti Polar Lipids) into 200 μL aliquots using a Drummond pipette. Evaporate chloroform using a gentle stream of nitrogen gas. Remove the remaining solvent in a vacuum chamber for 30 min. For long-term storage keep at -80°C. Prior to use, warm to room temperature and add 100 μL of octane (Sigma-Aldrich). Cap and refrigerate at 4°C for repeated short-term use.
3. Red Blood Cells (RBCs): 10% washed rabbit red blood cells (Lampire).
4. 1 M Tris stock solution: Weigh out 7.88 g of Tris-HCl and transfer to a 50 mL conical tube. Add water to 35 mL. Adjust to pH 7.5 using NaOH solution. Fill to 50 mL with water. Sterile filter buffer through a 0.22 μm cellulose acetate filter before use.
5. Low salt Tris buffer—5 T (used to make erythrocyte membranes): Prepare a 5 mM Tris solution by adding 250 μL of 1 M Tris pH 7.5 to water, adjusting the final volume to 50 mL. Sterile filter through 0.22 μm cellulose acetate filter before use.
6. High salt Tris buffer—10 T/2 N (used to make erythrocyte membranes and to isolate heptamers): 10 mM Tris pH 7.5, 2 M NaCl. Weigh 5.844 g NaCl and transfer to a 50 mL conical tube. Add 500 μL of 1 M Tris pH 7.5 and water to 35 mL. If necessary, adjust solution to pH 7.5. Fill to 50 mL with

water. Sterile filter buffer through a 0.22 μm cellulose acetate filter before use.

7. 10% SDS solution (used to make 1% SDS solution): Weigh 0.5 g SDS and transfer to a 15 mL conical tube. Add 5 mL of water and gently agitate until completely dissolved.
8. Luria-Bertani broth with Ampicillin—LB/Amp: In the presence of a flame, transfer the desired amount of Luria-Bertani solution into a sterile receptacle. Add Ampicillin to a concentration of 100 μg/mL (e.g., 200 mL Luria-Bertani broth + 200 μL of 100 mg/mL Ampicillin).
9. Isopropyl-β-D-thiogalactopyranoside (IPTG): Prepare a 1 M stock by dissolving 2.38 g IPTG in 10 mL water. Small aliquots (~1 mL) were stored at −20°C.
10. Lysis buffer (for bacterial lysis): Combine 10 mL Bacterial Protein Extraction Reagent (BPER; Thermofisher) with the protease inhibitors leupeptin (25 μL of 10 mg/mL stock in water) and phenylmethylsulfonyl fluoride, PMSF (10 μL of 1 M stock in ethanol).
11. 20 mM Tris pH 7.5, 100 mM KCl (20 T/100 K): Combine 1 mL of 1 M Tris pH 7.5 (described above) and 5 mL of 1 M KCl (3.73 g of KCl in 50 mL water). Dilute to 50 mL with water.
12. Column Elution Buffer, 400 mM Imidazole in 20 T/100 K: To 9 mL of 20 T/100 K, add 6 mL of 1 M Imidazole in 20 T/100 K (3.4 g Imidazole dissolved and diluted to 50 mL with 20 T/100 K).
13. Column Resin: HisPur Ni^{2+}-NTA (Thermofisher) as 1:1 slurry (resin:solvent).
14. 10 mM Tris (2-carboxyethyl) phosphine hydrochloride—TCEP: prepared fresh (e.g., 0.0112 g of TCEP in 3.94 mL of water).
15. Sulfo-Cy5 maleimide—Cy5 (BroadPharm; MW 803), prepared as 4 mg/mL in water. Stored at −20°C as small aliquots, well-sealed in an N_2 environment.

2.2 Control and Data Processing Software

1. Video capture of optical data: Nikon NIS-Elements v5.11.03.
2. Orbit mini control: Elements Data Reader v3.8.23 (EDR).
3. Electrical data analysis: Elements Data Analyzer v1.4.7 (EDA).
4. Video file display and processing: Fiji(ImageJ) v2.16.0/1.54p; Java 1.8.0_322 (64-bit).
5. Single particle tracking (SPT) algorithm: Fiji (ImageJ) plugin TrackMate v7.14.0 [32].
6. Data processing template in Excel (see preformatted supplemental file).

3 Methods

3.1 αHL Protein Production Methods

1. Expression of αHL protein in *E. coli* cells
   - A. A stab of *E. coli*(BL21(DE3) cells, stored at −80 °C; transformed with the pET plasmid carrying the *Staphylococcus aureus* αHL gene N244C point mutation with an N-terminal His_6 tag) is added to each of two culture tubes containing 4 mL LB/Amp.
   - B. The bacterial cultures are grown at 37°C with shaking (250 rpm) for approximately 18–20 h until bacterial growth is saturated.
   - C. Two 4 mL saturated cultures are added to ~190 mL of LB/Amp and grown in the shaker at 37°C and 250 rpm.
   - D. Bacterial growth is monitored via OD_{600} measurements using UV-Vis Spectrometry (calibrated with LB/Amp as a blank) until absorbance at 600 nm reaches 0.6–0.7. The OD measurement is made by taking a small sample (~1 mL) of the bacterial culture, placing it in a sterile cuvette, reading the OD_{600}, then returning the sample to the growing culture.
   - E. When the optimal OD_{600} is reached, 1 μL of 1 M IPTG per mL of culture is added to the culture flask to reach 1 mM IPTG for induction of protein expression (e.g., 198 μL 1 M IPTG is added to 198 mL culture).
   - F. After the addition of IPTG, the culture is incubated at 37°C and 250 rpm for 2 h to allow for protein expression.
   - G. After expression, the bacterial pellets are isolated by centrifugation. The culture is transferred to four 50 mL conical tubes and spun in a centrifuge (Eppendorf tabletop model 5810R) for 30 min at 4000 rpm.
   - H. After centrifugation, the supernatant is decanted into a waste beaker without disrupting the cell pellets. Tubes containing cell pellets are stored at −20°C (at least overnight prior to purification).
2. Purification of αHL from bacterial lysate
   - A. Two (50 mL) tubes containing the bacterial cell expression pellets are obtained from the freezer and thawed.
   - B. Lysis buffer is prepared (as described in the materials section).
   - C. Lysis buffer (3 mL) is added to each 50 mL conical tube containing cells. With a clean transfer pipette, the solution is carefully drawn up and down to resuspend the cell pellet in the buffer. Tip: Excess bubbling is avoided by retaining some solution in the pipette at all times during resuspension.

D. Resuspended pellets are combined into one tube, which is placed on a rocker for 30 min at room temperature to mix thoroughly and allow the cells time to lyse.

E. After this lysis period, the cell mixture (now called *lysate*) is evenly divided into four clean 1.7 mL microcentrifuge tubes and centrifuged at 15,000 rpm (Eppendorf bench-top model 5425) for 30 min.

F. After centrifugation, the supernatant (soluble fraction of the lysate; contains αHL) is carefully recovered and transferred into a 15 mL tube labeled *load on*. The insoluble fraction of the lysate (the pellet) is discarded.

3. IMAC Column Purification

A. Background: This purification method (IMAC: Immobilized Metal Ion Affinity Chromatography) utilizes the N-terminal His_6 tag on αHL, which interacts with high affinity with the immobilized Ni^{2+} ions on the HisPur resin. High concentrations of imidazole effectively compete with αHL for binding to Ni^{2+} ions and allow removal of αHL from the affinity resin.

B. Prior to loading onto an affinity column, the bacterial lysate supernatant is adjusted to contain 40 mM imidazole. Based on the amount of supernatant, the following calculation determines the volume of imidazole (1 M stock solution) needed to obtain a final working concentration of 40 mM imidazole (Im) (e.g., for 6 mL supernatant: [(6 + x) mL * 40 mM Im = x mL * 1 M Im] where x is the volume of 1 M Im needed to bring the final solution to 40 mM Im. In this example, 0.25 mL of 1 M imidazole is added to 6 mL of bacterial lysate supernatant).

C. After addition of Im, thoroughly mix using gentle inversions of the tube. A small sample (10 μL) of this solution now called *load on* can be removed for SDS PAGE analysis.

D. The IMAC column is prepared as follows: A small (~10 mL maximum capacity) plastic column is obtained and fixed to a metal ring stand using a utility clamp. HisPur affinity resin (1:1 slurry of beads in H_2O/ethanol) is mixed to homogeneity, and 200 μL is placed into the column. (This volume provides a 100 μL resin volume.) Rinse and equilibrate the light blue Ni^{2+} charged resin with two column volumes (200 μL) of water followed by two column volumes (200 μL) of lysis buffer.

E. The *load on* is transferred onto the column in 1–2 mL aliquots. The solution that flows through is collected in a 15 mL conical tube labeled flow-through, *FT*.

F. After collection of the FT is complete, rinse the column with two washes (W1 and W2). First, 300 μL of lysis buffer is applied to the column with the collected eluate labeled *W1*. Second, 300 μL of 20 T/100 K is applied to the column with the collected eluate labeled *W2*. Contaminating bacterial proteins are evident in *W1* and *W2* when analyzed using SDS-PAGE.

G. To elute the αHL protein, 300 μL 20 T/100 K/400 mM Im is applied to the column, and the elution is collected in a microcentrifuge tube labeled *E1*. A second elution is produced by application of 200 μL of 20 T/100 K to the column and is collected in a microcentrifuge tube as *E2*. All the samples from column purification are stored at 4°C.

H. SDS-PAGE analysis (using 4%–20% acrylamide) of column purification samples reveals high levels of expression of α HL compared to all other bacterial proteins, the presence of all contaminating bacterial proteins in the W1 and W2 samples, and purified αHL in E1 and E2. This preparatory procedure yields homogeneously pure αHL (see Fig. 1, Lane 2).

3.2 Cysteine Mutant Labeling

1. Fluorescent Labeling of αHL N244C with Cy5

   A. The concentration of pooled E1-E2 fractions containing pure αHL is estimated using SDS-PAGE with a BSA standard curve. Typical concentrations are ~0.6–1.0 mg/mL.

   B. A ten-fold molar excess of TCEP (8 μL of 10 mM stock; 80 nmol) is combined with aHL protein (8 nmol; e.g., 460 μ L of 0.63 mg/mL αHL) and placed on a rocker for 10 min

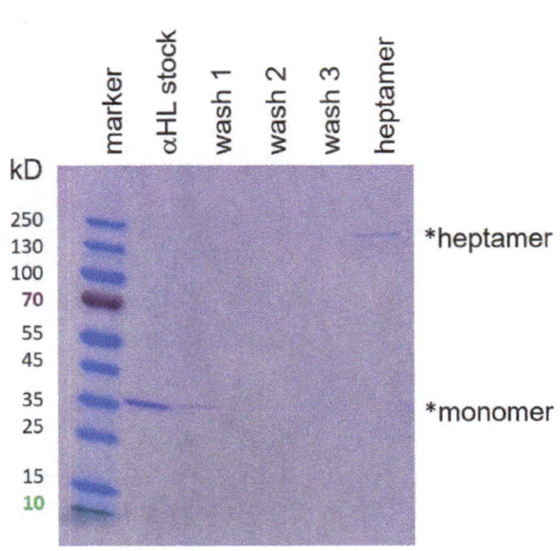

Fig. 1 αHL samples were analyzed using SDS-PAGE. A typical stock of monomeric αHL purified using IMAC (Lane 2) contains a small but detectable amount of heptamer (faint band below 250 kDa). Consecutive washes of αHL-exposed erythrocyte membranes reveal monomeric αHL in wash 1 (Lane 3) and the absence of contaminating monomers in washes 2 and 3 (Lanes 4 and 5). Isolated αHL heptamers migrate at a molecular weight consistent with their oligomerized form. Similar profiles are observed for αHL variants with and without His_6 tags or Cy5 labels

at room temperature followed by 10 min at 4°C to allow reduction of any disulfide bonds between αHL monomers.

C. A 10- to 20-fold molar excess of Cy5 maleimide dye (e.g., 24 μL of 4 mg/mL Cy5) is added to the TCEP/ αHL mixture and placed on a rocker at 4°C for 2.5 h.

D. The concentration of [Im] in the reaction mixture is calculated. The αHL elution pool is typically 3/5 by volume E1 and 2/5 by volume E2, so the [Im] of the E1/E2 pool is ~240 mM Im (sample calculation: 300 μL E1*400 mM Im/500 μL total volume = 240 mM Im). The example reaction mixture would contain 224 mM Im (sample calculation: 460 μL * 240 mM Im/492 μL total volume).

E. After the incubation at 4°C, the reaction mixture is diluted with 20T/100K to bring the [Im] below 40 mM (e.g., Dilute the reaction mixture 5.6-fold with 20T/100K to bring the [Im] from 224 mM to 40 mM; add 2.75 mL 20T/100K).

F. The diluted reaction mixture is then applied to a fresh HisPur column (prepared as described above). In order to remove excess Cy5 dye that has not reacted with αHL's cysteine residue, at least seven consecutive washes (150 μL of 20T/100K) are applied to the column until no hint of blue color from Cy5 dye is evident and the eluate is completely colorless.

G. After the washes are complete, immobilized Cy5-αHL is removed from the HisPur column with three elutions of 340 μL 20T/100K/400 M Imidazole (E1-3). Samples are stored at 4°C and wrapped in aluminum foil to protect them from light exposure.

H. SDS-PAGE analysis of Cy5-αHL followed by illumination with a 655 nm laser reveals Cy5 fluorescence comigrating with αHL. Washes contain free Cy5 dye (migrating near the bromophenol blue dye front of the polyacrylamide gel) while elutions of Cy5-αHL are devoid of contaminating free Cy5 dye.

I. The N-terminal His_6 tag is removed from Cy5-αHL by specific cleavage with TEV protease (from Invitrogen; use 100 units TEV per mg αHL protein) in dialysis (20T/100K/1 mM DTT). A second round of dialysis with DTT-free buffer is required prior to IMAC purification. Cleaved Cy5-αHL protein is collected as HisPur column flow-through while the His_6 tag, any uncleaved α HL protein, and TEV protease are retained on the HisPur column. SDS-PAGE analysis reveals 90% of N244C αHL monomers were labelled with Cy5.

3.3 αHL Nanopore Heptamer Formation

1. Erythrocyte Membranes

   A. Place 7 mL of 5 mM Tris pH 7.4 (5 T) in a 15 mL conical tube. Add a total of 50 μL 10% RBCs incrementally (20 μL, 20 μL, and then 10 μL) carefully inverting the tube between each addition to promote hypotonic lysis of the cells.

   B. Divide the lysed red blood cell solution into four 1.7 mL microcentrifuge tubes by pipetting ~1.60 mL into each tube (discard any leftover solution) and centrifuge at 20,000 × *g* for 20 min at 4°C to isolate the RBC membranes in the pellet.

   C. Using a micropipette, carefully remove all the supernatants without disrupting the tiny red pellets that form at the bottom of the tubes.

   D. Add 200 μL of 5T buffer to each tube and resuspend the pellets by drawing solution up and down with a pipettor. Combine all resuspended pellets into one tube and then centrifuge again at 20,000 × *g* for 10 min at 4°C.

   E. Remove all of the supernatant (~800 μL) and replace with 500 μL of 5T buffer to allow resuspension and washing of the pellet. Centrifuge again at 20,000 × *g* for 10 min at 4°C.

   F. Repeat the previous step one more time. With each wash, resuspension, and centrifugation, the red color of the pellet becomes less apparent.

   G. Again, remove all of the supernatant without disrupting the pellet, add 800 μL of 10T/2N buffer (10 mM Tris pH 7.5, 2 M NaCl) and resuspend the pellet. Centrifuge the tube at 20,000 × *g* for 10 min at 4°C. This high-salt wash will remove cytoskeletal proteins from the erythrocyte membranes.

   H. To wash the pellet again, remove all of the supernatant without disrupting the pellet, and add 750 μL of 10T/2N buffer and resuspend the pellet. Centrifuge at 20,000 × *g* for 10 min at 4°C.

   I. Repeat the previous step one more time. The pellet should be transparent at this point.

   J. Lastly, remove all of the supernatant without disturbing the pellet, and add 150 μL of 10T/2N buffer to resuspend the pellet. Store the tube containing resuspended erythrocyte membranes at 4°C. Prior to use, gently tap the tube to resuspend the membranes.

2. Isolating Heptamers

   A. Prepare 2× concentrated erythrocyte membranes by placing 100 μL of erythrocyte membranes into a 1.7 mL microcentrifuge tube and centrifuging at 20,000 × *g* for 10 min at 4°C. Remove and discard half (50 μL) of the supernatant and resuspend the pellet with the remaining 50 μL.

   B. To the 2× concentrated erythrocyte membranes, add 120 μL of αHL protein monomers (at ~0.6–1 mg/mL in 20 T/100 K).

   C. Incubate the erythrocyte membranes and αHL protein monomers for 15 min at room temperature on a rocker.

   D. Centrifuge the tube at 20,000 × *g* for 5 min at room temperature.

   E. Remove all the supernatant avoiding the clear pellet at the bottom and place this solution into a tube labeled *excess monomer*. Note: Multiple rounds of heptamer formation can be accomplished using the recovered *excess monomer* (120 μL) with 50 μL of fresh 2× concentrated erythrocyte membranes. Repeat the entire procedure for *isolating heptamers*.

   F. To the pellet, add 100 μL of 10T/2N buffer and resuspend. Centrifuge at 20,000 × *g* for 5 min at room temperature. Remove all the supernatant without disrupting the pellet and place it in a sample tube labeled *W1* (wash 1).

   G. Repeat the previous step two more times in order to complete three rounds of washes (*W1*, *W2*, *W3*) with 10T/2N buffer. These washes allow removal of peripherally bound αHL monomers. The pellet contains the αHL heptamers.

   H. Prepare a 1% SDS solution by placing 90 μL of 10T/2N into a 0.6 mL microcentrifuge tube and then adding 10 μ L 10% SDS to the surface of the 10T/2N buffer. Mix the solution by gently flicking the tube. A white, goopy precipitate will form on the surface. Avoid this precipitate in the next step.

   I. Use 50 μL of the 1% SDS/10T/2N to resuspend the pellet. Centrifuge at 20,000 × *g* for 20 min at room temperature.

   J. Remove the ~50 μL supernatant (avoiding the pellet) and place it into a sample tube labeled *heptamers*. This is the final product containing preassembled protein heptamers. Store at 4°C.

   K. Resuspend the pellet (stripped membranes) with 30 μL of 20 T/100 K if sample analysis is desired.

L. Formation of the preassembled heptamers is confirmed through gel electrophoresis and electrophysiology on a lipid membrane.

   i. SDS-PAGE: using a 4%–20% Tris-Glycine gel, the protein samples travel through an SDS-buffered environment according to their molecular weights. We observe the heptamer bands slightly below the 250 kD marker (see Fig. 1, Lane 6).

   ii. Electrophysiology: create a stable lipid membrane, apply a voltage, and add small amounts of preassembled heptamer in order to track jumps in the current that signify insertion of a heptamer through the membrane.

3.4 Apparatus

1. Electrophysiology

   A. Microelectrode cavity array (MECAopto-inv) with ring-shaped electrode (Nanion Technologies). This chip provides four microwells for forming bilayers and four independently addressable Ag/AgCl electrodes. The top chamber volume is ~400 μL. The microwell dimensions are 150 μm diameter × 25 μm (~440 nL).

   B. Orbit mini MECAopto-inv extension adapter (Nanion Technologies).

   C. Orbit mini amplifier (Nanion Technologies): four-channel amplifier and electrophysiology recording apparatus.

   D. Capacitance and resistance calibration test cell (Nanion Technologies).

   E. Orbit mini microscope stage adapter (3D printed using ABS).

   F. Gravity-driven external perfusion system. Fluid controller (Nanion Technologies). Angled suction and delivery tubes (Warner Instruments).

2. Fluorescence Microscope

   A. Ti-E inverted fluorescence microscope (Nikon) with x–y motorized stage & joystick controller.

   B. CFI Apo TIRF 100XC Oil objective (Nikon).

   C. Dichroic mirror (647 custom TIRF cube, Chroma Technologies), Emission filter (ET690/50 m, Chroma Technologies).

   D. Single-mode fiber-optically coupled laser (637 nm, 70 mW) with a distal collimation lens, field diaphragm (iris), and aperture lens.

   E. EMCCD camera (Andor, DU-897-CSO-#BV).

3. Configure all components as shown in Fig. 2A. The Orbit mini is mounted on the Nikon Ti-E microscope stage using a 3D-printed adapter designed for the footprint of the Orbit mini. The perfusion delivery and extraction tubes are mounted using removable magnetic holders positioned next to the MECAopto-inv chip. The laser input is adjusted with an iris to avoid illumination of the ring-shaped Ag/AgCl electrode within the microwell (Fig. 2B). The laser illuminates a 500–7000 μm^2 portion of the freestanding lipid bilayer that is suspended above the cover glass by the SU8 polymer layer. Fluorescent spots are detected in the bilayer using an EMCCD camera mounted in the microscope camera port. The dual electrical and optical recording system uses Elements Data Reader 3 (EDR) to monitor the formation of lipid bilayers and record nanopore ion currents. NIS-Elements imaging software controls the camera and all microscope imaging optics. Figure 2C illustrates the open/closed nanopore distribution under study in this protocol. All spots detected on the bilayer arise from multimeric structures (presumed heptamers) possessing one Cy5 fluorescent label per monomer (i.e., seven fluorescent labels per nanopore).

3.5 Bilayer Formation and Electrical Monitoring

1. Ensure all solutions are warmed to room temperature before using in the MECAopto-inv chip.
2. Rinse the MECAopto-inv chip thoroughly using ethanol and distilled water. It can be gently wiped dry with a lint-free cloth. A disposable canister of compressed air can be used to fully dry the chip.
3. Insert the calibration test cell into the Orbit mini. Compensate the voltage offset for all channels to obtain a reading of 0 nA. Remove the test cell.
4. Under the EDR *View* > *Analysis* > *RC estimation* tab, toggle OFF the resistance monitoring. Set the Amplitude and Tperiod to 100 mV and 40 ms, respectively. Press *Start analysis* (green arrow) to activate triangular wave voltage. Under *View* > *Settings* > *Controls* tab, set Range to 2 nA, Sampling Rate to 5 kHz, and Final Bandwidth to SR/2.
5. Add 400 μL of electrophysiology recording buffer into the MECAopto-inv chip. Use the hair of a sterile paint brush to gently remove any air bubbles trapped in the microcavities (view under a dissection microscope).
6. Insert the MECAopto-inv extension adapter into the Orbit mini (Fig. 2A), then gently insert the chip into the extension adapter. Off-scale square waves should immediately appear. This indicates electrical continuity between the top and bottom chambers (*see* **Note 1**).

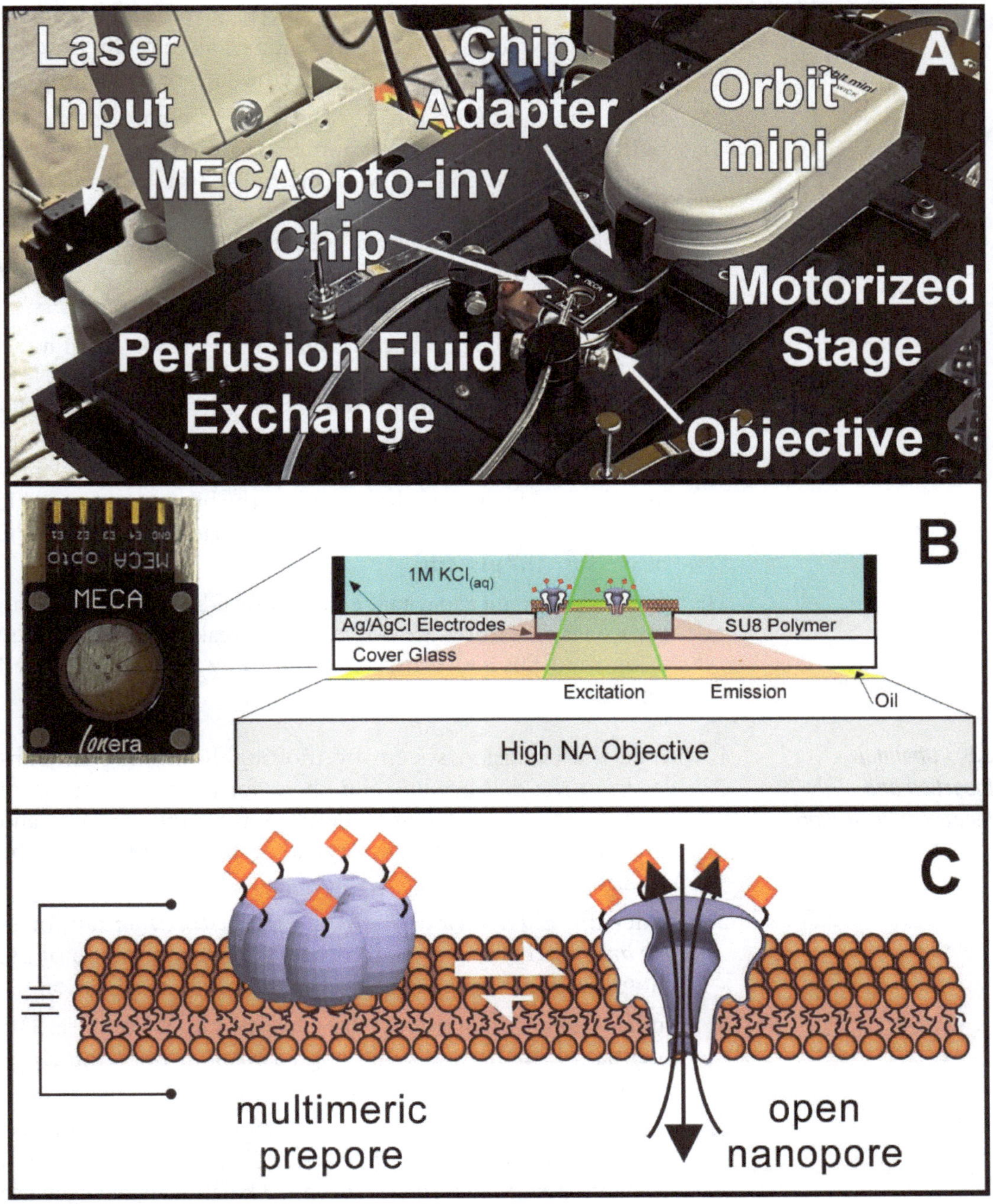

Fig. 2 Set up for simultaneous optical and electrical recording in free-standing lipid bilayers formed on MECAopto-inv chips. (**A**) Inverted microscope configuration with a mounted Orbit mini on a 3D motorized translation stage. A gravity-driven perfusion system enables solution exchange. (**B**) Expanded schematic of a single microwell within the chip and the optical interface. The excitation beam (green) is adjusted so that light propagates inside the ring-shaped Ag/AgCl electrode. The conically shaped excitation beam illuminates a portion (500–7000 μm^2) of the total bilayer area. Fluorescence emission (red) is collected by the objective (N.A. = 1.49) over a solid angle that is larger than the angle of the excitation beam. (**C**) Generalized equilibrium dynamics of bilayer-associated prepores and electrically conducting nanopores. For the open αHL nanopore, the equilibrium is effectively irreversible

7. Dip the tip of a small disposable pipette into the lipid solution (do not draw up an appreciable volume). A thin film of residual lipid solution should adhere to the surface of the tip.
8. Insert the pipette tip into the buffer-filled chip. Depress the pipette plunger until a small bubble is formed at the tip. Bring the bubble into contact with the SU8 layer and move the bubble back and forth over microcavities until long-lasting small-amplitude square waves (10–30 pF) are observed. This indicates lipid solution deposition over the microcavity and electrical isolation between the top and bottom chambers.
9. Wait 2–3 min for spontaneous thinning of the solvent annulus as indicated by growth of the square wave amplitude (*see* **Note 2**). The authors consider a reading of ~45–90 pF to be adequate. This corresponds to a bilayer diameter of 90–130 mm. Record the stabilized capacitance reading and save for entry into the Excel template that estimates the bilayer area available for nanopore insertion. The specific capacitance of DPhPC bilayers is ~0.7 μF/cm^2.
10. If the membrane capacitance never exceeds ~35 pF and clean square wave features are not evident, break the membrane(s) using the EDR *Zap* feature and return to 3.5.8 or 3.5.2 and repeat.

3.6 Channel Insertion and Electrical Recording

1. Prime the perfusion system by thoroughly flushing air bubbles from the delivery line and check for proper operation of the vacuum extraction line. Position the fluid delivery and extraction tubes from the gravity-fed perfusion system into the MECAopto-inv chip (Fig. 2A).
2. Under the EDR *View* > *Analysis* > *RC estimation* tab, press *Stop analysis* (black square) to deactivate the application of the triangular wave voltage used for capacitance determination.
3. Under *View* > *Settings* > *Voltage Protocols* tab > *Vhold*, set DC potential to the desired value (e.g., 100 mV). Press the *Apply* button.
4. Under *View* > *Controls*, press the *Setup Save Settings* button to activate the *EDR Save Preferences* dialog. Set the file path and name desired for data recording. The authors typically use 32-bit binary format with files split every 2 min. Start recording when ready, prior to the addition of nanopore protein.
5. Inject the nanopore proteins into the MECAopto-inv chip and wait for insertion into the bilayer.
6. Insertions appear as discrete current jumps, which enable counting of open nanopores in the bilayer.
7. Following the desired number of nanopore insertions (Fig. 3), rinse the MECAopto-inv chip with 10–20 chamber volumes of

recording buffer (5–10 mL) to remove labeled proteins from solution. Because the adsorption of αHL to the lipid bilayer is effectively irreversible, bilayer-associated heptamers remain intact. The surface density of nanopores is not appreciably reduced by rinsing. Removal of the protein from solution slows and prevents additional nanopore insertions, allowing optical recordings to be performed with an unchanging number of inserted nanopores. Figure 3 shows data from an experiment where a 16-s optical recording was performed from 50–66 s, with only one open nanopore in the bilayer.

8. Continue electrical recording until the optical measurements are complete.

3.7 Single-Molecule Optical Recording of the Bilayer

1. Activate the NIS-Elements Imaging software.
2. Under *View* > *Acquisition Controls* > *DU-897 Settings* > *Commands* > *Advanced Camera Settings*, Cool the camera to −80°C. Under *View* > *Acquisition Controls* > *DU-897 Settings* set *Format for Capture* > *Binning 2 × 2* and *Exposure Time* to 40 ms.
3. With no EM gain applied and a low white light transverse illumination intensity, position the MECAopto-inv chip to image the edge of the plus-shaped fiducial mark. Use the fiducial mark as a reference to navigate to a microwell that has an intact bilayer with inserted nanopores (Fig. 2B).
4. Center the microwell in the camera field of view.
5. Raise the Z position of the objective ~25–30 μm, turn off transverse light, set the *EM Gain Multiplier* to 600 or greater. Briefly activate laser illumination (50–200 W/cm²) to view individual fluorescent spots diffusing in the membrane. Quickly refine the

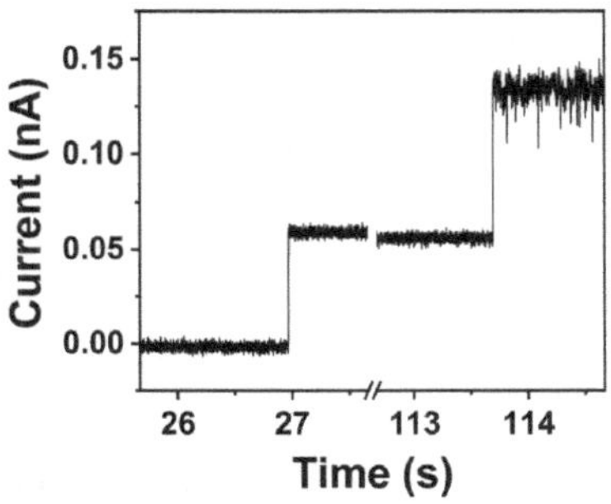

Fig. 3 Single nanopore insertions occur shortly after protein addition and reach a steady state following a chamber rinse. In this recording (performed at 100 mV), a single nanopore was present in the bilayer from ~27 to 113.7 s. A corresponding optical recording was acquired from 50 to 66 s (Fig. 6). The slight drop in current overtime (visible at the break) is likely due to minor Ag/AgCl electrode polarization

focus to sharply resolve moving fluorescent spots. Verify membrane planarity by ensuring that the spots remain in focus over the entire illuminated area. If spots cannot be focused from edge to edge, a bilayer with a concave or convex curvature may have formed. If all intact bilayers exhibit curvature, the experiment will need to be restarted from Step 3.5.1. Avoid premature photobleaching of spots by keeping the exposure time minimal and the laser intensity low. Deactivate the laser illumination.

6. Under *Acquire* > *Fast Timelapse*, specify the path and filename for the video recording, as well as the number of frames to acquire.
7. Save the file in *.nd2 format to retain microscope-specific metadata, including the frame interval (*see* **Note 3**).
8. Press *Apply* and *Run Now* to acquire the video. Figure 4 shows a frame from a fluorescent spot recording along with spot trajectories (added after processing) from a nanopore titration experiment. The video file was collected after an initial low-concentration addition of heptameric protein. The surface density of nanopores adsorbed to the bilayer is relatively low and no open nanopores are present (i.e., all spots are prepores). Low surface coverage produces fewer linking errors and yields more accurate diffusion constants.

3.8 Spot Detection and Tracking with Fiji (ImageJ) and TrackMate

1. Open the video file (*.nd2) of the optical recording in Fiji (ImageJ).
2. Enhance the image quality using *Image* > *Adjust* > *Brightness/Contrast* (*see* **Note 4**).
3. If the camera has been previously calibrated in NIS-Elements, the calibration settings will be loaded automatically with the file header. If not, input the pixel calibration values and appropriate units using *Image* > *Properties*. Set the voxel depth to a

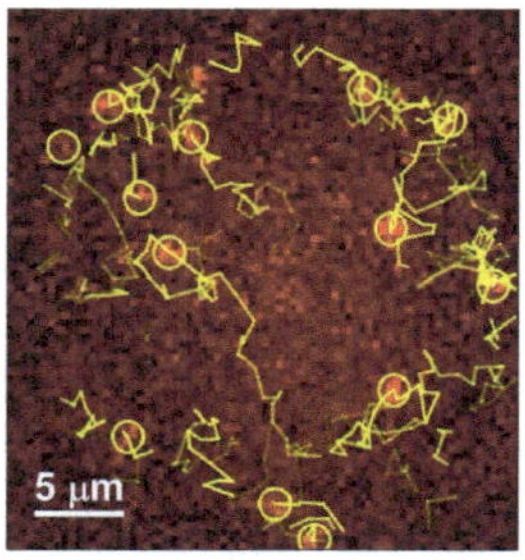

Fig. 4 Fluorescent spots from nanopores in the lipid bilayer and their corresponding diffusion trajectories. The trajectory intensity fades for spot locations in earlier frames

value of 1. If needed, adjust the frame interval to match the experimental acquisition conditions (*see* **Note 5**).

4. Identify a region of interest (ROI) that matches the illuminated area.

   A. Select the oval drawing tool from the ImageJ toolbar. Draw an ROI around the illuminated area. Right-click on the ROI > *Add to ROI Manager* > *More* > *Specify* to control specific ROI settings.

5. Determine the illuminated area using the *Measure* tool (*Analyze* > *Measure*). The area will be reported in calibrated units. Save this value for entry into the Excel template.

6. Crop the video to the dimensions of the ROI (*Image* > *Crop*).

7. Open TrackMate for the cropped video using *Plugins* > *Tracking* > *TrackMate.*

   A. Select *LoG detector.*

   B. Use an estimated object diameter of 1.5 to 2.0 μm. Activate median filter pre-processing and sub-pixel localization.

   C. Select Preview. Adjust the Quality threshold setting so that the number of detected spots visually matches the perceived number of particles in the video (*see* **Note 6**).

   D. Under *Initial thresholding*, set the threshold so that all spots are selected.

   E. Under *Set filters on spots*, adjust the slider to keep all spots.

   F. Under *Select a tracker*, select *Simple LAP tracker.*

   G. Under *Simple LAP tracker*, set a *linking max distance* of 1.8–2.1 μm (*see* **Note** 7). Set both gap-closing parameters to 0.

   H. Under *Set filters on tracks*, adjust the slider setting to keep all tracks. Figure 4 shows typical trajectories arising from the fluorescent spots recorded over time.

   I. Under *Display options*, select *Tracks.* In the *Track tables*, select *Spots.*

      i. Click the *Frame* column heading to sort by frame.

         a. Count the number of tracks in the initial frame.

         b. Calculate the initial particle density by dividing the number of tracks by the measured illumination area. If initial density is above 0.05 particles/μm^2, allow photobleaching to naturally reduce the spot density. Crop out the beginning frames of the video file where the particle density is high and repeat Step 3.8.7 before diffusion coefficient analysis (*see* **Note 8**).

         c. Also keep a processed copy of the full video file (i.e., not cropped in time) for accurate particle density analysis using the supplemental Excel template.

ii. Click the *Track ID* column heading to sort by track ID (*see* **Note 9**).

iii. Select *Export to CSV*. Save as a *.csv file.

3.9 Analysis Using Excel

1. Open a new Excel workbook. Excel computations can be developed from a blank file using the following instruction set. Alternatively, a pre-formatted template file is provided as a supplement at: Supplementary File 1. The template accommodates up to 60,000 detected spots from the TrackMate *.csv file. Example data from the TrackMate *.csv file with processed results are provided in Sheet 2 of the workbook.

2. Open the TrackMate.csv trajectory file in Excel. Isolate the block of columns (A-I) from the *.csv file: LABEL, ID, TRACK_ID, QUALITY, POSITION_X, POSITION_Y, POSITION_T, FRAME. If using the supplement template file, simply copy and paste columns A-I into the template. Add additional data values in the yellow-coded cells as described in Sects. 3.9.3, C, D, J, P, S, U. If developing a blank workbook file, delete all other columns and add formulas and data as directed below.

3. If developing an Excel workbook from scratch, add the following columns. *See* **Note 10** for exact Excel formulas.

   A. Insert column J and label it "dr." Calculate the step displacement (dr) between consecutive particle positions within the same trajectory.

      i. Compute the stepwise displacement between successive frames only when FRAME increases by +1 and TRACK_ID is unchanged from the previous row. Leave rows that do not satisfy this condition as 0 (*see* **Note 10a**).

      ii. The stepwise lateral displacement for every valid frame interval is calculated as:

$$dr = \sqrt{(x_{t+1} - x_t)^2 + (y_{t+1} - y_t)^2}$$

   B. Calculate the squared displacement (dr^2)

      i. Insert column K. Label it "dr^2." This contains squared displacement (SD) values.

      ii. Square the values obtained in column J to determine the two-dimensional SD for each trajectory step (*see* **Note 10b**).

$$SD = (dr)^2$$

C. Enter measured localization uncertainty for the microscope into column L, row 5. Use units of μm^2. Label the column "σ^2" (*see* **Note 11**).

D. Enter the camera exposure time. The authors recommend using 40 ms when the laser illumination is continuous and set to ~200 W/cm². Label column M as "τ_{exp}." Enter data in units of seconds.

E. Determine the motion-blur correction factor (R). This correction factor is necessary if the exposure time is a significant fraction of the frame interval.

   i. Insert column N and label it "Blur Factor (R)."

   ii. Calculate the correction value where τ_{exp} is the camera exposure time and Δt is the frame interval. This factor compensates for motion blur when τ_{exp} approximates Δt (*see* **Note 10c**).

$$R = \frac{\tau_{\exp}}{6\Delta t}$$

F. Calculate the corrected SD

   i. Insert column O and label it "r^2_{corr}" (corrected squared displacement).

   ii. Compute the corrected SD, where σ^2 equals the localization uncertainty and R is the blur correction factor.

   iii. Leave rows without valid dr^2 value as 0 (*see* **Note 10d**).

$$r^2_{corr} = \frac{(dr^2 - 4\sigma^2)}{(1 - 2R)}$$

G. Compute single-step diffusion coefficients

   i. Insert column P and label it "Single-step diffusion (D_{step})."

   ii. Divide the corrected SD by $4\Delta t$ to yield the apparent diffusion coefficient for each step, where Δt equals the frame interval (*see* **Note 10e**).

$$D_{step} = \frac{r^2_{corr}}{4\Delta t}$$

H. Remove zeros and negatives

   i. Some measured SDs are smaller than the localization uncertainty. This occasionally produces negative SD values. In addition, the first spot in every trajectory does not produce a valid SD. All negative and zero values need removal.

ii. Insert column Q and label it, "Remove zeros & negs"

iii. *See* **Note 10f**.

I. Count the number of valid SDs

i. Determine the total number of valid SDs to use in a Complementary Cumulative Squared Displacement (cCSD) plot by processing column Q with the COUNT function.

ii. Insert column R and label it, "Total # of SDs (N)." The first spot in each trajectory and all negative values produced by correction are excluded.

iii. *See* **Note 10g**.

J. Sort the single-step diffusion values.

i. Insert column S and label it "Sorted D_{step}." Sort all D_{step} values in ascending order (*see* **Note 10h**).

K. Enter the linking maximum distance used in TrackMate into column T, row 5. Use units of μm. Label the column Link Max. Dist. (MLR). This value is used to eliminate data points that are biased. Under representation at large D_{step} occurs because large displacements get preferentially eliminated near the edge of the field of view and also by the MLR cutoff.

L. Truncate the sorted D_{step} column to exclude biased data points.

i. Exclude all values greater than or equal to $MLR^2/(4\Delta t)$.

ii. Insert column U and label it "MLR Truncated D_{step}."

iii. *See* **Note 10i**.

M. Calculate the Complementary Cumulative Squared Displacement (cCSD)

i. Insert column V and label it "c*CSD*."

ii. For each sorted value, compute the cCSD probability distribution using the following equation, where *i* is the sorted rank index. This time-normalized distribution represents the fraction of apparent diffusion constants that are larger than the corresponding D_{step} value. The plot can be conveniently used to extract diffusion coefficients (*see* **Note 10j**).

$$cCSD_i = 1 - \frac{(i-1)}{N}$$

N. Compute the natural logarithm of the cCSD.

   i. Insert column W and label it "ln(cCSD)."
   ii. *See* **Note 10k**.

O. Create cCSD plot

   i. Select columns U and W and insert a scatter chart. For a single homogeneous diffusing species, typical of αHL prepores and open nanopores, the plot should appear linear. Nonlinearity indicates multiple diffusing species.
   ii. Add a linear trendline with the intercept fixed at zero.
   iii. The slope from the trendline on the plot can be used to compute the diffusion constant using

$$D\left(\mu m^2/s\right) = -1/\text{slope}$$

Figure 5 shows an example cCSD plot produced using the Excel template. Replicate trials (N=6) give a diffusion constant of 8.1 ± 0.3 $\mu m^2/s$ for the type of αHL heptamer described in this report.

P. Compute and display the diffusion constant directly within a cell in the template table.

   i. Insert column X and label it "Diffusion Constant."
   ii. Compute the homogeneous diffusion constant using the linear slope and y-intercept fixed at 0 (*see* **Note 10l**).

Q. Enter the bilayer capacitance recorded from the EDR.

   i. Insert column Y and label it "Bilayer Capacitance." Enter capacitance value in units of picofarads (pF).

R. Assuming a specific capacitance of 0.7 $\mu F/cm^2$, estimate the total bilayer area in μm^2. This estimate generally represents the nanopore-permeable portion of the membrane only, excluding the thick solvent annulus that surrounds the bilayer.

   i. Insert column Z and label "Total Bilayer Area."
   ii. *See* **Note 10m**.

S. Compute the bilayer diameter, assuming a planar membrane geometry.

   i. Insert column AA and label it "Bilayer Diameter."
   ii. *See* **Note 10n**.

T. Enter the measured number of nanopores inserted in the bilayer during the optical recording. This is determined by the number of stepwise current increases at the time of the optical recording (Fig. 3).

   i. Insert column AB and label it "Open Nanopores."

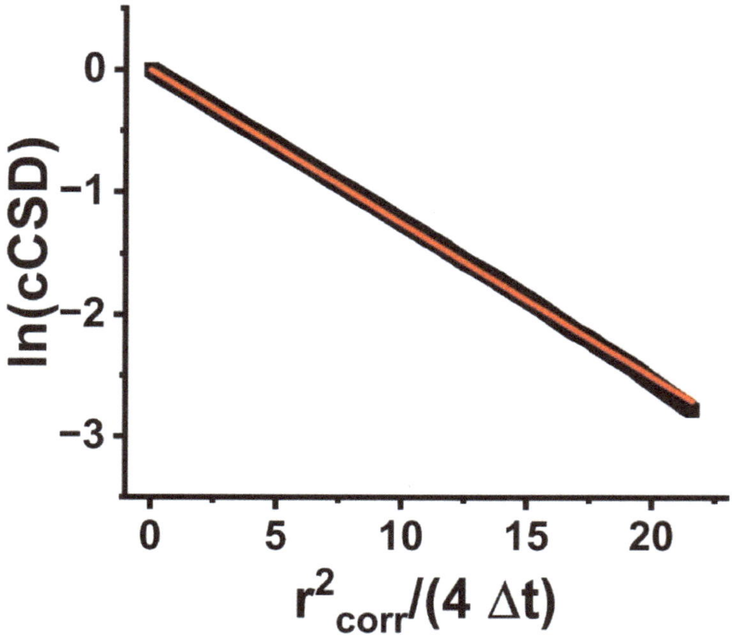

Fig. 5 Complementary cumulative squared displacement analysis of trajectories from heptameric N244C-Cy5 αHL adsorbed to the bilayer yields a monodisperse diffusion constant of 8.1 ± 0.3 μm²/s (N = 6 trials)

U. Compute the density of open nanopores in the bilayer based on the Total Bilayer Area.

   i. Insert column AC and label it "Open Nanopore Density."

   ii. *See* **Note 10o**.

V. Enter the bilayer area illuminated by the laser, as determined in 3.8.5. This will typically be a smaller portion of the total bilayer area than is available for nanopore insertion.

   i. Insert column AD and label it "Optically Probed Area."

   ii. Enter the ROI value determined with the Fiji *Measure* tool (Step 3.8.5).

W. Compute the average number of nanopore trajectories per frame from the first five frames.

   i. Insert column AE and label it "Average # Traj."

   ii. *See* **Note 10p**.

X. Compute the total nanopore density on the bilayer determined by the optical measurement.

i. Insert column AF. Label it "Meas. Total Nanopore Density."

ii. The TrackMate output gives trajectory information for the total (i.e., both open and closed) number of nanopores in the bilayer. Figure 6 indicates that photobleaching can rapidly reduce the number of detected spots, if the laser intensity is high. Therefore, use the average number of trajectories per frame from the beginning of the video recording (i.e., first five frames) to minimize bias introduced by photobleaching (*see* **Note 10q**).

Y. Correct the Total Nanopore Density for spatially overlapping spots, if the surface density is high.

i. When the spatial density of nanopores grows large, TrackMate cannot identify and count unique spots. The number of measured spots (ρ_{meas}) generally needs correction if the spot density exceeds ~0.05 spots/μm^2. The authors use a two-dimensional diffusion simulation [33] set to the diffusion constant of the αHL heptamer (i.e., 8.1 μm^2/s) to generate a surface density saturation curve of the form

$$\rho_{meas} = a\rho_{cor} / \left(b + \rho_{cor}\right)$$

where ρ_{meas} is the particle density determined from TrackMate and ρ_{cor} is the simulated particle density. After determining values for the fitting parameters (a and b), the corrected particle density is calculated by entering ρ_{meas} and solving for ρ_{cor}. The supplemental Excel template uses values of a and b from a particle simulation of 8.1 μm^2/s within a circularly illuminated area having a diameter of 35 μm.

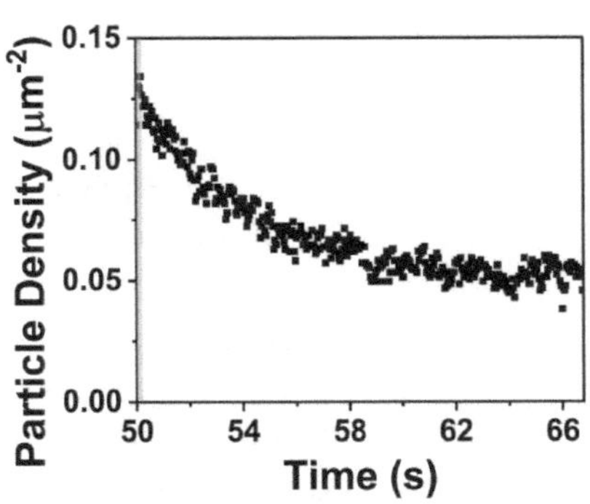

Fig. 6 Photobleaching reduces the observed particle density over time. Only data from the initial frames (green) are used to count the number of heptamers on the bilayer. The total number of N244C-Cy5 αHL prepores covering the membrane is typically much larger than the number of open nanopores

ii. Insert column AG. Label it "Corr. Total Nanopore Density."

iii. *See* **Note 10r**.

Z. Estimate the total number of nanopores over the entire bilayer.

i. Use the corrected total nanopore density to estimate the total number of nanopores associated with the Total Bilayer Area. This calculation accounts for nanopores outside the region of the laser illumination by assuming a uniform spatial density. Spatial uniformity can be verified by observing the spot density close to the annulus with movements of the motorized stage. For αHL in DPhPC bilayers, the spatial density of diffusing spots appears generally uniform in all directions.

ii. Insert column AH. Label it "Est. Total # of Nanopores."

iii. *See* **Note 10s**.

AA. Compute the Open/Closed Nanopore Ratio.

i. For N244C-Cy5 αHL, very few bilayer-associated heptamers insert to form open nanopores. Most are in a nonconducting prepore state. The authors find that the Open/Closed ratio for this heptamer type is typically ~10^{-3}-10^{-4} under conditions that produce single-nanopore insertions. Other mutant types and tag locations can produce a much higher Open/Closed ratio.

ii. Insert column AI, label it "Open/Closed Nanopore Ratio."

iii. *See* **Note 10t**.

4 Notes

1 If off-scale square waves do not appear, carefully manipulate the chip until proper electrical contact is established with the MECAopto-inv adapter.

2 No thinning is an indication of a thick lipid deposit. Thin out the deposit by repeating Step 3.5.8 with a fresh pipette tip without additional lipid, until annulus thins. Alternatively, use a small piece of clean Teflon film to spread the deposited lipid solution over a larger area. An ideal membrane is characterized by a symmetric square wave possessing a nearly flat top. A triangular shape indicates leakage of the membrane or an uneven deposition of lipid and solvent across the microcavity.

3 For the DU-897 camera used in this experiment, the frame interval is 42 ms when the readout mode is 10 MHz at 14-bit resolution.

4 Adjusting image *Brightness/Contrast* is a visual aid and does not affect intrinsic intensity values. It is important to be able to visualize spots clearly in order to provide an accurate Estimated Object Diameter.

5 Frame interval equals the sum of exposure and camera read-out times.

6 The Quality parameter takes into account both the intensity and diameter of a spot (following normalization and filtering). It gets larger for brighter spots and for spots with a diameter that matches the object diameter. In cases where the difference between particles and background fluorescence is not distinct, the authors prefer a high Quality threshold to avoid linking erroneous spots into tracks.

7 The Linking max distance, alternatively called the maximum linking radius (MLR), sets the maximum distance between two spots in successive frames that will be linked into a single trajectory. It is crucial to set an appropriate value. An MLR that is too large will cause erroneous linking of unrelated spots, while an MLR that is too small will result in artificially truncated trajectories.

8 High particle densities cause increased error due to incorrect linkage of spots. To minimize the formation of incorrect linkages for diffusion measurements, the authors keep the particle density below 0.05 particles/μm^2.

9 Sorting first by frame, then by track ID prepares the export *.csv file for use with the Excel template. If this combination of sorting steps is not performed in the proper sequence, the template will not function properly.

10 Excel Formulas for Direct Implementation

   a Step displacement (dr): `=IF((I5=I4+1)*AND(C5=C4),SQRT((E5-E4)^2+(F5-F4)^2),)`

   b Squared displacement (dr^2): `=J5^2`

   c Motion blur factor (R): `=$M$5/(6*($H$6-$H$5))`

   d Corrected squared displacement (r^2_{corr}): `=IF(K5<>0,((K5-4*$L$5)/(1-2*$N$5)),)`

   e Single-step diffusion coefficients ($r^2_{corr}/4\Delta t$): `=O5/(4*($H$6-$H$5))`

   f Remove zeros and negatives: `=IF(P5<=0,"“,P5)`. Use the fill handle to copy the formula down to row 60004.

   g Total # of SDs (N): `=COUNT(Q5:Q60004)`

h Sorted D_{step} values: =SORT(Q5:Q60004,1)

i Truncate sorted D_{step} values: =FILTER(S5:S60004, (S5:S60004<(($T$5^2/(4*($H$6-$H$5))))))

j Complementary cumulative squared displacement (cCSD): =IF(ISNUMBER(U5),(1-((ROW(S5)-ROW(S$5))/$R$5)),”“)

k Natural Logarithm of cCSD: =IF(ISNUMBER(V5), LN(V5),”“)

l Compute diffusion constant from slope of cCSD plot: =-1/INDEX((LINEST (FILTER (W5:W60004, ISNUMBER(U5:U60004)*ISNUMBER(W5:W60004)), FILTER(U5:U60004,ISNUMBER(W5:W60004)*ISNUMBER(U5:U60004)),FALSE,FALSE)),1)

m Compute total bilayer area: =((Y5*0.000001)/0.7)*100000000

n Compute bilayer diameter: =2*SQRT(Z5/PI())

o Compute open nanopore density: =AB5/Z5

p Compute the average number of nanopore trajectories per frame from the first 5 frames of the video recording: = AVERAGE(COUNTIF(I5:I60004,0),COUNTIF(I5:I60004,1),COUNTIF(I5:I60004,2),COUNTIF(I5:I60004,3),COUNTIF(I5:I60004,4))

q Compute the density of both open and closed nanopores (i.e., total nanopores) as determined by the optical recording: =AE5/AD5

r Correct the measured total nanopore density for spatial overlap of spots: =0.25974/((0.25856/AF5)-1)

s Estimated total number of nanopores over the total bilayer area of the membrane: =Z5*AG5

t Open/Closed Nanopore Ratio: =AB5/(AH5-AB5)

11 Optical imaging systems cannot determine the exact spatial location of single molecules. Even stationary particles exhibit apparent motion due to camera readout noise and nanoscale-level equipment oscillation. Localization uncertainty is typically estimated by measuring positional fluctuations of fixed dye molecules on a cover glass. This value must be determined for each apparatus individually. The value for our system is 0.007 μm^2.

Acknowledgments

This work was supported by the National Science Foundation, Grant# 2105892, and the Wheaton College Alumni Association.

References

1. Zhang M, Chen C, Zhang Y, Geng J (2022) Biological nanopores for sensing applications. Proteins 90(10):1786–1799. https://doi.org/10.1002/prot.26308
2. Mayer SF, Cao C, Dal Peraro M (2022) Biological nanopores for single-molecule sensing. iScience 25(4):104145. https://doi.org/10.1016/j.isci.2022.104145
3. Zhang T, Li H, Jiang M, Hou H, Gao Y, Li Y, Wang F, Wang J, Peng K, Liu Y-X (2024) Nanopore sequencing: flourishing in its teenage years. J Genet Genom 51(12):1361–1374. https://doi.org/10.1016/j.jgg.2024.09.007
4. Wang Y, Zhang S, Jia W, Fan P, Wang L, Li X, Chen J, Cao Z, Du X, Liu Y, Wang K, Hu C, Zhang J, Hu J, Zhang P, Chen H-Y, Huang S (2022) Identification of nucleoside monophosphates and their epigenetic modifications using an engineered nanopore. Nat Nanotechnol 17(9):976–983. https://doi.org/10.1038/s41565-022-01169-2
5. Meng F, Li X, Zou N, Wang X (2025) Protein profiling by nanopore-based technology. Anal Chem 97(19):10110–10125. https://doi.org/10.1021/acs.analchem.5c00992
6. Yi Y, Li Z, Liu L, Wu H (2025) Towards next generation protein sequencing. ChemBioChem 26(6):e202400824. https://doi.org/10.1002/cbic.202400824
7. Das N, Chakraborty B, RoyChaudhuri C (2022) A review on nanopores based protein sensing in complex analyte. Talanta 243:123368. https://doi.org/10.1016/j.talanta.2022.123368
8. Wiswedel R, Bui ATN, Kim J, Lee M-K (2024) Beta-barrel nanopores as diagnostic sensors: an engineering perspective. Biosensors 14(7):345. https://doi.org/10.3390/bios14070345
9. Ying Y-L, Hu Z-L, Zhang S, Qing Y, Fragasso A, Maglia G, Meller A, Bayley H, Dekker C, Long Y-T (2022) Nanopore-based technologies beyond DNA sequencing. Nat Nanotechnol 17(11):1136–1146. https://doi.org/10.1038/s41565-022-01193-2
10. Crnković A, Srnko M, Anderluh G (2021) Biological nanopores: engineering on demand. Life 11(1):27. https://doi.org/10.3390/life11010027
11. Fragasso A, Schmid S, Dekker C (2020) Comparing current noise in biological and solid-state nanopores. ACS Nano 14(2):1338–1349. https://doi.org/10.1021/acsnano.9b09353
12. Ide T, Yanagida T (1999) An artificial lipid bilayer formed on an agarose-coated glass for simultaneous electrical and optical measurement of single ion channels. Biochem Biophys Res Commun 265(2):595–599. https://doi.org/10.1006/bbrc.1999.1720
13. Ide T, Takeuchi Y, Yanagida T (2002) Development of an experimental apparatus for simultaneous observation of optical and electrical signals from single ion channels. Single Mol 3(1):33–42.
14. Ide T, Takeuchi Y, Aoki T, Yanagida T (2002) Simultaneous optical and electrical recording of a single ion-channel. Jpn J Physiol 52(5):429–434. https://doi.org/10.2170/jjphysiol.52.429
15. Borisenko V, Lougheed T, Hesse J, Füreder-Kitzmüller E, Fertig N, Behrends JC, Woolley GA, Schütz GJ (2003) Simultaneous optical and electrical recording of single gramicidin channels. Biophys J 84(1):612–622. https://doi.org/10.1016/S0006-3495(03)74881-4
16. Harms GS, Orr G, Montal M, Thrall BD, Colson SD, Lu HP (2003) Probing conformational changes of gramicidin ion channels by single-molecule patch-clamp fluorescence microscopy. Biophys J 85(3):1826–1838. https://doi.org/10.1016/S0006-3495(03)74611-6
17. Harms G, Orr G, Lu HP (2004) Probing ion channel conformational dynamics using simultaneous single-molecule ultrafast spectroscopy and patch-clamp electric recording. Appl Phys Lett 84(10):1792–1794. https://doi.org/10.1063/1.1652228
18. Chandler EL, Smith AL, Burden LM, Kasianowicz JJ, Burden DL (2004) Membrane surface dynamics of DNA-threaded nanopores revealed by simultaneous single-molecule optical and ensemble electrical recording. Langmuir 20(3):898–905. https://doi.org/10.1021/la035728i
19. Heron AJ, Thompson JR, Cronin B, Bayley H, Wallace MI (2009) Simultaneous measurement of ionic current and fluorescence from single protein pores. J Am Chem Soc 131(5):1652–1653. https://doi.org/10.1021/ja808128s

20. Bartsch P, Walter C, Selenschik P, Honigmann A, Wagner R (2012) Horizontal bilayer for electrical and optical recordings. Materials 5(12):2705–2730. https://doi.org/10.3390/ma5122705
21. Weatherill EE, Wallace MI (2015) Combining single-molecule imaging and single-channel electrophysiology. J Mol Biol 427(1):146–157. https://doi.org/10.1016/j.jmb.2014.07.007
22. Huang S, Romero-Ruiz M, Castell OK, Bayley H, Wallace MI (2015) High-throughput optical sensing of nucleic acids in a nanopore array. Nat Nanotechnol 10(11):986–991. https://doi.org/10.1038/nnano.2015.189
23. Perez-Mitta G, Sezgin Y, Wang WW, Mackinnon R (2024) Freestanding bilayer microscope for single-molecule imaging of membrane proteins. Sci Adv 10(25):eado4722. https://doi.org/10.1126/sciadv.ado4722.
24. Ide T, Takeuchi Y, Noji H, Tabata KV (2010) Simultaneous Optical and Electrical Single Channel Recordings on a PEG Glass. Langmuir 26(11):8540–8543. https://doi.org/10.1021/la9045594
25. Rajapaksha SP, Wang X, Lu HP (2013) Suspended lipid bilayer for optical and electrical measurements of single ion channel proteins. Anal Chem 85(19):8951–8955. https://doi.org/10.1021/ac401342u
26. Thompson JR, Cronin B, Bayley H, Wallace MI (2011) Rapid assembly of a multimeric membrane protein pore. Biophys J 101(11):2679–2683. https://doi.org/10.1016/j.bpj.2011.09.054
27. Gu L-Q, Braha O, Conlan S, Cheley S, Bayley H (1999) Stochastic sensing of organic analytes by a pore-forming protein containing a molecular adapter. Nature 398(6729):686–690. https://doi.org/10.1038/19491
28. Rosen CB, Rodriguez-Larrea D, Bayley H (2014) Single-molecule site-specific detection of protein phosphorylation with a nanopore. Nat Biotechnol 32(2):179–181. https://doi.org/10.1038/nbt.2799
29. Rosen CB, Bayley H, Rodriguez-Larrea D (2020) Free-energy landscapes of membrane co-translocational protein unfolding. Commun Biol 3(1):160. https://doi.org/10.1038/s42003-020-0841-4
30. Yan S, Wang L, Du X, Zhang S, Wang S, Cao J, Zhang J, Jia W, Wang Y, Zhang P, Chen H-Y, Huang S (2021) Rapid and multiplex preparation of engineered mycobacterium smegmatis Porin A (MspA) nanopores for single molecule sensing and sequencing. Chem Sci 12(27):9339–9346. https://doi.org/10.1039/D1SC01399H
31. Soskine M, Biesemans A, Maglia G (2015) Single-molecule analyte recognition with ClyA nanopores equipped with internal protein adaptors. J Am Chem Soc 137(17):5793–5797. https://doi.org/10.1021/jacs.5b01520
32. Tinevez J-Y, Perry N, Schindelin J, Hoopes GM, Reynolds GD, Laplantine E, Bednarek SY, Shorte SL, Eliceiri KW (2017) TrackMate: an open and extensible platform for single-particle tracking. Methods 115:80–90. https://doi.org/10.1016/j.ymeth.2016.09.016
33. Bailey DJ, Kindt JT, Taylor MM, Paulson AR, Jones BH, Hubbell KL, Keranen-Burden LM, Burden DL (2010) Post-hoc vibration mitigation for single-molecule tracking and diffusion measurements in lipid membranes. Spectrosc Lett 43(7–8):586–596. https://doi.org/10.1080/00387010.2010.510734

Chapter 9

Simulation of pH-Dependent, Loop-Based Membrane Protein Gating Using Pretzel

Alan Perez-Rathke, Monifa A. V. Fahie, Christina M. Chisholm, Min Chen and Jie Liang

Abstract

Bacterial porins often exhibit ion conductance and gating behavior, which can be modulated by pH. However, the underlying control mechanism of gating is often complex, and direct inspection of the protein structure is generally insufficient for full mechanistic understanding. Here we describe Pretzel, a computational framework that can effectively model loop-based gating events in membrane proteins. Our method combines Monte Carlo conformational sampling, structure clustering, ensemble energy evaluation, and a topological gating criterion to model the equilibrium gating state under the pH environment of interest. We discuss details of applying Pretzel to the porin Outer membrane protein G (OmpG).

Keywords Computer simulation, Protein loop modeling, Sequential Monte Carlo sampling, Clustering, Protein conformation, Ion channel gating, Bacterial outer membrane proteins, OmpG protein

1 Introduction

Bacterial membrane porins, such as outer membrane protein G (OmpG) [1], show promise for use as nano-sensor devices [2–12]. Typically, these devices measure the flow of ionic current through the porin channel to detect and quantify the presence of a specific molecular species in solution. However, environmental factors, such as pH, can alter the gating behavior of these proteins, as is the case of OmpG, which exhibits pH-dependent gating behavior [1].

Here, we describe how the computational framework, called Pretzel (Protein Topology of Zoetic Loops) [13], can be applied to investigate loop-driven gating behavior of membrane proteins under different pH conditions. Our method allows atomic-level study of loop-based conformational gating events that may occur over time scales not amenable to conventional computational methods, such as

Fanjun Li (ed.), *Nanopore Technology*, Methods in Molecular Biology, vol. 3072,
https://doi.org/10.1007/978-1-0716-5527-6_9,

molecular dynamics and Brownian dynamics simulations. Specifically, we describe detailed application of Pretzel to OmpG, a porin in which loop invasion into the lumen is thought to be a central component of the observed pH-dependent gating process [2].

Application of the Pretzel framework results in energy and gating state summaries for an ensemble of 3-D protein conformations. These summaries may be used for further analysis to identify residues and regions, which drive the gating process [13]. Although here we discuss only details in applying Pretzel to OmpG, one may modify the provided shell scripts to model other membrane proteins in which loop regions are intrinsic to the gating mechanism.

2 Materials

Pretzel is a C++ framework with source code and scripts available from the public git repository: https://github.com/uic-lianglab/ompg-public. In addition, several data processing steps (e.g., clustering) employ the SCOBY suite of utilities available at: https://bitbucket.org/aperezrathke/scoby.

2.1 Additional Equipment

1. Linux HPC cluster or workstation, equipped with at least 2 GB RAM per computational node.
2. G++ compiler version ≥4.7 (https://gcc.gnu.org/).
3. CMake software version ≥4.8 (https://cmake.org/).
4. Protein structure as a PDB atom coordinates file, ideally membrane aligned from the OPM database [14]. In particular, membrane-aligned OmpG PDBs 2iww and 2iwv [1] may be downloaded from: https://opm.phar.umich.edu/ (*see* **Note 1**).
5. Optional membrane PDB atom coordinates file from CHARMM-GUI [15] (http://www.charmm-gui.org/) (*see* **Note 1**).
6. Git software, for obtaining Pretzel and SCOBY source code, version ≥1.7.1 (https://git-scm.com).
7. CASTp for detecting cavities and channels within proteins [16] (http://sts.bioe.uic.edu/castp/).
8. BOOST C++ libraries, required by SCOBY, version ≥1.58 (https://www.boost.org/)
9. NAMD [17] molecular dynamics software (*see* **Note 12**).

2.2 Equipment Setup

We assume access is available to a Linux terminal operating a bash shell. Download and install all listed software. It is recommended to obtain the Pretzel and SCOBY source codes via the *git clone* terminal command; unless otherwise stated, all relative file paths are relative to the Pretzel repository folder retrieved by the *clone* operation (./).

To compile the Pretzel and SCOBY source codes:

1. Navigate to Pretzel's *./scripts* folder and run the shell scripts *cmake_init.sh* followed by *cmake_build.sh*.
2. Copy the compiled *pretzel* binary from *./CMakeBuild* to *./* as all demonstration shell scripts provided assume ./ to be the location of the compiled binary.
3. Follow an analogous procedure to compile the SCOBY source code. Similarly, copy the *scoby* binary into Pretzel's ./ directory.

If additional details are desired for Pretzel and SCOBY compilation, please refer to the *readme.md* files within the respective git repositories.

3 Methods

The application of the Pretzel method consists of the following general stages: fragment (single-loop) building, multi-loop generation, protonation of ionizable residues, relaxation of structures and energy evaluation, and gating state assignment (Fig. 1). Briefly, the fragment building stage consists of producing 3-D conformation libraries for the backbones of each individual loop region to be modeled. The

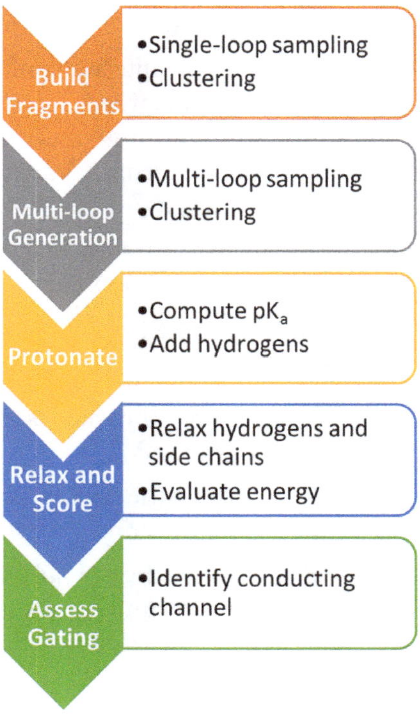

Fig. 1 Stages of the Pretzel framework

next stage, multi-loop generation, utilizes the single-loop fragment libraries to efficiently generate an ensemble of 3-D conformations of the protein with multiple loops (*see* **Note 2**). Next, the pK_a value is computed for each ionizable residue and hydrogens are placed according to the pH level for each multi-loop conformation. Subsequently, the multi-loop conformations have hydrogens and side chains relaxed using an implicit solvent molecular dynamics simulation; the energy of each conformation is captured after the last relaxation step. Finally, we assess the gating state at each conformation using a topological criterion.

3.1 Single-Loop Fragments

1. Generate the raw (i.e., unclustered), single-loop, backbone fragments by running the shell script *./scripts/a_frag_lib/a_queue_raw.sh* (*see* **Notes 3 and 4**).
2. Combine the raw fragments into a single PDB file per loop by running the shell script *./scripts/a_frag_lib/b_cat_raw.sh.*
3. Convert the raw PDB fragments into comma-separated values (CSV) format by running the shell script *./scripts/a_frag_lib/c_pdb_2_cluster_csv.sh.*
4. Convert the raw, plain-text CSV fragments into binary format by running the shell script *./scripts/a_frag_lib/d_cluster_csv_2_cluster_bin.sh* (*see* **Note 5**).
5. For clustering efficiency and to limit the effect of noise, perform PCA dimensionality reduction on the raw fragments by running the shell script *./scripts/a_ frag_lib/e_pca_reduce.sh* (*see* **Note 6**).
6. To help mitigate conformational sampling bias, cluster the raw fragments at each loop by running the shell script *./scripts/a_frag_lib/f_cluster_shwap.sh;* then resample from the clusters by running the shell script *./scripts/a_frag_lib/g_resample_ids.sh* followed by *./scripts/a_frag_lib/h_resample_pdbs.sh* (*see* **Notes 7 and 8**).
7. Copy the refined fragment libraries to the location expected by the multi-loop generation stage by running the shell script *./scripts/a_frag_lib/i_build_lib_dir.sh.*

3.2 Multi-Loop Generation

1. Generate the raw (i.e., unclustered), multi-loop ensemble by running the shell script *./scripts/b_multi_loop/a_gen_pdbs.sh* (*see* **Note 3**).
2. Convert the raw, multi-loop PDBs into CSV format by running the shell script *./scripts/c_cluster/a_pdb_dirs_2_cluster_csv.sh.*
3. Convert the raw, plain-text, multi-loop CSVs into binary format by running the shell script *./scripts/c_cluster/b_cluster_csv_2_cluster_bin.sh.*
4. For multi-loop clustering efficiency, perform PCA dimensionality reduction by running the shell script *./scripts/c_cluster/c_pca_reduce.sh* (*see* **Note 6**).

5. To help mitigate multi-loop sampling bias, cluster the raw ensemble by running the shell script *./scripts/c_cluster/d_cluster_shwap.sh*; then resample the ensemble by running the shell script *./scripts/c_cluster/e_resample_ids.sh* followed by *./scripts/c_cluster/f_resample_pdbs.sh (see* **Note 7**).

3.3 Protonate

1. Compute the pK_a of each ionizable residue among the multi-loop ensemble by running the shell script *./scripts/b_multi_loop/b_calc_pka.sh* (*see* **Note 9**).
2. After pK_a computation has completed, it is optional but recommended to remove unneeded files generated by PROPKA3 [18], as this will free up disk space that may be used by other applications. To remove these unneeded files, run the shell script *./scripts/b_multi_loop/c_clean_propka.sh.*
3. Next, for each multi-loop conformation, place hydrogens at all residues and assign protonation states according to the pH level by running the shell script *./scripts/c_cluster/h_gen_NAMD_topo.sh* (*see* **Note 10**).
4. Similarly, after hydrogens have been placed, it is optional but recommended to remove unneeded files generated by NAMD [17]. To remove these unneeded files, run the shell script *./scripts/i_psf_clean.sh.*

3.4 Relax and Score

1. To relax hydrogens and side chains in implicit solvent, run the shell script *./scripts/d_relax/a_NAMD_minimize.sh* (*see* **Note 11**).
2. To evaluate the potential energy at each relaxed conformation, run the shell script *./scripts/d_relax/b_NAMD_capture.sh* (*see* **Notes 12 and 13**).
3. For easier downstream analysis, concatenate (i.e., combine) individual conformation energy data files into aggregated CSV files by running the shell script *./scripts/d_relax/c_1_cat_NAMD_energy.sh* (*see* **Note 14**) followed by *./scripts/d_relax/c_2_munge_NAMD_interactions.sh* (*see* **Note 15**).

3.5 Assess Gating

1. Convert the NAMD [17] relaxed atomic coordinates at each conformation back to a hydrogen-free PDB format expected by CASTp [16] by running the shell script *./scripts/d_relax/d_1_coor_to_castp_pdb.sh.*
2. Compute topological properties at each relaxed conformation by running the shell script *./scripts/d_relax/e_call_castp.sh* (*see* **Note 16**).

   1. To assign "open" (conducting) vs "closed" (nonconducting) labels to each relaxed conformation, run the shell script *./scripts/d_relax/f_open_closed_classify.sh* (*see* **Note 17**).

4 Notes

1. For OmpG modeling, the membrane-aligned PDBs [14] 2iww and 2iwv are available in the Pretzel git repository folder: ./ompg/templates/. In addition, corresponding PDBs containing atom coordinates for DMPC lipid bilayer membranes, generated using CHARMM-GUI [15], are located in the folder: ./ompg/membrane/. All demonstration shell scripts for OmpG modeling assume the relevant PDB files reside in these folders.
2. The mDiSGro method [19, 20] is used for sampling protein loop 3-D conformations.
3. The bash shell scripts ./scripts/a_frag_lib/a_queue_raw.sh and ./scripts/b_multi_loop/a_gen_pdbs.sh generate backbone fragments and multi-loop conformations, respectively, by spawning multiple concurrent (i.e., parallel) pretzel processes. To control the number of concurrent pretzel processes, edit the corresponding script to assign the desired value to the MAX_NPROC variable; it is recommended to not exceed the number of available logical cores. For detailed pretzel argument descriptions, please refer to the comments in https://github.com/uic-lianglab/ompg-public/blob/master/source/params.h.
4. Other noteworthy variables in ./scripts/a_frag_lib/a_queue_raw.sh include START (a list of start residue numbers for each loop region of interest) and END (a list of end residue numbers for each loop region). To modify the loop regions simulated, simply change the corresponding entries in the START and END lists.
5. The purpose of the shell scripts ./scripts/a_frag_lib/c_pdb_2_cluster_csv.sh and ./scripts/a_frag_lib/d_cluster_csv_2_cluster_bin.sh is to convert the fragment data into a format usable by the SCOBY clustering utilities.
6. The shell script ./scripts/a_frag_lib/e_pca_reduce.sh has two noteworthy variables that may be modified through a text editor. The first variable, MAX_NPROC, controls the number of concurrent scoby instances used for processing each loop's fragment library; if RAM or CPU availability is limited, it is recommended to decrease this value. The second variable, PERC_KEEP, is a value in range (0.0, 1.0) and controls the percentage of variance retained by the principal components; the corresponding multi-loop ensemble script ./scripts/c_cluster/c_pca_reduce.sh also has an analogous PERC_KEEP variable. For detailed descriptions of arguments used by the SCOBY PCA reduce utility, please refer to

the script example_pca_reduce.sh at https://bitbucket.org/aperezrathke/scoby/src.

7. The mDiSGro [20] sampling algorithm used by Pretzel has inherent sampling bias. Clustering the conformational geometries followed by resampling can help mitigate this bias by achieving a more uniform coverage of the structural landscape. SCOBY provides an implementation of a streaming affinity propagation clustering algorithm [21, 22]; for detailed argument descriptions, please refer to the script example_shwap.sh at https://bitbucket.org/aperezrathke/scoby/src.
8. A representative clustering of a large set of raw fragments is shown in Fig. 2. The resampled fragment libraries for OmpG loops 1 and 6 are shown in Fig. 3.
9. The PROPKA3 method [18] is used for computing acid dissociation constants (pK_a). Similar to other scripts, ./scripts/b_multi_loop/b_calc_pka.sh launches multiple PROPKA (python) instances in parallel; to adjust RAM and CPU usage, one may modify the MAX_NPROC variable to increase or decrease the number of instances used.
10. The shell script ./scripts/c_cluster/h_gen_NAMD_topo.sh uses the NAMD [17] psfgen utility with a custom Tcl script to append hydrogen atoms and to protonate ionizable residues according to the parameter pH and computed pK_a. For demonstration OmpG modeling, the pH levels used are 5 and 7. To simulate different pH levels, use a text editor to change the PH_LEVELS variable within the shell script. Also, similar

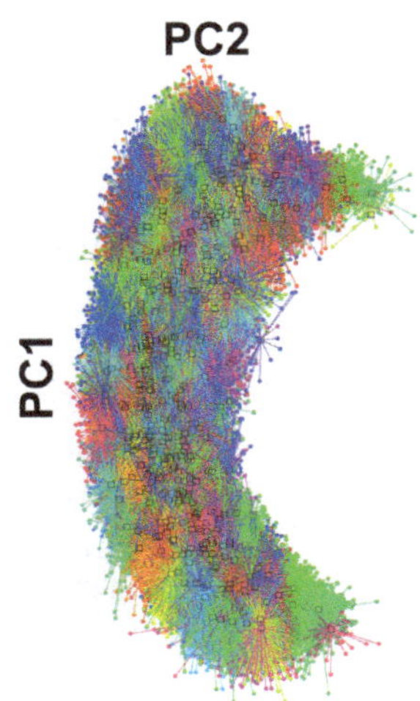

Fig. 2 StrAP [13] algorithm clustering of 278,009 single-loop (fragment) conformations for OmpG loop 6, resulting in 398 clusters (black boxes). Only the first two principal components (PC1, PC2) for each fragment's atomic coordinates are shown

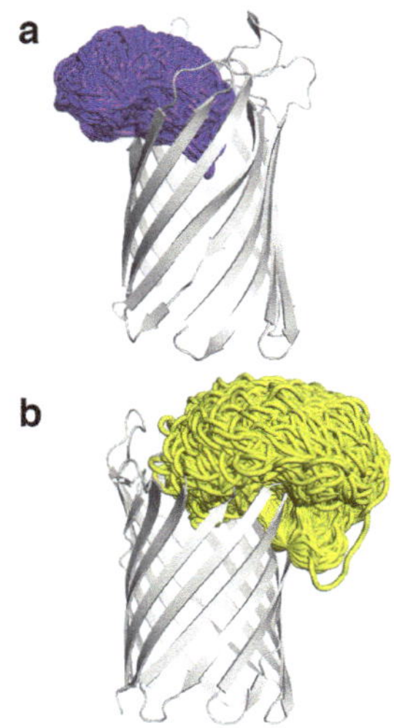

Fig. 3 OmpG single-loop fragment libraries after clustering and resampling: (**a**) loop 1 fragment library and (**b**) loop 6 fragment library. Each library consists of 10,000 loop conformations

to other scripts, the shell script's MAX_NPROC variable can be modified to adjust for CPU and RAM usage.

11. Each protonated 3-D conformation undergoes a modicum of relaxation primarily to resolve hydrogen and side chain steric clashes. For each conformation, the "relaxation" consists of a separate NAMD [17] molecular dynamics simulation in implicit solvent using the CHARMM36 [23] force field. Experience suggests that this phase is the most computationally intensive. To address this, multiple NAMD processes are spawned concurrently. Similar to other scripts, to control the number of spawned process (to adjust for CPU and RAM usage), simply edit the MAX_NPROC variable within ./scripts/d_relax/a_NAMD_minimize.sh.
12. The potential energy is captured at each relaxed protein conformation using a 0-Step NAMD [17] simulation. The NAMD C++ source code has been modified to provide detailed reporting of all nonbonded potential energy terms at each residue. Pre-compiled binaries are provided in the Pretzel git repository folder: ./tools/namd/bin/. However, if these binaries are not system compatible, then the modified NAMD source code (provided in the Pretzel git repository folder: ./tools/namd/source/) must be compiled and placed in the corresponding ./tools/namd/bin/ subfolder. See the "Compiling NAMD" section within ./tools/namd/source/notes.txt for detailed instructions on how to compile the NAMD source code; the shell script ./tools/namd/source/full_build.sh also provides a compilation reference.

13. Evaluation of the potential energy at each relaxed conformation is computationally intensive. To address this, multiple NAMD processes are spawned concurrently. Similar to other scripts, to control the number of spawned processes (to adjust for CPU and RAM usage), simply edit the MAX_NPROC variable within ./scripts/d_relax/b_NAMD_capture.sh.
14. The shell script ./scripts/d_relax/c_1_cat_NAMD_energy.sh generates a plain-text, comma-separated values (CSV) energy summary file, ener.csv, for each simulated pH level. Each CSV energy summary file may be opened within any spreadsheet editor or serve as input for downstream analysis in statistical computing environments, such as R, Matlab, or Python. The energy summary has the following columns: NAME, PH, POTENTIAL, ELECT, VDW, BOND, IMPRP, BOUNDARY, and MISC. The first column, NAME, contains the file name of each relaxed conformation that had its energy evaluated. The next column, PH, is the pH at which the energy was captured. The remaining column entries are the corresponding potential energy terms at each conformation: POTENTIAL is the total potential energy, ELECT is the electrostatic potential energy, VDW is van der Waals potential energy, etc. See the NAMD 2.11 user guide, Sect. 5.1 (available at https://www.ks.uiuc.edu/Research/namd/2.11/) for descriptions of the potential energy terms used by the molecular dynamics force field.
15. The shell script ./scripts/d_relax/c_2_munge_NAMD_interactions.sh generates a plain-text, comma-separated values (CSV) interactions summary file, ints_electro.10000.csv, of the electrostatic potential energies at each residue–residue pair. The 10,000 in the file name indicates that the summary is limited to the 10,000 lowest energy conformations; to change this limit, modify the MAX_ENERGY_RANK variable within the shell script. The first two columns of the CSV interactions summary file are the PDB residue numbers of the two electrostatically interacting residues; the remaining columns are the observed electrostatic energy values for the corresponding row's residue pair at each of the evaluated conformations.
16. The CASTp method [16] is used to compute topological properties, such as pockets and mouths [24, 25]. These properties will be used to heuristically identify possible conducting channels within each relaxed conformation [13]. The shell script ./scripts/d_relax/e_call_castp.sh assumes the newcast executable binary is available through the user's $PATH environment variable; for queries regarding batch usage of newcast, please contact uic.lianglab@gmail.com. In addition, the shell script models conductance by setting the

PROBE_RADIUS variable equal to 2.75, where this value in Angstroms is the van der Waals radius of a potassium (K^+) ion [26]. To modify the radius of the conducting ion, simply change the PROBE_RADIUS variable in a text editor. Similar to other shell scripts, multiple newcast instances are spawned to run in parallel; to adjust CPU and RAM usage, modify the MAX_NPROC variable in a text editor.

17. The shell script ./scripts/d_relax/f_open_closed_classify.sh will process the CASTp data for each relaxed conformation to identify possible conducting channels. The output is a comma-separated values (CSV) summary file, oc.2.75.csv, with two columns. The first column is the file name of each conformation. The second column is a "1" or "0" entry indicating if that row's conformation is in an "open" (conducting) or "closed" (nonconducting) state, respectively. The 2.75 in the name of output CSV file is to denote the radius of the conducting K^+ ion (*see also* **Note 16**); this may be changed by modifying the PROBE_RADIUS variable within the shell script.

Acknowledgments

This research was supported by the US National Institutes of Health, grants R01-GM126558, R01-GM079804, and R35GM127084.

References

1. Yildiz Ö, Vinothkumar KR, Goswami P, Kühlbrandt W (2006) Structure of the monomeric outer-membrane porin OmpG in the open and closed conformation. EMBO J 25:3702–3713
2. Chen M, Khalid S, Sansom MS, Bayley H (2008) Outer membrane protein G: engineering a quiet pore for biosensing. PNAS 105:6272–6277
3. Fahie M, Yang B, Mullis M, Holden MA, Chen M (2015) Selective detection of protein homologues in serum using an OmpG nanopore. Anal Chem 87:11143–11149
4. Fahie M, Chen M (2015) Electrostatic interactions between OmpG nanopore and analyte protein surface can distinguish between glycosylated isoforms. J Phys Chem B 119:10198–10206
5. Fahie M, Chisholm C, Chen M (2015) Resolved single-molecule detection of individual species within a mixture of anti-biotin antibodies using an engineered monomeric nanopore. ACS Nano 9(2):1089–1098
6. Pham B, Eron SJ, Hill ME, Li X, Fahie MA, Hardy JA, Chen M (2019) A nanopore approach for analysis of caspase-7 activity in cell lysates. Biophys J 117(5):844–855
7. Pham B, Chisholm CM, Foster J, Friis E, Fahie MA, Chen M (2020) A pH-independent quiet OmpG pore with enhanced electrostatic repulsion among the extracellular loops. Biochim Biophys Acta (BBA)—Biomembr 1863(1):183485. https://doi.org/10.1016/j.bbamem.2020.183485
8. Fahie MA, Candido J, Andree G, Chen M (2021) Tuning protein discrimination through altering the sampling interface formed between the analyte and the OmpG nanopore. ACS Sens 6(3):1286–1294. https://doi.org/10.1021/acssensors.0c02580
9. Fahie MAV, Li F, Palmer C, Yoon C, Chen M (2022) Modifying the pH sensitivity of OmpG nanopore for improved detection at acidic pH. Biophys J 121(5):731–741. https://doi.org/10.1016/j.bpj.2022.01.023

10. Kim M, Foster JC, Moore MD, Chen M (2023) Improving single-molecule antibody detection selectivity through optimization of peptide epitope presentation in OmpG nanopore. ACS Sens 8(7):2673–2680. https://doi.org/10.1021/acssensors.3c00528
11. Foster JC, Pham B, Pham R, Kim M, Moore MD, Chen M (2023) An engineered OmpG nanopore with displayed peptide motifs for single-molecule multiplex protein detection. Angew Chem—Int Ed 62(7):e202214566. https://doi.org/10.1002/anie.202214566
12. Foster JC, Pham B, Pham R, Ryan P, Tong N, Sharp J, Inoue S, Liang J, Beis K, Chen M (2025) Barrel expansion of outer membrane protein G nanopore through β-hairpin duplication. Prot Sci 34(8):e70203. https://doi.org/10.1002/pro.70203
13. Perez-Rathke A, Fahie MA, Chisholm C, Liang J, Chen M (2018) Mechanism of OmpG pH-dependent gating from loop ensemble and single channel studies. J Am Chem Soc 140:1105–1115
14. Lomize MA, Pogozheva ID, Joo H, Mosberg HI, Lomize AL (2012) OPM database and PPM web server: resources for positioning of proteins in membranes. Nucleic Acids Res 40:D370–D376
15. Jo S, Kim T, Iyer VG, Im W (2008) CHARMM-GUI: a web-based graphical user interface for CHARMM. J Comput Chem 29:1859–1865
16. Tian W, Chen C, Lei X, Zhao J, Liang J (2018) CASTp 3.0: computed atlas of surface topography of proteins. Nucleic Acids Res 46:W363–W367
17. Phillips JC, Braun R, Wang W, Gumbart J, Tajkhorshid E, Villa E, Chipot C, Skeel RD, Kalé L, Schulten K (2005) Scalable molecular dynamics with NAMD. J Comput Chem 26:1781–1802
18. Olsson MHM, Søndergaard CR, Rostkowski M, Jensen JH (2011) PROPKA3: consistent treatment of internal and surface residues in empirical pKa predictions. J Chem Theory Comput 7:525–537
19. Tang K, Zhang J, Liang Jie (2014) Fast protein loop sampling and structure prediction using distance-guided sequential chain-growth Monte Carlo method. PLoS Comput Biol 10:e1003539
20. Tang K, Wong SW, Liu JS, Zhang J, Liang J (2015) Conformational sampling and structure prediction of multiple interacting loops in soluble and β-barrel membrane proteins using multi-loop distance-guided chain-growth Monte Carlo method. Bioinformatics 31:2646–2652
21. Zhang X, Furtlehner C, Sebag M (2008) Data streaming with affinity propagation. In: Joint European Conference on machine learning and knowledge discovery in databases. Springer, Heidelberg, pp 628–643
22. Frey BJ, Dueck D (2007) Clustering by passing messages between data points. Science 315:972–976
23. Huang J, MacKerell AD Jr (2013) CHARMM36 all-atom additive protein force field: validation based on comparison to NMR data. J Comput Chem 34:2135–3145
24. Liang J, Edelsbrunner H, Fu P, Sudhakar PV, Subramaniam S (1998) Analytical shape computation of macromolecules: I. Molecular area and volume through alpha shape. Proteins 33:1–17
25. Liang J, Edelsbrunner H, Fu P, Sudhakar PV, Subramaniam S (1998) Analytical shape computation of macromolecules: II. Inaccessible cavities in proteins. Proteins 33:18–29
26. Bondi A (1964) van der Waals volumes and radii. J Physical Chem 68:441–451

Part II

Solid-State Nanopores

Chapter 10

Fabrication and Characterization of Glass Nanopores

Xiaoyu Chen, Zhongfei Liu, Aiyuan Li, Yanmei Ma, Weidi Chen, Rui Hu and Kaikai Chen

Abstract

Nanopore technology has emerged as a transformative tool for single-molecule detection, enabling label-free, real-time analysis of nucleic acids, proteins, and other biomolecules. Among various nanopore platforms, glass nanopores fabricated by laser-assisted pulling offer distinct advantages, including exceptionally low electrical noise, excellent chemical stability, and rapid, cost-effective production. However, achieving reproducible nanopore fabrication with precise diameter control remains challenging, often limiting their widespread adoption. This chapter presents a detailed protocol for fabricating high-quality glass nanopores with diameters ranging from approximately 4 nm to 14 nm, suitable for diverse single-molecule detection applications. By following this standardized workflow, researchers can reproducibly generate size-controlled glass nanopores and achieve enhanced success rates in biomolecule detection experiments. This protocol is designed to be accessible to laboratories without specialized nanofabrication facilities, democratizing access to nanopore technology for the broader research community.

Keywords Glass nanopores, Nanopore fabrication, Laser pulling, Nanopore characterization, Single-molecule detection

1 Introduction

The ability to identify and characterize individual biomolecules in solution remains a central challenge in analytical chemistry and molecular diagnostics. Nanopore technology has emerged as a promising solution, offering a label-free platform that monitors biomolecules through characteristic ionic current signatures as they traverse nanoscale pores under an applied electric field [1, 2]. This approach has enabled breakthroughs in DNA sequencing [3, 4], protein fingerprinting [5–7], glycan profiling [8–10], and disease biomarker detection [11, 12], with each application demanding precise control over nanopore diameters.

Solid-state nanopores are fabricated from inorganic materials, including silicon nitride (SiN_x) membranes, two-dimensional (2D)

Fanjun Li (ed.), *Nanopore Technology*, Methods in Molecular Biology, vol. 3072,
https://doi.org/10.1007/978-1-0716-5527-6_10,

materials, and glass capillaries [13]. Among these solid-state nanopores, glass nanopores offer a compelling combination of practical advantages. They exhibit low capacitive noise, intrinsic mechanical and chemical stability, ensuring robust performance across diverse experimental conditions [14, 15]. Importantly, glass nanopores can be fabricated in minutes using a standard laser-assisted puller, bypassing the complex instrumentation and specialized expertise required for TEM drilling or FIB milling [16]. This accessibility has positioned glass nanopores as an increasingly popular choice for laboratories seeking to implement nanopore sensing without substantial infrastructure investment.

Despite their growing adoption, the successful fabrication of glass nanopores requires substantial practical knowledge. Building on our previous work [17–19], we present here a comprehensive protocol that guides researchers through the complete workflow from nanopore fabrication to characterization. The protocol details the laser-assistedpulling process for fabricating glass nanopores in two distinct size regimes: approximately 14 nm diameter for general single-molecule studies and approximately 4 nm diameter for high-resolution applications. We describe the preparation of custom fluidic flow cells using PDMS molding and the fabrication of Ag/AgCl electrodes for stable electrochemical measurements. Step-by-step instructions are provided for integrating nanopores into the fluidic system. Finally, we outline characterization methods, including scanning electron microscopy (SEM) imaging for morphological assessment, current–voltage (*I–V*) curve measurements for electrical validation, and geometric modeling approaches for accurate nanopore diameter estimation. This protocol aims to lower the barrier to entry for nanopore research while preserving the flexibility needed for application-specific customization.

2 Materials

2.1 Capillaries

Quartz capillary for fabrication of glass nanopores: inner diameter = 0.2 mm, outer diameter = 0.5 mm, length = 7.0 cm (Sutter Instrument, QF050-20-7).

2.2 Solutions and Reagents

1. 4 M LiCl electrolyte (pH 9.0): dissolve 8.478 g LiCl powder and 0.5 mL of 1 M Tris-HCl buffer (10×, pH 8.0) in ultrapure water to 50 mL total volume, adjust pH to 9.0 with LiOH, and filter twice through a 0.22 μm sterile syringe filter.
2. Saturated $FeCl_3$ solution: $FeCl_3$ powder dissolved in ultrapure water to saturation.
3. Polydimethylsiloxane (PDMS) gel: prepared by mixing the silicone elastomer base with the curing agent at a 10:1 volume ratio and stored at −20°C until use (SYLGARD 184 Silicone Elastomer Kit).

3 Methods

3.1 Fabrication of Glass Nanopores with Diameters of ~14 nm

1. Turn on the laser-assisted puller (P-2000, Sutter Instrument, USA) and set parameters, including "Heat," "Filament," "Velocity," "Delay," and "Pull" (details shown in Table 1) (*see* **Notes 1 and 2**).
2. Place a quartz capillary (QF050-20-7) into the puller bars of the laser-assisted puller.
3. Run the preset pulling program; the laser heats the middle section of the capillary to a molten state, and then a mechanical force is applied, breaking the capillary into two conical glass nanopores at the tips (*see* **Note 3**). Figure 1a presents a schematic illustration of the glass nanopore fabrication process, and Fig. 1(b shows an SEM image of the fabricated nanopore with a diameter of ~14 nm.

Table 1
Parameters for pulling glass nanopores with diameters of ~14 nm

Nanopore diameters (nm)	Heat	Filament	Velocity	Delay	Pull
~14	480	0	25	170	200

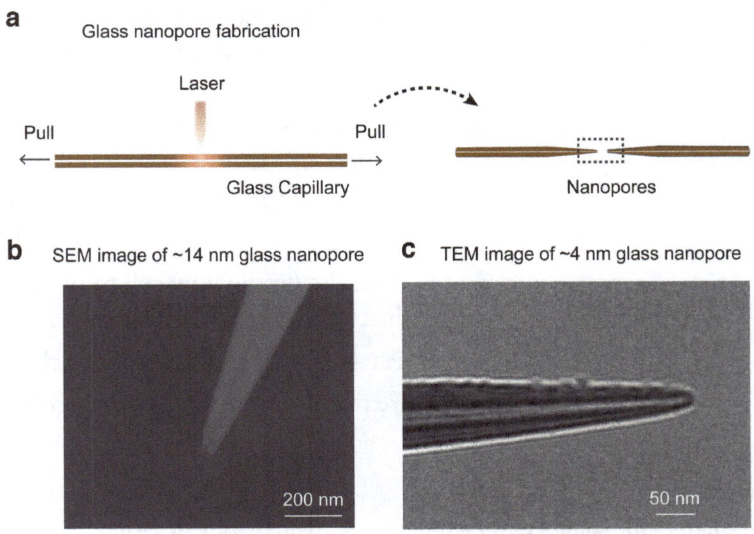

Fig. 1 (**a**) Schematic illustration of the laser-assisted pulling process of glass nanopores. A quartz capillary is heated by laser to a molten state and mechanically pulled to form a pair of conical nanopores. (**b**) SEM image of the nanopore tip with a diameter of ~14 nm. (**c**) TEM image of the nanopore tip with a diameter of ~4 nm (reproduced from [19] with permission from John Wiley and Sons)

3.2 Fabrication of Glass Nanopores with Diameters of ~4 nm

In this section, nanopores are fabricated using a method similar to that in Sect. 3.1; the parameters of the puller are shown in Table 2. Figure 1c shows TEM image of a glass nanopore with a diameter of ~4 nm.

3.3 Fabrication of Fluidic Flow Cell

1. Inject mixed polydimethylsiloxane into the custom-made mold of the fluidic flow cell and bake for 15 min at 130°C.
2. After curing, demold the PDMS fluidic flow cell from the mold.

3.4 Preparation of Ag/AgCl Electrodes

1. Cut a pair of silver wires (1 mm diameter and about 2 cm length, or as required).
2. Immerse silver wires into saturated $FeCl_3$ solution to a depth of approximately 1 cm for 30 min, and a uniform AgCl film will be coated on the surface of silver wire to obtain Ag/AgCl electrodes.
3. Remove Ag/AgCl electrodes from saturated $FeCl_3$ solution and rinse them thoroughly with deionized water.
4. Polish the ends of the electrodes with sandpaper until the Ag layer is exposed to establish an electrical connection. This process yields the final Ag/AgCl electrodes.
5. Connect the electrodes to the head stage of the current amplifier (e.g., Axopatch 200B, Molecular Devices) (*see* **Note 4**).

3.5 Integration of Fluidic Flow Cell with ~14 nm Nanopores

The workflow is shown in Fig. 2.

1. Assemble the nanopores into the fluidic flow cell.
2. Activate the surfaces of the glass slide and the assembled fluidic flow cell for bonding by plasma treatment (e.g., plasma surface cleaning system: Zepto, Diener electronic GmbH & Co.KG) for 30 s.
3. Inject PDMS into the sealed zone to separate the *trans* and *cis* chambers of the fluidic flow cell.
4. Bake the fluidic flow cell at 130°C to cure the PDMS for 15 min.
5. Expose the fluidic flow cell to plasma for 10 min, increasing the hydrophilicity of glass nanopore.
6. Inject 4 M LiCl electrolyte into the *trans* and *cis* chambers.
7. Store the assembled fluidic flow cell at 4°C until the experiment.

Table 2
Parameters for pulling glass nanopores with diameters of ~4 nm

Nanopore diameters (nm)	Heat	Filament	Velocity	Delay	Pull
~4 nm	580	0	25	170	225

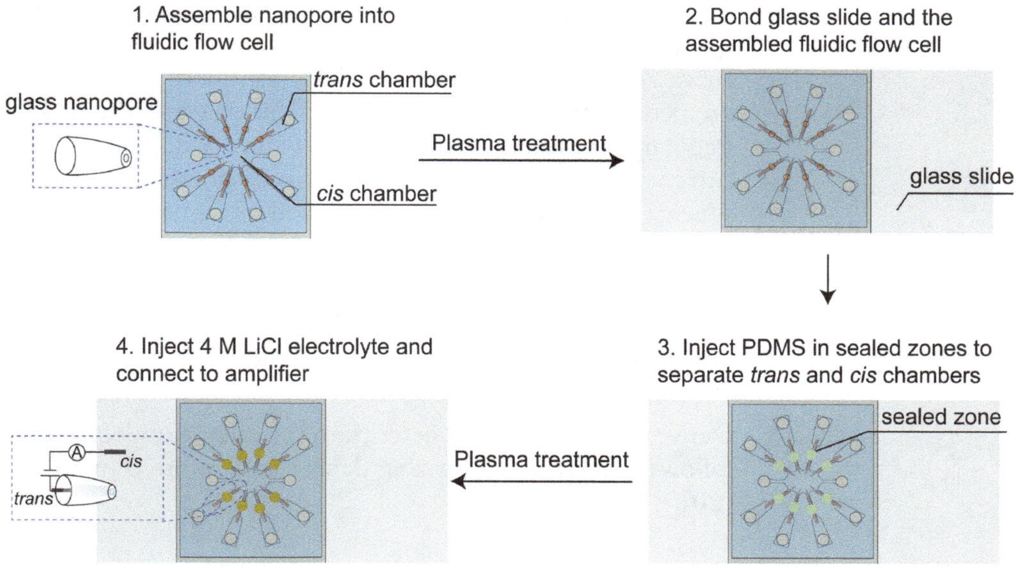

Fig. 2 Schematic workflow for nanopore device assembly. First, assemble eight nanopores into a custom-made fluidic flow cell. Next, plasma treatment of the glass surfaces to activate bonding between glass and PDMS. Then inject PDMS into sealed zones to isolate the *trans* and *cis* chambers, curing at 130°C for 15 min. Finally, perform an additional plasma treatment for 10 min to increase the hydrophilicity of glass nanopores. Fill both the *trans* and *cis* chambers with 4 M LiCl electrolyte and connect the fluidic flow cell to the amplifier

3.6 Integration of Fluidic Flow Cell with ~4 nm Nanopores

The fabrication follows the method described in Sect. 3.5, with the only modification being a 20-min plasma exposure of the pulled nanopores at Step 5 in Sect. 3.5.

3.7 Characterization of Nanopores

Both SEM and TEM are suitable for nanopore imaging. Here, we introduce the SEM characterization method.

3.7.1 Image Characterization of Nanopores

1. Clean the glass nanopores using plasma cleaner for 5 min.
2. Mount the glass nanopore on an SEM stub and secure with conductive tape or silver paint.
3. Locate and align the tip zone of glass nanopore at low magnification.
4. Increase the magnification and adjust the working distance to focus sharply on the tip region for a high-resolution image of glass nanopore (*see* **Note 5**).

3.7.2 Characterization of Nanopores Using I–V Curves

1. Insert the Ag/AgCl electrodes into the *trans* and *cis* chambers, respectively.
2. Apply a voltage (*V*) sweep to the electrodes. Set the range from −600 mV to +600 mV with a 50 mV step size.

3. Record ionic current (I) and plot the data points with voltage on the x-axis and ionic current on the y-axis to generate the I–V curve and calculate resistance of the nanopore (in our measurement system, the sample rate is 250 kHz and the filtering frequency is 50 kHz) (*see* **Note 6**). Figure 3 displays a typical current–voltage (I–V) curve and the corresponding root-mean-square (RMS) noise of the nanopore measurement system.

3.7.3 Estimate the Nanopore Diameter

1. Define a geometric model from the SEM image, as shown in Fig. 4.
2. Calculate the electrical resistance of the open pore from the I–V curve data.
3. Derive the nanopore diameter from the calculation equation as follows (*see* **Note 7**): variable definitions are listed in Table 3 [20]:

$$R = \frac{4L}{\sigma \pi d_i (d_i + 2L \tan\theta_i')} + \frac{2}{\sigma \pi (d_i + 2L \tan\theta_i') \tan\theta_i''} + \frac{1}{2\sigma d_i} I.$$

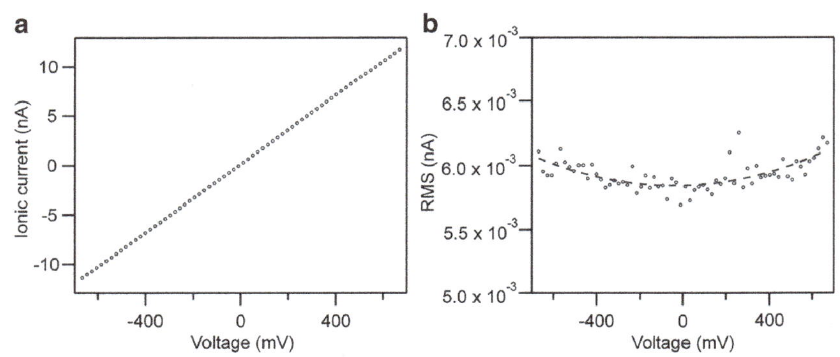

Fig. 3 Electrical characterization of a glass nanopore. (**a**) Current–voltage (I–V) curve demonstrating the linear and symmetric response of an open nanopore under applied voltage sweep from −600 mV to +600 mV. (**b**) Root-mean-square (RMS) noise profile of the nanopore measurement system, showing the baseline noise level

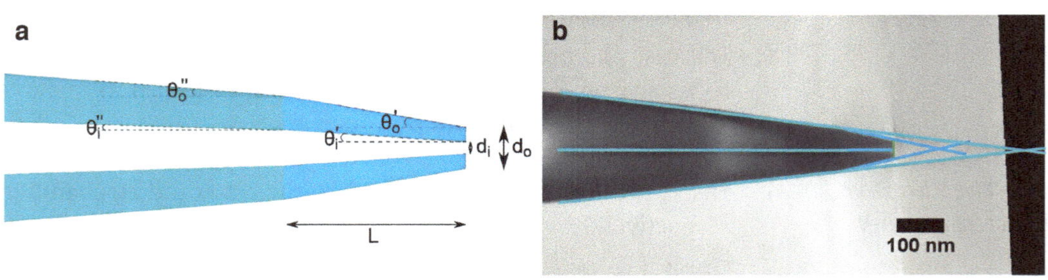

Fig. 4 Nanopore diameter estimation using geometric modeling. (**a**) The geometric model and (**b**) a TEM image of a glass nanopore. Those parameters indicated in a are key factors used to calculate the nanopore diameter from electrical resistance measurements (reproduced from [20] with permission from American Chemical Society)

Table 3
Variable definitions in the equation used for nanopore diameter calculation

Symbol	Description	Unit
R	The measured electrical resistance of the open nanopore	Ω
σ	The conductivity of the electrolyte solution	S/m
d	The diameter of the nanopore tip	m
L	The length of the first conical section of the nanopore	m
θ_i'	The semi-angle of the first conical section	rad
θ_i''	The semi-angle of the second conical section	rad

4 Notes

1. To ensure stable and reproducible nanopore diameters, warm up the laser-assisted puller for at least 10 min before pulling the first capillary. Variations in laser power during warm-up can result in inconsistent nanopore sizes.
2. Keep the laser-assisted puller clean and free from dust. Wipe the optical lenses regularly (every 1–2 months of active use) with lens paper and appropriate lens cleaning solution (isopropanol or specialized optical cleaner). Do not use regular tissue paper or cloth, which may scratch the lenses. Dust and debris on the lenses can scatter or attenuate laser power, affecting heating uniformity and pulling reproducibility.
3. The pulling parameters are specific to the puller and are very sensitive to the experimental environment, such as the humidity and room temperature. It is recommended to optimize parameters for your specific setup by testing with several capillaries.
4. Fresh Ag/AgCl electrodes should be prepared for each experiment to ensure stable electrochemical potential. The quality of the AgCl coating is critical for low-noise measurements. The coating should appear uniform and dark gray after treatment in the $FeCl_3$ solution.
5. When characterizing nanopores by TEM or SEM, use the minimum beam power necessary to obtain clear images. High beam power can cause nanopore deformation, collapse, or enlargement due to heating and charging effects.
6. Effective electromagnetic shielding (e.g., Faraday cage) is crucial for the measurement system to reduce noise and achieve high-resolution signal measurement. The entire measurement setup, including the head stage, flow cell, and electrodes, should be enclosed within the Faraday cage.

7. Prior to detection of biomolecules, measure the *I–V* curve to assess nanopore performance and calculate the effective nanopore diameter. For 4 M LiCl electrolyte, the curve should be linear and symmetric across positive and negative voltage ranges, indicating an open nanopore without blockages, defects, or rectification effects. The slope of the *I–V*curve provides nanopore conductance, which can be used to estimate the diameter using the equation in Sect. 3.7.3.

References

1. Branton D, Deamer DW, Marziali A et al (2008) The potential and challenges of nanopore sequencing. Nat Biotechnol 26:1146–1153
2. Ying Y-L, Hu Z-L, Zhang S et al (2022) Nanopore-based technologies beyond DNA sequencing. Nat Nanotechnol 17:1136–1146
3. Clarke J, Wu H-C, Jayasinghe L et al (2009) Continuous base identification for single-molecule nanopore DNA sequencing. Nat Nanotech 4:265–270
4. Wang Y, Zhao Y, Bollas A et al (2021) Nanopore sequencing technology, bioinformatics and applications. Nat Biotechnol 39:1348–1365
5. Ritmejeris J, Chen X, Dekker C (2024) Single-molecule protein sequencing with nanopores. Nat Rev Bioeng 3:303–316
6. Ahmad M, Ha J-H, Mayse LA et al (2023) A generalizable nanopore sensor for highly specific protein detection at single-molecule precision. Nat Commun 14: 1374
7. Lu C, Bonini A, Viel JH, Maglia G (2025) Toward single-molecule protein sequencing using nanopores. Nat Biotechnol 43:312–322
8. Yao G, Ke W, Xia B, Gao Z (2024) Nanopore-based glycan sequencing: state of the art and future prospects. Chem Sci 15:6229–6243
9. Li M, Xiong Y, Cao Y et al (2023) Identification of tagged glycans with a protein nanopore. Nat Commun 14: 1737
10. Zhang S, Cao Z, Fan P et al (2024) Discrimination of disaccharide isomers of different glycosidic linkages using a modified MspA nanopore. Angew Chem, Int Ed 63:e202316766
11. Ying Y-L, Long Y-T (2019) Nanopore-based single-biomolecule interfaces: from information to knowledge. J Am Chem Soc 141:15720–15729
12. Chen Y, Davidson NM, Wan YK et al (2025) A systematic benchmark of nanopore long-read RNA sequencing for transcript-level analysis in human cell lines. Nat Methods 22:801–812
13. Fried JP, Swett JL, Nadappuram BP et al (2021) In situ solid-state nanopore fabrication. Chem Soc Rev 50:4974–4992
14. Fragasso A, Schmid S, Dekker C (2020) Comparing current noise in biological and solid-state nanopores. ACS Nano 14:1338–1349
15. Guan X, Li H, Chen L et al (2023) Glass capillary-based nanopores for single molecule/single cell detection. ACS Sens 8:427–442
16. Bell NAW, Thacker VV, Hernández-Ainsa S et al (2013) Multiplexed ionic current sensing with glass nanopores. Lab Chip 13:1859
17. Alawami MF, Bošković F, Zhu J et al (2022) Lifetime of glass nanopores in a PDMS chip for single-molecule sensing. iScience 25:104191
18. Chen K, Bell NAW, Kong J et al (2017) Direction- and salt-dependent ionic current signatures for DNA sensing with asymmetric nanopores. Biophys J 112:674–682
19. Chen K, Choudhary A, Sandler SE et al (2023) Super-resolution detection of DNA nanostructures using a nanopore. Adv Mater 35: 2207434
20. Bell NAW, Keyser UF (2015) Specific protein detection using designed DNA carriers and nanopores. J Am Chem Soc 137:2035–2041

Chapter 11

Preparation and Detection of Designed DNA Carriers for Nanopore Sensing

Yaxian Liu, Chengwei Luo, Ran Ji, Yufei Li, Yuxia Nie, Zemin Zhou, Rui Hu and Kaikai Chen

Abstract

Single-molecule detection is a powerful strategy for early disease diagnosis, allowing the specific identification of nucleic acids and proteins. While nanopore technology enables fluorescent label-free, high-resolution analysis of individual biomolecules, its lack of intrinsic molecular selectivity limits performance in complex biological samples. The DNA carrier strategy addresses this limitation by introducing target recognition capability, incorporating molecular barcodes, and enhancing signal clarity in complex backgrounds. When combined with low-noise, high-resolution glass nanopores, these carriers enable specific and multiplexed biomarker detection for early-stage cancer and other diseases. This chapter presents a detailed and standardized protocol for preparing and detecting assembled DNA carriers using nanopores, providing an accessible workflow for broader adoption in single-molecule sensing and molecular diagnostics.

Keywords DNA carrier assembly, Single-molecule sensing, Nanopore detection, Molecular barcoding, Glass nanopore, Biomarker, Molecular diagnostics

1 Introduction

Early-stage disease detection increasingly relies on the ability to identify low-abundance biomarkers with high specificity and sensitivity. Single-molecule nanopore detection, which involves the direct identification of individual nucleic acids and proteins, offers a powerful approach for such diagnostic applications [1–5]. Nanopore sensing has emerged as a promising platform for this purpose, enabling fluorescent label-free, real-time detection of single molecules by monitoring ionic current modulations as analytes pass through a nanoscale pore [6]. However, a key limitation of conventional nanopores is their lack of intrinsic selectivity, making it difficult to distinguish target molecules from abundant background species in complex clinical samples.

Supplementary Information The online version contains supplementary material available at https://doi.org/10.1007/978-1-0716-5527-6_11

Fanjun Li (ed.), *Nanopore Technology*, Methods in Molecular Biology, vol. 3072,
https://doi.org/10.1007/978-1-0716-5527-6_11,

To address this limitation, the DNA carrier strategy has been developed to improve the specificity, sensitivity, and interpretability of nanopore-based detection [2, 7–9]. These carriers are rationally designed DNA nanostructures that enable selective target binding, assembled from a single-stranded DNA scaffold hybridized with complementary oligonucleotides. Through programmable design, they can incorporate structural barcodes and molecular recognition sites like aptamers or hybridization probes at defined positions along the DNA backbone (Fig. 1). These functional elements enable improved detection accuracy, signal amplification, and the potential for multiplexed biomarker identification from clinical biofluids.

When coupled with DNA carriers, glass nanopores form the basis of a portable and robust sensing platform for specific and multiplexed detection [10, 11]. This system is particularly suitable for detecting disease-associated nucleic acids and proteins under clinically relevant conditions [12]. In this chapter, we present a fully standardized and reproducible protocol for the synthesis of DNA carriers and their readout using glass nanopores. Our aim is to lower the technical entry

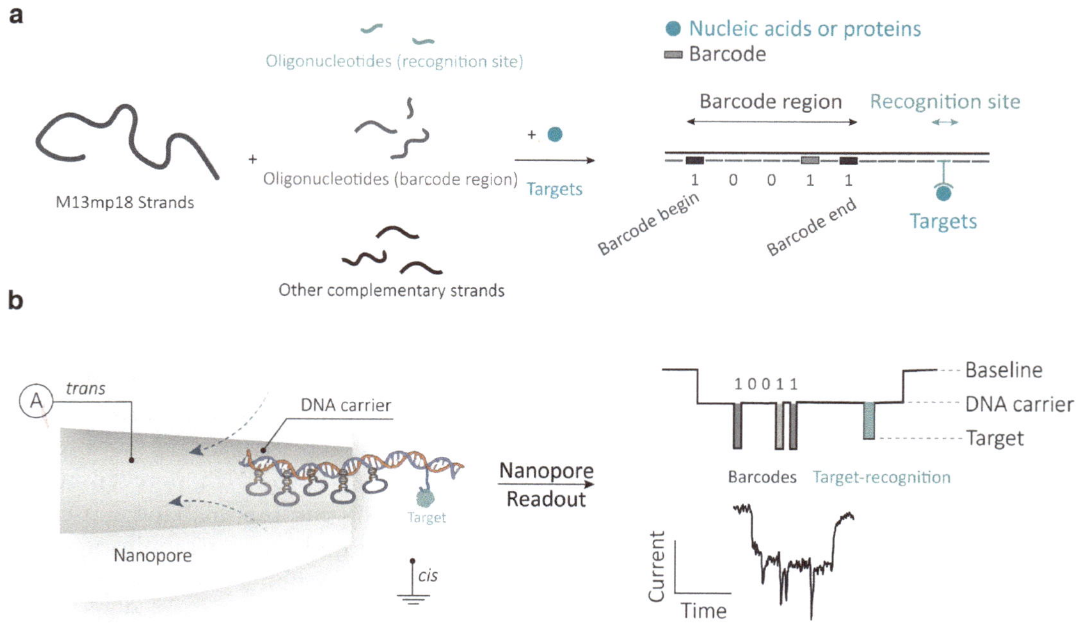

Fig. 1 Programmable DNA carriers for nanopore sensing. (**a**) The M13mp18 single-stranded DNA scaffold is hybridized with two sets of functional oligonucleotides: one forming a barcode region for positional encoding and the other introducing a recognition site for specific target binding, such as nucleic acid or proteins. The remaining complementary oligonucleotides complete the duplex structure, generating a linearized double-stranded DNA carrier. (**b**) The DNA carrier is electrophoretically driven through a glass nanopore from the *cis* to the *trans* chamber. As the DNA carrier translocates the nanopore, barcode features and bound biomolecules (e.g., nucleic acids and proteins) generate distinguishable current signatures. The resulting ionic current trace reflects both structural markers and molecular recognition information (Event adapted from [1])

barrier, ensure experimental consistency, and promote wider adoption of carrier-assisted nanopore sensing in molecular diagnostics.

2 Materials

Prepare all solutions using ultrapure double-distilled water. All buffers and reagents are stored at room temperature unless otherwise indicated. Here, we introduce DNA carriers designed with two functional domains as an example: a barcode region formed by short oligonucleotides at one end for digital signal encoding, and a binding site for target recognition at the other, which can be designed as an aptamer, complementary DNA sequence, or antibody-linked motif (*see* **Note 6**). To construct these features, a set of dumbbell-shaped oligonucleotides is designed to replace staple strands and form these markers within the carrier.

2.1 DNA Carrier Samples and Reagents

1. M13mp18 single-stranded DNA.
2. Barcode and target-recognition strands mixture: Add 1 μL each of the designed oligonucleotides in the coding region and 1 μL of each of the designed oligonucleotides in the target recognition region to a tube, each at 100 μM in 10 mM Tris–HCl (pH 8.0). Adjust the total volume to 250 μL with 10 mM Tris–HCl (pH 8.0), resulting in a final concentration of 400 nM for each oligonucleotide. See Supplementary Table S1 for the sequences used.
3. M13mp18 complementary oligonucleotide mixture: Add 1 μL each of the remaining complementary oligonucleotides (from a total pool of 190), corresponding to positions not replaced in the previous step, each at 100 μM in 10 mM Tris–HCl (pH 8.0), Adjust the total volume to 250 μL with 10 mM Tris–HCl (pH 8.0), resulting in a final concentration of 400 nM for each oligonucleotide. See Supplementary Table S2 for the sequences used.
4. Dilution buffer: 10 mM Tris–HCl, pH 8.0. Add 400 μL of 1 M Tris–HCl to a 50 mL falcon tube, add water to a volume of 40 mL, vortex thoroughly, and filter through a sterile syringe filter.
5. Hybridization buffer: 10 mM Tris–HCl, 10 mM $MgCl_2$, pH 8.0. Add 400 μL of 1 M Tris–HCl and 400 μL of 1 M magnesium chloride ($MgCl_2$) to a 50 mL falcon tube, add water to a volume of 40 mL, vortex thoroughly, and filter through a sterile syringe filter.
6. Wash buffer: 10 mM Tris–HCl, 0.5 mM $MgCl_2$, pH 8.0. Add 400 μL of 1 M Tris–HCl and 20 μL of 1 M magnesium chloride ($MgCl_2$) to a 50 mL falcon tube, add water to a volume of

40 mL, vortex thoroughly, and sterilize by filtration through a sterile syringe filter.

7. DNA purification kit.
8. Filter: 100 kDa Amicon Ultra centrifugal filter.
9. Tubes: 0.2, 1.5, 2.0 mL.
10. PCR thermocycler.
11. High Speed Centrifuge.
12. 0.22 µm sterile syringe filter.

2.2 Glass Nanopore Fabrication

1. Quartz capillaries with filaments: Inner diameter = 0.2 mm, outer diameter = 0.5 mm, and length = 7.5 cm.
2. CO_2 laser-assisted puller (e.g., P-2000, Sutter Instrument).
3. Plasma cleaner.
4. Heater.
5. Vacuum pump.
6. Polydimethylsiloxane (PDMS): the base components and curing agent are mixed at a 10:1 weight ratio.
7. Tweezers: fine-tip and anti-static tweezers.
8. 3D-printed PDMS moulds: custom moulds (see Chap. 10 for design and fabrication details).

2.3 Nanopore Measurement Equipments and Reagents

1. Patch-clamp amplifier: for the detection of ionic currents (e.g., Axopatch 200B, Molecular Devices).
2. Data acquisition system: digitizer (e.g., NI PCIe-6251) and customized LabVIEW software for data acquisition and analysis, or a Digidata 1440 A digitizer with pCLAMP software (Molecular Devices). Both platforms support real-time current recording.
3. Low-pass Bessel filter (50 kHz): reduce high-frequency noise while preserving signal integrity.
4. Faraday cage: minimize electrical interference.
5. Anti-vibration table.
6. Electrodes: Ag/AgCl wires for working and reference electrodes.
7. 1 M Tris (pH 8.0).
8. Measuring buffer (4 M LiCl, 10 mM Tris, pH 9.0): Weigh out 8.47 g LiCl, add 500 µL of 1 M Tris (pH 8.0), dissolve in water to a final volume of 50 mL, and adjust pH to 9.0 using LiOH (*see* **Note 5**).

2.4 Data Acquisition and Analysis Tools

1. Data acquisition software: Custom LabVIEW-based program for real-time ionic current and amplitude measurements. Other

tools include pCLAMP software. The software is highly adaptable and can be tailored to specific experimental needs.

2. Data analysis program: Custom LabVIEW-based program for trace recording, event detection, and current peak analysis. The software is highly adaptable and can be tailored to specific experimental needs. For users seeking alternative options, several open-source tools, such as OpenNanopore, Transalyzer, and others, are available for nanopore data acquisition and analysis.
3. Additional tools: OriginPro, Igor, and Python.

3 Methods

3.1 Preparation of DNA Carriers

The preparation of DNA carrier barcoding structures, including M13mp18 linearization, oligonucleotide assembly, purification, target binding, and nanopore measurement, is schematically summarized in Fig. 2.

1. DNA scaffold linearization. In two separate 0.2 mL PCR tubes, combine 40 μL of 100 nM M13mp18 phage DNA (7249 nt), 8 μL CutSmart buffer, 28 μL nuclease-free water, and 2 μL of 100 μM complementary oligonucleotides (5‘- TCTAGA GGATCCCCGGGTACCGAGCTCGAATTCGTAATC-3’). Perform thermal annealing by first rapidly heating the mixture to 65°C, then gradually cooling to 25°C at a rate of 1°C/min. Add 1 μL each of EcoRI and BamHI restriction enzymes (100,000 units/mL) to each PCR tube and incubate at 37°C for 60 min to cut the circular M13mp18 Phage DNA.
2. Purification of the DNA scaffold. Combine every two PCR tubes and purify the nucleic acids by removing enzymes using the Omega Bio-tek D6492 DNA purification kit, following the manufacturer's instructions (*see* **Note 1**).
3. Quantification and quality assessment. Measure the concentration of the purified DNA using a NanoDrop spectrophotometer. Typical yields range from 80–120 nM. To evaluate purity, confirm that the A260/A280 ratio is approximately 1.8–2.0, indicating minimal protein contamination. An A260/A230 ratio above 2.0 further supports the absence of residual salts or organic solvents. Dilute to 100 nM using 10 mM Tris–HCl (pH 8.0) buffer and store at 4°C until further use.
4. Assembly of DNA scaffold with oligonucleotides (1:5 ratio). In a 0.2 mL PCR tube, combine 8 μL of 100 nM linearized M13mp18 single-stranded DNA, 10 μL of 400 nM barcode and target-recognition mixture, 10 μL of 400 nM M13mp18

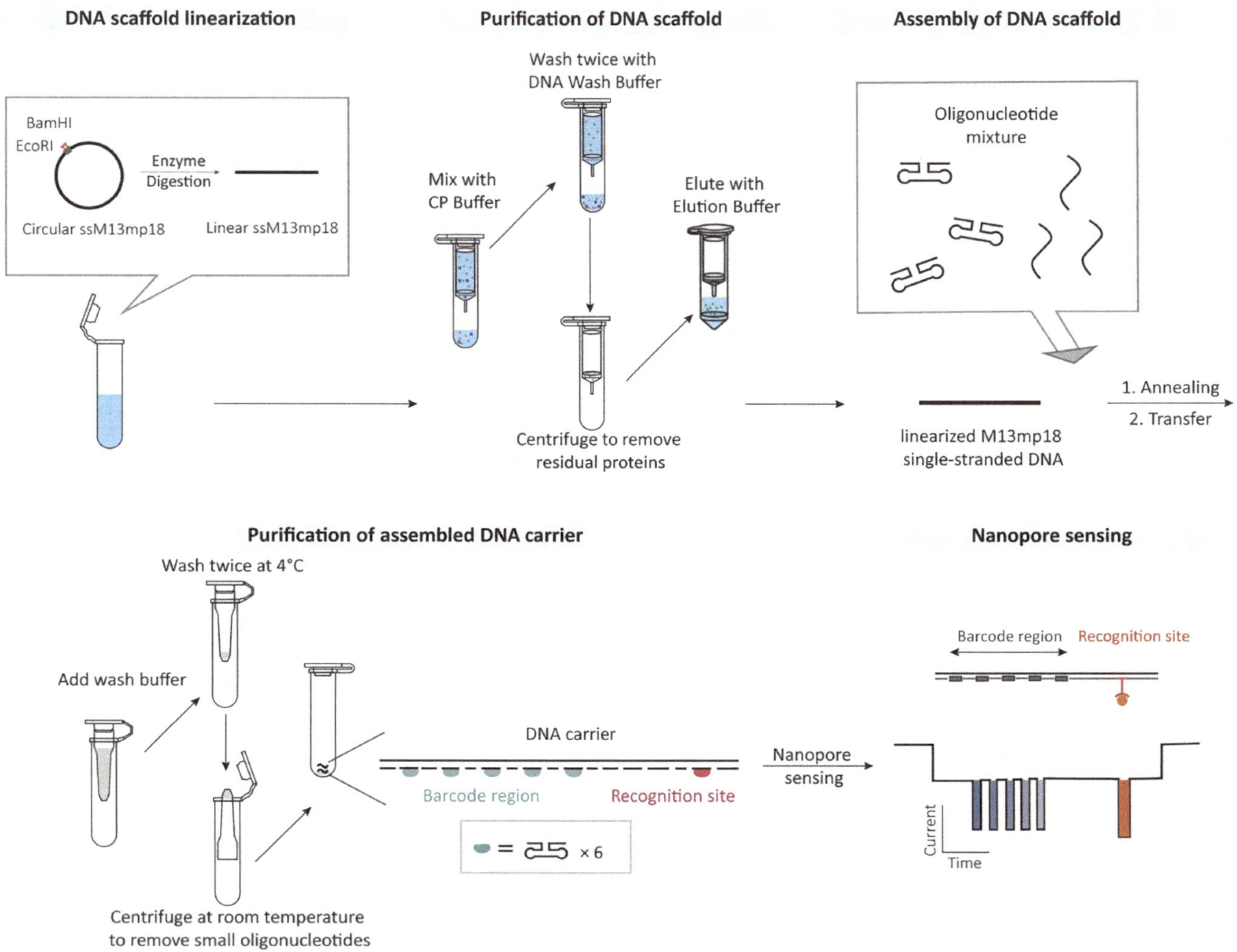

Fig. 2 Workflow for the preparation of DNA carriers. M13mp18 single-stranded DNA is first annealed with complementary strands and linearized by restriction digestion. After purification, the linearized scaffold is hybridized with a designed oligonucleotide mixture. The resulting DNA carrier contains defined barcode regions and recognition sites, enabling multiplexed signal readout in nanopore experiments

complementary oligonucleotide Mixture, 1.2 μL of 100 mM Tris–HCl (pH 8.0), 4 μL of 100 mM $MgCl_2$, and 6.8 μL nuclease-free water. Perform thermal annealing by first rapidly heating the mixture to 70°C, then gradually cooling to 25°C at a rate of 1°C/min (*see* **Note 3**).

5. Purification of assembled DNA carrier. Dilute the annealed products with 460 μL wash buffer (10 mM Tris–HCl, 0.5 mM $MgCl_2$, pH 8.0) and transfer the mixture to a 100 kDa Amicon Ultra centrifugal filter. Centrifuge at 11,000 × *g* for 10 min at 4°C, discard the filtrate, and repeat the wash once. Then invert the filter unit into a clean collection tube and centrifuge at 1000 × *g* for 2 min at room temperature to collect the eluate, yielding the synthesized and purified DNA carrier barcoding structures and recognition sites (*see* **Note 4**).
6. Nanodrop quantification assessment. Measure the concentration and assess the purity of the final product using NanoDrop. The typical concentration of DNA carrier is 10–20 nM (*see* **Note 2**). Store the purified DNA carriers at 4°C until use (*see* **Note 7**).
7. Target binding and nanopore measurement. Before the nanopore measurement, the target is incubated with DNA carrier under appropriate binding conditions (*see* **Note 8**).

3.2 Preparation of Glass Nanopore

A detailed description of the glass nanopore fabrication procedure is provided in Chap. 10. Here, nanopores with diameters of approximately 10 nm were fabricated using the following puller settings: Heat = 530, Velocity = 25, Filament = 0, Delay = 170, and Pull = 225.

3.3 Nanopore Measurement

A detailed description of the nanopore sensing experimental set-up can be found in Chap. 10 (*see* **Notes 9–11**).

3.4 Data Analysis Workflow

We use customized software for data analysis. The workflow illustrates the data processing steps used to extract and analyze single-molecule nanopore translocation events, shown in Fig. 3. The analysis consists of three main steps: (1) identifying valid translocation events and extracting their key features, (2) filtering unfolded events (*see* **Note 12**), and (3) analyzing barcodes and recognition site signals to decode molecular identity and binding information [10, 13].

1. Event selection and feature extraction. Raw ionic current traces are first screened to identify valid translocation events, defined by transient current blockades relative to the baseline. For each event, key features such as dwell time, event charge deficit, and current amplitude are extracted.
2. Filtering of unfolded events. Detected events are then categorized as either unfolded or folded based on the current levels and peak morphology. Unfolded events display a single, stable

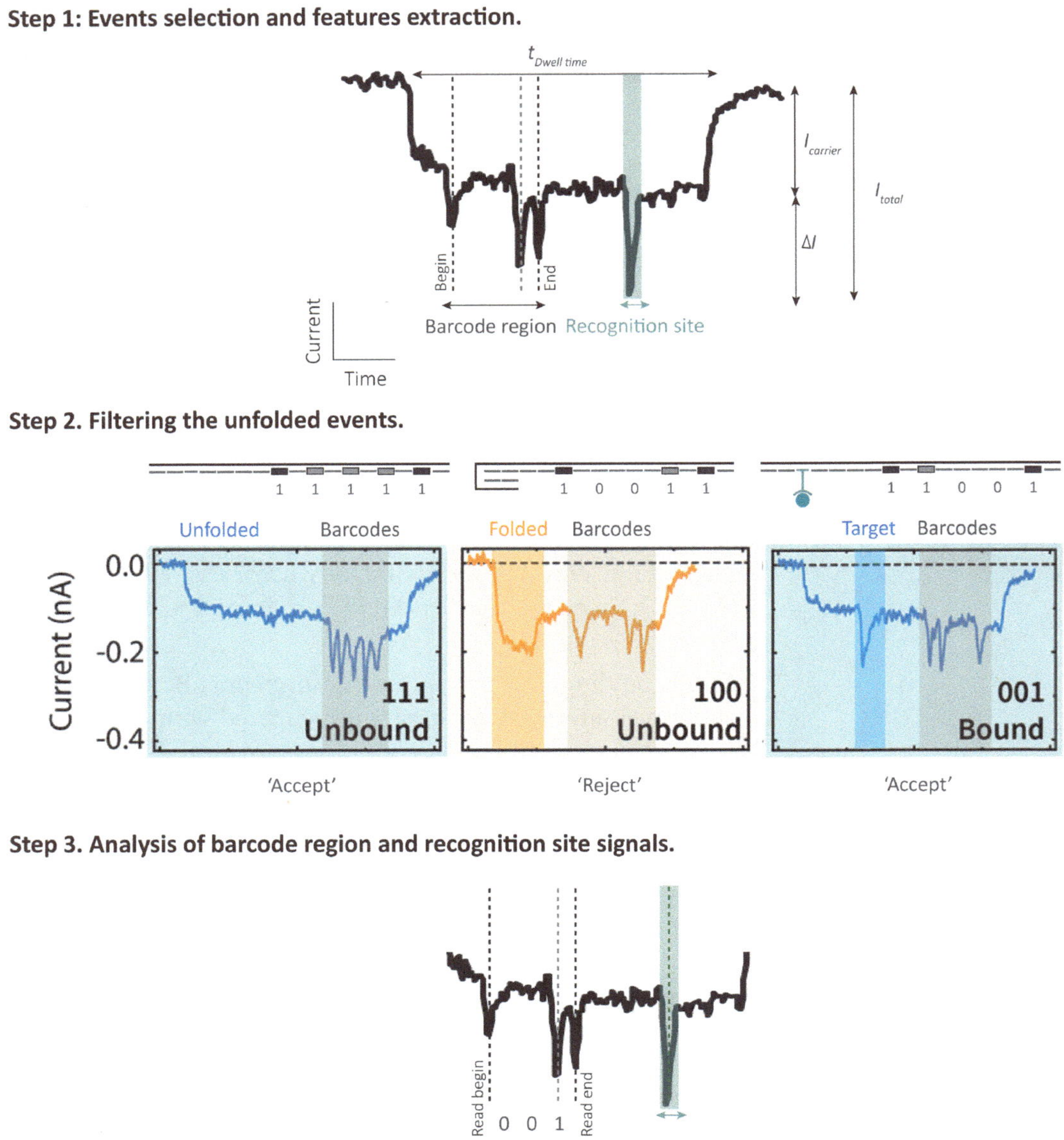

Fig. 3 Data analysis workflow for DNA carrier nanopore sensing signals. The analysis consists of three main steps. Step 1, events selection and feature extraction—raw ionic current traces are screened to identify events, followed by extracting key features, such as dwell time ($t_{Dwell\ time}$), current blockade depth (ΔI). The barcode region and recognition site are indicated along the translocation trace. Step 2, filtering of unfolded events. Detected events are categorized as unfolded (left, showing barcode "111" without target binding, accepted), folded (middle, showing barcode "100" with irregular current pattern, rejected), or target-bound (right, showing barcode "001" with characteristic blockade at the recognition site, accepted). Step 3, analysis of barcode region and recognition site signals. The selected unfolded events are decoded to identify the carrier's digital barcode (here shown as "011") and assess molecular binding status at the recognition site (Events adapted from [1, 13])

current blockade representing a correctly extended DNA carrier and are accepted for further analysis. Folded events, typically showing multiple or prolonged current blockades, often at the beginning of the traces, are rejected at this step. Notably, folded signals typically show blockades approximately twice as deep as those of unfolded DNA carriers, while target-binding events appear as sharper, localized blockades within the recognition site region.

3. Analysis of barcode region and recognition site signals. the selected unfolded events are further analyzed to decode the carrier's digital barcode and assess molecular binding. Distinct current peaks corresponding to the programmed barcode positions (e.g., "001") are used to identify the specific carrier type, whereas an additional blockade at the recognition site confirms the presence of a bound target. This final analysis step integrates structural (barcode) and functional (recognition-site) information, providing single-molecule resolution of binding events across multiple carrier designs.

4 Notes

1. Immediately wash the restriction enzyme away to minimize the risk of nonspecific cleavage.
2. Occasionally, the measured concentration of the purified sample may be higher than expected (by 10–20 nM), which is likely due to insufficient washing during purification due to the presence of remaining complementary oligonucleotides. Increasing the number of washes with 10 mM Tris–HCl (pH 8.0) and 0.5 mM $MgCl_2$ can improve purity, but sample loss may also increase correspondingly.
3. Maintain a strict molar ratio between carrier DNA and oligonucleotides (commonly 1:3 or 1:5) to achieve complete and uniform hybridization.
4. Include $MgCl_2$ (typically 10 mM) in the hybridization buffer to stabilize double-stranded DNA and secondary structures. Avoid using EDTA in subsequent steps, as it chelates Mg^{2+} and interferes with hybridization efficiency.
5. Ensure measuring buffer is prepared at the specified pH range. A pH range from 9.0 to 10.0 is recommended to enhance the surface charge of the glass nanopore and avoid DNA clogging.
6. When designing the barcode region and detection region on the DNA carrier, introduce intentional asymmetrical design between the two ends. This enables unambiguous determination of the carrier orientation during nanopore translocation

and facilitates accurate identification of different targets in multiplexed detection [2].

7. The assembled DNA carrier is chemically stable and can be stored at 4°C for at least 3 months without noticeable degradation.
8. When incubating DNA carriers with different targets, carefully optimize and determine the appropriate buffer composition for each target type, as binding efficiency can vary with ionic strength, pH, and temperature conditions.
9. For nanopore experiments, the carrier concentration in 4 M LiCl is typically maintained at 100–500 pM. A higher sample concentration increases the risk of nanopore clogging.
10. Monitor the ionic current baseline and noise level during experiments. If the noise increases substantially or the baseline drops below threshold, promptly apply a reverse bias or voltage alternation (a "kick-out" pulse).
11. For measurement with a long time for more than several hours, increase in electrolyte concentration and decrease in DNA capture rate may occur, even in the same nanopore. These changes can affect event features, such as current blockage and dwell time. To mitigate variability, we recommend normalizing signal parameters during data analysis.
12. DNA translocation events often exhibit complex folding patterns. Folded signals are typically more difficult to interpret. We recommend analyzing linear translocation events for more reliable quantification.

References

1. Bell NA, Keyser UF (2015) Specific protein detection using designed DNA carriers and nanopores. J Am Chem Soc 137:2035–2041
2. Bell NAW, Keyser UF (2016) Digitally encoded DNA nanostructures for multiplexed, single-molecule protein sensing with nanopores. Nat Nanotechnol 11:645–651
3. Sze J, Ivanov A, Cass A et al (2017) Single molecule multiplexed nanopore protein screening in human serum using aptamer modified DNA carriers. Nat Commun 8:1552–1552
4. Bošković F, Zhu J, Tivony R et al (2023) Simultaneous identification of viruses and viral variants with programmable DNA nanobait. Nat Nanotechnol 18:290–298
5. Ren R, Cai S, Fang X et al (2023) Multiplexed detection of viral antigen and RNA using nanopore sensing and encoded molecular probes. Nat Commun 14:7362
6. Xue L, Yamazaki H, Ren R et al (2020) Solid-state nanopore sensors. Nat Rev Mater 5:952–952
7. Keyser UF (2016) Enhancing nanopore sensing with DNA nanotechnology. Nat Nanotechnol 11:106–108
8. Chen K, Jou I, Ermann N et al (2021) Dynamics of driven polymer transport through a nanopore. Nat Phys 17:1043–1049
9. Wang X, Liu Y, Ren R et al. (2025) Single-molecule protein profiling using nanopores and dimeric aptamer-modified DNA carriers. Angew Chem Int Ed 64:e202505902
10. Chen K, Juhasz M, Gularek F et al. (2017) Ionic current-based mapping of short sequence motifs in single DNA molecules using solid-state nanopores. Nano Lett 17:5199–5205
11. Chen K, Choudhary A, Sandler SE et al (2023) Super-resolution detection of DNA nanostructures using a nanopore. Adv Mater 35: 2207434
12. Liu Y, Wang X, Campolo G et al (2023) Single-molecule detection of alpha-synuclein oligomers in Parkinson's disease patients using nanopores. ACS Nano 17:22999–23009
13. Misiunas K, Ermann N, Keyser UF (2018) QuipuNet: convolutional neural network for single-molecule nanopore sensing. Nano Lett 18:4040–4045

Chapter 12

Fixed Aptamer-Functionalized DNA Carriers for Protein Detection with Movable Nanopores

Helena Miljkovic, Wayne Yang and Aleksandra Radenovic

Abstract

Nanopore sensing has shown great potential; however, one of its major challenges remains the fast and uneven translocation of analytes in free translocation nanopore measurements. We recently introduced Scanning Ion Conductance Spectroscopy (SICS), a technique that enables precise control of translocation speed by immobilizing the analyte on the surface and regulating the speed through controlled capillary movement. This methodology requires a unique preparation of the analyte over conventional free-translocation experiments, as the analyte is captured by and immobilized along a DNA carrier that is attached to a glass surface. This chapter describes the preparation of the essential components required for experiments, including the capillary, buffer, measurement system, and, most importantly, the preparation of DNA carriers that can be functionalized with aptamers.

Keywords Capillary, DNA carrier, Nanopore, SICS, Aptamer, Protein

1 Introduction

Single-molecule techniques have emerged as powerful tools for the identification and quantification of proteins for use in a broad range of applications [1–3]. However, traditional nanopore sensing is limited by the fast translocation of analytes [4], leading to a low signal-to-noise ratio (SNR) of protein translocation events and complicating data analysis—signal can be hidden in noise, averaged out, or masked by electronic noise [5]. Additionally, analyte capture and detection in traditional nanopore experiments rely on stochastic diffusion [6]; low-charged proteins would lead to low capture efficiencies. To address these limitations, we attach the analyte onto the functionalized DNA carriers with user-defined aptamer targets before approaching the analyte with a movable nanopore for sensing by using Scanning Ion Conductance Spectroscopy (SICS) [7, 8].

The following components are part of our measurement systems: capillary, buffer, electrodes, and DNA carrier (Fig. 1). The glass

Fanjun Li (ed.), *Nanopore Technology*, Methods in Molecular Biology, vol. 3072,
https://doi.org/10.1007/978-1-0716-5527-6_12,

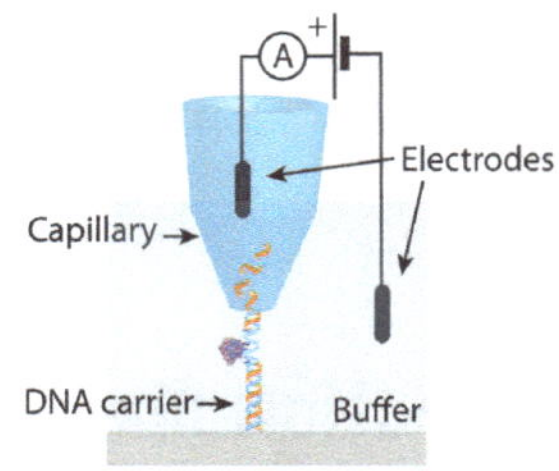

Fig. 1 Description of the SICS system. Schematic illustration of the setup, showing the capillary inserted into the buffer. There are two electrodes—one positioned inside the capillary and the other in the bulk solution. The DNA carrier is immobilized on the glass surface within the buffer

nanocapillary is filled and placed in a chamber containing the same buffer conditions. Silver chloride ($AgCl_2$) electrodes are respectively inserted inside the capillary and in the reservoir of the chamber to define the potential drop across the pore. The DNA carrier is immobilized on the glass surface of the Petri dish in the buffer. When a voltage is applied between the two electrodes, the movable capillary captures and moves along the DNA carrier; the ionic current passing through the pore is read out to give insight on the topography along the analyte.

1.1 DNA Carrier for Sensing

The DNA carrier is a long double-stranded DNA molecule that provides easily programmable sites for aptamer attachment [7, 9–11]. The uniform negative charge of the DNA carrier provides a sufficient electrophoretic driving force for the analyte, including weakly charged proteins, to be driven through the nanopore upon application of a voltage [7–10, 12].

1.2 Nanopore Measurement (Free Translocation)

Two reservoirs are filled with an ionic buffer and separated by a small hole, the nanopore. When electrodes are placed in each reservoir and a voltage is applied across the pore, ions flow through it, producing an ionic current [2, 3]. When a molecule such as DNA, RNA, or a protein enters the pore, it blocks a portion of the ion flow, causing a drop in current. These current blockades reveal information about the molecule's size, structure, and interactions [11, 13]. Because each translocation event creates a distinct electrical signature, nanopores enable real-time, label-free detection of single molecules.

1.3 Scanning Ion-Conductance Spectroscopy (SICS)

Scanning ion-conductance spectroscopy is a movable nanopore sensing technique that detects fixed, surface-bound molecules. The advantage of this method is that it allows precise control over the analyte's translocation speed, spatial multiplexing, and the ability to revisit the same molecules even after several days [8]. This also enables measurements on the same molecules under different conditions or at different time points, allowing for time-dependent studies. Furthermore, individual molecules can be measured up to 100 repetitions, leading to an improved signal-to-noise ratio and more reliable statistical analysis [8]. The analyte attached to the surface is a DNA carrier that can be functionalized in various ways, allowing us to push the limits of precision and reliability in nanopore measurement systems [7].

2 Materials

To control the speed of molecular translocation, we used glass capillaries mounted on a stepper motor in combination with a piezo stage holding the sample for fine adjustment. Together, this setup allows precise and reproducible regulation of the translocation velocity as defined by the movement of the capillaries along the analyte down to 0.1 nm. This section covers the preparation of the capillaries, the most commonly used buffers, and the preparation and attachment strategies for the DNA-based carriers and the ionic sensing methodology.

2.1 Capillary Fabrication

1. Borosilicate or quartz glass capillaries (available from Hilgenberg GmbH and Sutter Instruments). The parameters of the capillary depend on the desired pore diameter at the tip. To fabricate capillaries with pore sizes in the range of 5–300 nm, use the following capillary: Length $L = 70 \pm 1$ mm, outer diameter $Da = 0.5 \pm 0.1$ mm, inner diameter $Di = 0.20 - 0.13 + 0.15$ mm, spacing $S = 0.15$ mm.
2. Ethanol 70%.
3. Sutter Instruments puller, model P-2000.
4. Capillary holder (Fig. 2C), lab-made.
5. Oxygen plasma-cleaner.
6. Scanning electron beam microscope (SEM)—Zeiss Microscope.
7. Desiccator (low-pressure conditions, 1–10 mbar).
8. Petri dish.
9. Double-sided polyimide (Kapton) tape.

2.2 Preparation of Buffer

1. The choice of buffer and its concentration depends on the protein of interest. In our experiments, the best signal-to-noise ratio is achieved using potassium chloride (KCl). Potassium chloride (KCl) supplied as a powder (Sigma-Aldrich).
2. Buffer filter 0.02 μm (Fisher Scientific).
3. Mili-Q water.

2.3 DNA Carrier Preparation

1. pPIC9K vector (Thermo Fisher Scientific, V17520).
2. Thermal cycler.
3. 1× PBS (pH 7.4, Gibco).

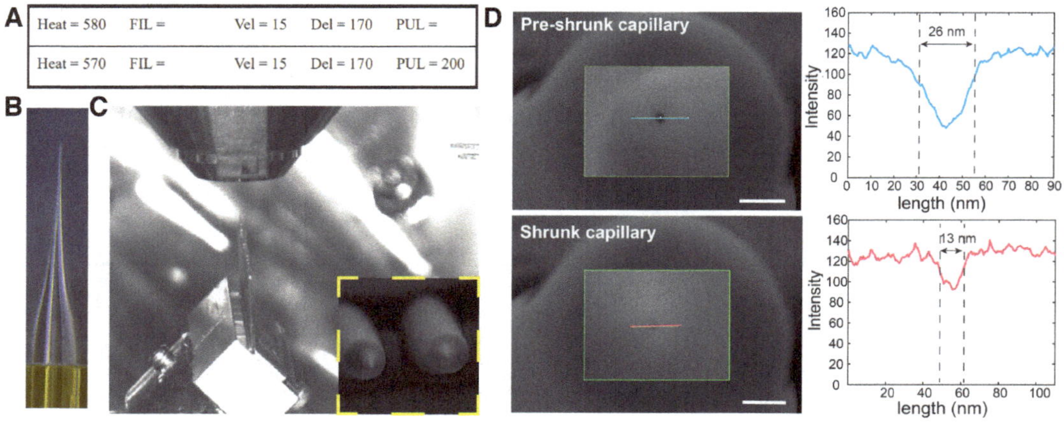

Fig. 2 Capillary fabrication. (**A**) Protocol that can be used on a quartzglass capillaries (length $L = 70 \pm 1$ mm, outer diameter $Da = 0.5 \pm 0.1$ mm, inner diameter $Di = 0.20 - 0.13 + 0.15$ mm, spacing $S = 0.15$ mm) to get a capillary pore size of diameter between 20 and 40 nm [7, 8]. (**B**) Pulled capillary viewed from the side. (**C**) SEM setup used for imaging the capillary, showing the 45° capillary holder. Inset: SEM image of the capillary tip. (**D**) SEM image of the capillaries before (top) and after shrinking (bottom), including the profile measurement of the capillary size, with their respective intensities across the pore. The pore edge is defined as the point where the intensity drops to 80% of the baseline intensity value. Scale bars = 100 nm [8]

4. Monarch PCR & DNA Cleanup Kit (New England Biolabs, T1030L).
5. Nt.BbvCI nicking enzymes (New England Biolabs, R0632) with rCutSmart Buffer.
6. EcoRV-HF nicking enzyme (New England Biolabs, R3195) with rCutSmart Buffer.
7. Gel extraction kit (example: Monarch DNA Gel Extraction Kit, T1020S).
8. Agarose TBA 1% gel, Tris–Glycine 16% gel, and electrophoresis system (Bio-Rad). Quick-Load 1 kb DNA Ladder (Sigma-Aldrich, D0428).
9. Extended aptamer sequence for TBA 1: 5′- TCAGCTTGCGG AGATGCGAT GGTTGGTGTGGTTGG ATCGAGAAAAA GGTACGCCC -3′ (Microsynth).

2.4 Electrodes

1. Silver Wire (Ag/AgCl, Advent, Purity 99.99%, GI4777).
2. Bleach/sodium hypochlorite solution.
3. HCl solution (Sigma-Aldrich, 0.1 M).
4. Constant low voltage source (Axopatch 200B or equivalent).

2.5 Nanopore (Free Translocation) Measurements

1. Teflon/PEEK chambers (custom-fabricated).
2. Chimera Instruments (Chimera Instruments VC100).
3. Faraday cage.

2.6 Controlled Translocation (Scanning Ion-conductance Spectroscopy—SICS)

1. Micro Petri dish with a bottom glass surface (Ibidi company; μ-Dish 35 mm, high glass bottom). To change the size of the coated area, we use a customized hollow cylinder (can be also formed by cutting the end of a 1000 μL pipette tip) that is glued on top.
2. Polydimethylsiloxane (PDMS; SYLGRAD 184 silicone elastomer kit).
3. Bovine serum albumin (BSA-Bt, Sigma-Aldrich, Albumin, biotin-labeled bovine, A8549).
4. Streptavidin (Sigma-Aldrich, Streptavidine from Streptomyces avidinii, 85878).
5. YO-PRO-1 dye (Thermo Fisher Scientific, Y3603).
6. Scanning ion conductance microscope (SICM).

3 Methods

3.1 Capillary Fabrication

Glass capillaries are typically made from either borosilicate or quartz glass. The capillaries are typically pulled using a Sutter Instruments puller. After pulling, the capillaries are transferred to a custom metal holder that can be placed in the scanning electron beam microscope (SEM) for inspection. Here, the nanopore size (as defined by the diameter across the tip) can be further tuned by focusing the electron beam at the capillary opening to induce controlled shrinking. As an example, we show the protocol for quartz glass capillaries from Hilgenberg GmbH—Length $L = 70 \pm 1$ mm, outer diameter $Da = 0.5 \pm 0.1$ mm, inner diameter $Di = 0.20 - 0.13 + 0.15$ mm, spacing $S = 0.15$ mm:

1. Quartz capillaries, in particular, are fabricated with the CO_2 laser-based puller P-2000. When using this instrument, careful positioning of the capillary is essential: it should be aligned precisely perpendicular to the beam path of the laser, i.e., centered along the capillary's length, and oriented at a 90° angle to the laser direction. These capillaries can be prepared using the pulling protocol shown in Fig. 2A. If you encounter issues with the pulling parameters, refer to **Note 1**. After pulling, the left and right capillaries (also known as the twin pair) should be symmetrical under electron beam inspection, i.e., with similar taper angles (Fig. 2B), pore sizes, and shape (typically 1.8 ± 0.3 degrees, 25–40 nm). It is also important to use capillaries with minimal initial dimensional variation so that the pulling protocol yields consistent results across batches. Using this protocol, we can fabricate nanopores with diameters typically ranging from 25 nm to 40 nm [7].

2. Visualization and shrinking with SEM—Before imaging, the nanopipettes are mounted in the conductive holder (Fig. 2C) and cleaned in an oxygen plasma cleaner to remove any organic contamination (30 sccm of O_2, 100 W, duration of 5 min). Because glass is not conductive, SEM visualization is challenging and can lead to significant charge build-up at the tip. Parameters used in this specific case are as follows: EHT = 3 kV, probe current = 370 pA, and working distance ≈ 6 mm (insertion in Fig. 2C). For shrinkage, the same area is imaged repeatedly, which leads to local melting and reforming during repeated scans of the capillary tip. For imaging tips, refer to **Note 2**. After this step, the capillaries are typically reduced to a diameter of 5–15 nm (Fig. 2D) [8, 14].
3. When the desired size is achieved, the capillaries are removed from the SEM and plasma-cleaned once more to remove carbon buildup. Next, capillaries are attached to the Petri dish with double-sided polyimide (Kapton) tape and then immediately immersed in the desired nanopore sensing buffer. To help remove bubbles and improve wettability, the submerged capillaries are then placed in a desiccator (low-pressure conditions, 1–10 mbar) for 1 h, followed by several cycles of brief microwave heating as reported in Navikas et al. [15]. These heating cycles help remove trapped air bubbles from the nanopipettes and promote complete filling with the buffer solution.

To accurately sense biomolecules, the dimensions of the nanopore must be carefully selected. The size and shape of the pore have a significant influence on its conductivity and on the rate of translocation events.

3.2 Preparation of Buffers

Solid-state nanocapillaries employ the use of ionic buffers typically ranging from the 100 mM to 4 M range [7, 9, 11, 16, 17]. Generally, the higher ionic strength, the slower the translocation time [18]. The concentration and type of salt also affect the SNR of the analyte with the highest SNR achieved at 4 M KCl [5, 12, 19]. Other important factors to consider include the stability of the analyte in the sensing buffer. For example, for DNA structures, stability is generally high across many buffer conditions, enabling broad flexibility in buffer choice. In contrast, protein measurements require greater care, as buffer composition must maintain protein stability, avoid charge screening effects, and preserve any interactions of interest. Hence, the ionic strength of the sensing buffer is generally lower and in the < 1 M range [7, 10, 11]. Similarly, when studying interactions between two molecules, the buffer must support binding conditions. For these reasons, tailoring the buffer composition to the specific analyte and experimental goal is essential (*see* **Note 3**).

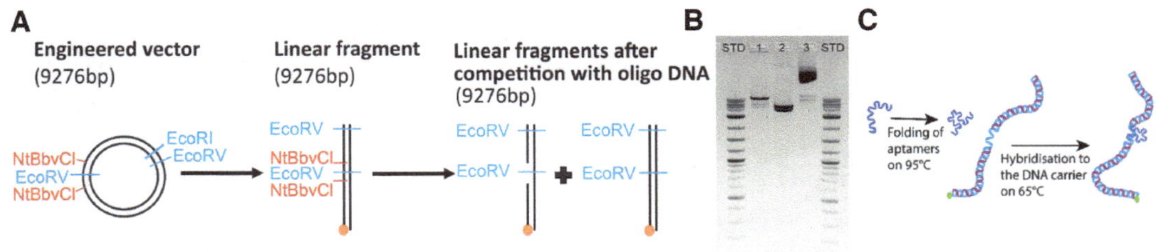

Fig. 3 Protocol for DNA carrier formation. (**A**) Circular plasmid vector that was first linearized and then gap was created by competition with the short DNA oligos. (**B**) Characterization of the construct with 1% agarose gel. **Lane 1:** Purified biotinylated, gapped DNA after EcoRV digestion. **Lane 2:** The engineered vector digested with EcoRV *before* gap formation, yielding two linear fragments that appear as a single thick band. **Lane 3:** Undigested engineered vector [8]. (**C**) Annealing of the aptamer to the DNA handle. In the first step, the aptamer is folded, and in the second step, it is hybridized to the DNA handle [7]

3.3 DNA Carrier Preparation

In our lab, we generated an 8.6 kb DNA carrier containing one to three single-stranded DNA regions, each 40 bp in length and spaced approximately 1 kb apart. We synthesized it from 9k plasmid [8] that was modified in order to introduce sequences for primers and restriction sites (Nt.BbvCI and EcoRV). To generate gaps (single-stranded region in the middle of double-stranded DNA), we introduced two Nt.BbvCI recognition sites positioned 40 bp apart, with an EcoRV restriction site located between them.

1. For SICS measurements, the plasmid was linearized by PCR using a forward primer modified with a 5′ Biotin tag, whereas for free translocation experiments, the forward primer does not have the biotin modification in order to avoid interactions. PCR is performed, and the product is purified using the clean-up kit Monarch PCR and DNA Cleanup Kit (New England Biolabs, T1030L) (*see* **Note 4**).
2. After purifying the PCR product, the DNA was digested with Nt.BbvCI nicking enzymes (ordered from New England Biolabs company) for 3 h at 37°C.
3. The nicked plasmid was incubated with a 100× molar excess (with respect to the strand concentration) of a single-stranded DNA competitor and thermally treated with the following thermal protocol: 90°C (5 min), 80°C (1 min), and −1°C cooldown (1 min, 30×), 50°C (30 min), and −1°C touchdown (1 min, 30×) (Fig. 3A).
4. Additional digestion with EcoRV for 2 h at 37°C allowed selective cutting of double-stranded DNA, leaving full size only the nicked constructs. The EcoRV site positioned between the Nt.BbvCI sites serves to eliminate ungapped molecules. After successful gap formation, the EcoRV site becomes single-stranded and is no longer recognized. In contrast, if the gap fails to form and the region remains double-stranded,

EcoRV digestion produces two smaller fragments that can be removed through a gel purification process. EcoRV cuts only double-stranded DNA, and the ~9 kb gapped fragment remains intact.

5. The DNA gap constructs are separated and extracted using Agarose gel (1%) electrophoresis (Fig. 3B). The fraction of ungapped molecules shows up as faint lower bands when running a gel, while the longest band in the gel is the DNA gap. The DNA gap construct was extracted by cutting a corresponding band in the gel and using a gel extraction kit (Monarch DNA Gel Extraction Kit, T1020S). This gel can be used as a quality check of the process (*see* **Note 5**).

For SICS experiments, the single-gap DNA construct had the gap positioned 3670 bp from the biotin-labeled end, while the double-gap DNA carried an additional gap located 1030 bp further along the molecule [7, 8]. When designing the DNA carrier, the first aptamer binding site should be placed at least 1500 bp away from the biotin end that is attached to the surface to ensure accessibility by the nanocapillary. The capture distance between the substrate surface and the nanopore, as well as the DNA capture rates, can vary between experiments. These parameters are influenced by factors, such as nanopore size, size of the sensing molecule, applied voltage, charge of the surface, and the concentration and type of ions present in the buffer.

3.4 Incubation of Aptamers to DNA Carriers

We can introduce the aptamer in the place of the DNA gap by extending the sequence of the aptamer with the complementary sequences of the gaps. Using the TBA1 aptamer (5′-GGTTGGTGTGGTTGG-3′) as an example, the sequence was extended on both sides with sequences complementary to half of the gap of the DNA carrier, resulting in 5′-TCAGCTTGCGGAGATGCGAT GGTTGGTGTGGTTGG ATCGAGAAAAGGTACGCCC - 3′ [7].

1. To ensure proper folding of the aptamer, we first fold the aptamer, together with the extended sequences, into its active tertiary structure (the conformation that enables binding to the target protein). Initially, the aptamer was folded by annealing from 95°C (Fig. 3C) and its folding was verified using circular dichroism spectroscopy (Chirascan V100) (*see* **Note 6**).
2. Afterward, we mixed the folded aptamers with the DNA carrier and performed a second annealing step at 65°C (Fig. 3C). Here, we hybridized the extended aptamer sequence to the DNA gap structure.

It is recommended to verify that the aptamer remains active after the extended region of the sequence binds to the complementary single-stranded DNA (ssDNA) region of the DNA carrier. For further details on the verification procedure, *see* **Note 7**. The salt concentration of the final sensing buffer may be increased for certain aptamers (*see* **Note 8**).

3.5 Preparation of Electrodes

The most commonly used electrodes for nanopore sensing experiments are Ag/AgCl (Silver Wire, Advent, Purity 99.99%, GI4777). This AgCl coating acts as a reversible redox interface, enabling stable ion-to-electron transduction and minimizing electrode drift during measurements. These electrodes can be prepared by chemical or electrochemical chlorination of a silver wire.

1. Chemical chlorination: Chemical chlorination is achieved by first sanding off the top layer of silver wire with sandpaper and then immersing the silver wire in a bleach or sodium hypochlorite solution— typically for 30 min. Household bleach can be used for this purpose. During this process, the surface of the silver wire reacts with chloride ions to form a uniform AgCl layer.
2. Electrochemical chlorination: Electrochemical chlorination uses two silver wires that are sanded with sandpaper before being placed in an HCl solution (Sigma-Aldrich, 0.1 M). A constant voltage (typically a few volts) is applied across them using an external battery (9 V). The silver at the anode terminal undergoes oxidation, releasing silver ions that combine with chloride ions in the solution to produce AgCl on the electrode surface. By adjusting parameters such as applied voltage, duration, and electrolyte concentration, the thickness and uniformity of the AgCl layer can be tuned. This process typically takes less than 30 s [20].

The wires are rinsed with deionized water to remove any residual chemicals before the measurements. A properly coated electrode has a uniform dark brown color absent of any grey spots or patches.

3.6 Nanopore (Free Translocation) Measurements

The setup includes two Teflon chambers (holding the nanopipette), electrodes, and an amplifier (Fig. 4). Electrodes, the chamber with the buffer, and the capillary are placed inside a grounded Faraday cage to minimize external electrical interference and noise. The cage shields the system from electromagnetic noise sources such as surrounding electronic devices and mechanical vibrations, which could introduce fluctuations in the ionic current signal.

To set up the experiment, we perform the following steps [2, 3]:

1. Two Teflon chambers with small nanopipette holders are placed in proximity. The chambers are filled with the desired buffer (typically 1 M LiCl or KCl); in our case, each contains 400 μL of a prefiltered 1 M KCl electrolyte solution.
2. The electrodes (prepared as described in the electrodes section) are placed into each of the respective chambers and

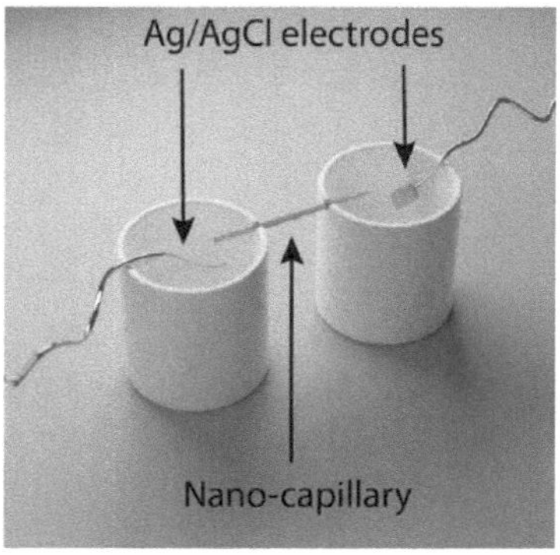

Fig. 4 Schematics of free translocation setup. The capillary is positioned between two Teflon wells to ensure proper insulation between the chambers, with both ends of the capillary immersed in one buffer-filled well each. An electrode is placed in each chamber and connected to the amplifier for signal recording. The entire setup is enclosed in a Faraday cage during measurements to minimize electrical noise [23]

connected to the amplifier—in our case, Chimera (Chimera Instruments VC100) or Axopatch 200B. One needs to ensure that the electrical connections leading up to the electrodes are properly secured and stable.

3. A wetted capillary is positioned between two chambers, connecting the buffers and forming the nanopore for analyte sensing. After positioning the capillary, close the Faraday cage and zero the voltage (correct the voltage until the current that the amplifier is measuring is zero). This configuration ensures a reference potential across the nanopipette.
4. Add the analyte of interest to the chamber (typically the one containing the capillary tip) and apply a voltage. The ionic current across the nanopore is then recorded and analyzed. The typical acquisition rate for the current measurement ranges between 100 KHz and 1 MHz. If excessive noise is observed, refer to **Note 9**.

The current data are acquired by software provided by a number of companies, including Chimera Instruments, Nanion Technologies, Oxford Nanopore Technologies, etc. Afterward, the nanopore data can be analyzed by software, such as OpenNanopore (Python) and Transanalyzer (Matlab) [21].

Nanopore (free translocation) measurements are typically done before SICS experiments to assess analyte stability, particularly by checking the DNA carrier length and its sticking to the capillary.

3.7 Controlled Translocation (Scanning Ion-conductance Spectroscopy—SICS)

Here, we will demonstrate the method we used to incorporate aptamer modifications that will enable the detection of the specific protein from the protein mixture within the buffer.

3.7.1 Immobilization of the Surface with DNA

1. Obtain the micro Petri dish with a bottom glass surface (from the Ibidi company; μ-Dish 35 mm, high glass bottom). On the top of the dish, we glue with polydimethylsiloxane (PDMS; SYLGRAD 184 silicone elastomer kit), an array of cylindrical cones making chambers that would contain the area we want to coat with the DNA carriers. The size of this chamber would determine the volume of the buffer we need to use, and it is in the range of 5–400 μL. Plasma cleaning of the Petri dish improves coverage of DNA binders and functionalization (30 sccm of O_2, 100 W, duration of 5 min).
2. Next, prefill the chamber with phosphate-buffered saline (PBS pH 7.4, Gibco) with half of the volume you would like to have at the end—make sure that this volume is completely covering the surface of the cover glass. For example if the desired final volume is 400 μL, place 200 μL of PBS followed by 200 μL (i.e., the other half of the final volume) with a 1 mg/mL concentration of biotin-conjugated bovine serum albumin (BSA-Bt, Sigma-Aldrich, Albumin, biotin labeled bovine, A8549). Incubate the solution in the chamber for at least 1 h at room temperature.
3. Rinse the chamber with PBS solution by pipetting half of the solution out and pipetting in fresh PBS in the same amount of volume that is taken out, for at least 10×.
4. To prepare the streptavidin binding step, we remove half of the volume in the chamber and pipette in 200 μL (or half of the final desired volume) with 0.1 mg/mL concentration of streptavidin (Sigma-Aldrich, Streptavidin from *Streptomyces avidinii*, 85878). Incubate for at least 1 h at room temperature.
5. Rinse the chamber with PBS solution by another 10× wash as described above.
6. Finally, pipette out 200 μL of solution (or half of the final desired volume) and pipette in the prepared DNA carrier solution as described in the DNA carrier section (typically at a concentration of 100–200 pM).
7. Incubate the DNA carriers in the Petri dish for at least 3 days at 4°C. To prevent evaporation, seal the sample and Petri dish well with parafilm.
8. After incubation and before the start of the experiment, remove unbound DNA carriers. The sample should be washed using the same half-dilution step as described above (repeated

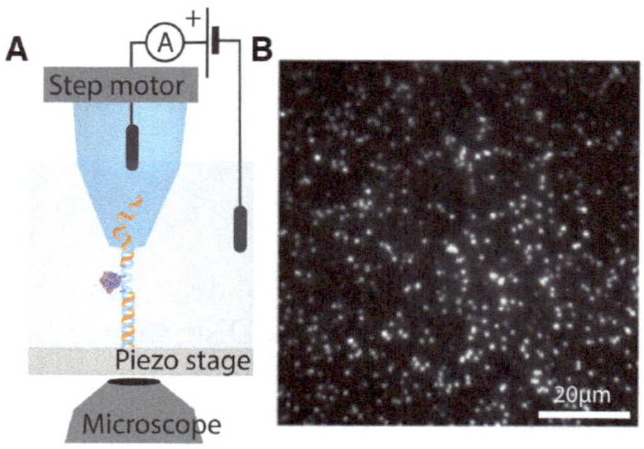

Fig. 5 Single-molecule DNA carrier fluorescence imaging. (**A**) Schematic illustration of the setup, showing that measurements can be performed simultaneously with fluorescence imaging. The diagram also shows that the capillary is mounted on a stepper motor and that the sample is positioned on a piezo stage [7]. (**B**) Imaging of tethered DNA carriers onto the glass surface with a widefield fluorescence microscope. The BSA-Bt tether concentration indicated is 1 pM, 0.1 mg/mL is the concentration of streptavidin, and the DNA carrier concentration is 100 pM [8]

~20 times). After washing, the buffer can be exchanged to match the desired experimental conditions/buffer again using the half-dilution method as the washing step. In the example shown here, the buffer is replaced with 1 M KCl containing a mixture of human serum proteins. Note that care must be taken to ensure that the sample does not fully dry out.

Optical imaging and check of bound DNA carriers: In order to check and optimize the protocol to maximize the surface density and distribution of attached DNA carriers, we used 5–30 μL of 5 nM YO-PRO-1 dye (Thermo Fisher Scientific, Y3603) to visualize the bound DNA carriers. After incubating the YO-PRO dye sample for 30 min with the sample, we imaged the surface in TIRF mode (see Fig. 5A) with a 488 nm laser (see Fig. 5B). If the surface is not properly coated, (*see* **Note 10**).

The prepared DNA carrier sample can be stored for several months at 4°C without degradation in a parafilm-sealed container in PBS buffer. The DNA carrier can be modified even after incubation on the surface. For details, *see* **Note 11**.

3.7.2 DNA Carrier Approach and Sensing

The instrument system that is adapted for nanopore sensing is called Scanning ion conductance microscope (SICM) [7, 8, 22]. Here, we will describe the protocol used to locate the molecule of interest and to scan over the area.

1. **Finding the surface**: To approach the surface, we used the characteristic decrease in ionic current that occurs when the nanopore comes into proximity to the surface (Fig. 6 Step 1). The capillary is inserted in the buffer and lowered toward

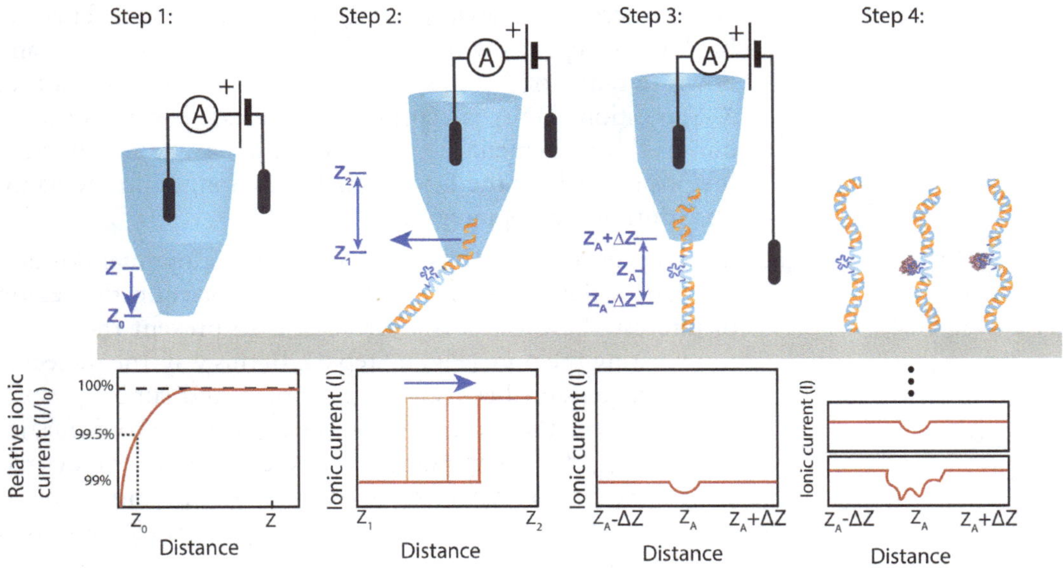

Fig. 6 Protocol for DNA carrier approach and sensing. Step 1: Approach of the capillary to the surface. **Step 2:** Centering the capillary over the path of the DNA carrier. **Step 3:** Reducing the ramp range to focus on the region of interest along the DNA carrier. **Step 4:** Move the capillary in the XY direction to analyze different molecules [7]

the surface with coordinated movement in the *z*-direction using both the stepper motor and the piezo stage. During the approach, a trigger is activated to stop the motion in case the measured ionic current decreases to 99.5% of its initial value. The approach starts with the piezo stage moving through its full range (for our system −20 μm; with a speed of 15–30 μm/s). If the threshold is not reached, the piezo retracts to its minimum position, and a step motor makes a step that is 60% of the piezo's full range. Alternating these two steps allows us to progressively assess proximity to the surface using the piezo stage, followed by stepwise advancement with the step motor. Once the threshold is reached, the capillary is in the approached position (with reference to the surface) and stops moving.

2. **Capturing and centering the DNA carrier within the nanopore**: After the nanopipette was brought close to the surface, a positive voltage was applied to the capillary, and the tip–surface distance was adjusted to enable efficient DNA capture. These parameters depend strongly on the salt concentration of the buffer (for buffers with lower ionic strength, the capillary has to be positioned further from the surface). The capillary is moved in the X–Y plane until the DNA is captured by the pore (marked by a stepwise decrease in current) (see Fig. 6 Step 2). After detecting the DNA signal, we center the pore above the

point where the analyte is attached to the surface. This can be done by repeatedly moving/hopping the nanopore up and down (z range of 2 μm, speed of 500 nm/s) as a function of X–Y position. DNA is captured and released from the pore during this movement. The center of the analyte is defined as the position where the DNA signal is the longest in the nanopore during the hopping step.

3. **Sensing along the DNA carrier:** Once the optimal position is reached, the ramp range is reduced to focus on the region of interest along the DNA carrier and to prevent the carrier from exiting the pore (Fig. 6 Step 3). In this way, the molecule is held in tension between the nanopore and surface, which would prevent the molecule from being lost due to diffusion. At this step, the voltage range can be changed to the desired condition, or to maximize SNR, the translocation (ramp) speed can be adjusted, and the same molecule can be re-sensed multiple times.
4. **Interrogating other molecules:** By moving the capillary in the XY-direction to capture other DNA carriers, one can interrogate independent DNA carriers and build up statistics on the sample (Fig. 6 Step 4).

To detect the protein in the target analyte solution, the preceding buffer (i.e., the one used for DNA attachment and storage) is exchanged with one containing the target protein. It is also possible to use buffers containing multiple proteins, such as human serum, and selectively capture the protein of interest using aptamers. Now, when approaching the DNA carrier with the captured protein, a deeper decrease in current at the aptamer sites along the DNA carrier indicates that an additional molecule is bound at that location. When working with proteins, the nanocapillary is more prone to contamination. Methods for addressing this issue are described in **Note 12**.

4 Notes

1. Pulling parameters for nanocapillary production: The pulling parameters may require adjustment depending on room temperature fluctuations or instrument drift. In such cases, the time delay (change in steps of 10 on the dial) and heat settings (change in steps of 5 on the dial) are typically the primary parameters to optimize.
2. SEM imaging tip to reduce charge buildup: Minimize the imaging time in order to minimize the charge buildup and use a lower working current and voltage. Typical imaging voltage and current are 1–3 kV and 50–370 pA. During imaging, the imaging window is aligned to the center of the pore

to ensure accurate visualization. If needed, a small amount of carbon can be sputtered onto the pipette prior to imaging to lower the charging effect while imaging.

3. Preparation of high-salt buffer: Due to the high salt concentration, even small amounts of evaporation can lead to the formation of salt crystals and change the ionic strength of the solution. It is recommended to always filter the buffer prior to the experiment.
4. Troubleshooting the DNA Carrier preparation protocol, tip 1: After performing the PCR amplification, leave one tube aside to verify that the first step was done correctly. This sample should be run on a 1% Agarose TBE gel.
5. Troubleshooting the DNA Carrier preparation protocol, tip 2: The extraction kit may need optimization if you use DNA carriers of different lengths or create gaps of different sizes. Aim to minimize the amount of purification, as excessive purification can reduce the final concentration and lower efficiency.
6. Hybridization buffer: The buffer may need to be adjusted to promote formation of the functional configuration of the aptamer. Options that worked for TBA 1 are 1× PBS (pH 7.4) and 150 mM NaCl, 10 mM $MgCl_2$, Tris-EDTA buffer at pH 7.4.
7. Aptamer activity after binding to the DNA carrier: To verify that the aptamer remains active after the extended region binds to the complementary ssDNA segment of the DNA carrier, we performed a functional assessment. The extended aptamer sequence was hybridized to a new, only ssDNA strand that matches the ssDNA region of the DNA carrier. This new ssDNA strand is labeled with a Cy5 fluorescent dye at one end to enable imaging. A gel shift assay was then performed using a 16% Tris–Glycine gel. After fluorescence imaging, a quick blue protein-staining dye was added to identify the regions containing thrombin proteins. Comparison of the two images showed that a large fraction of the proteins is bound [7].
8. Sensing buffer: The salt concentration in the annealing buffer is important for the stability of the DNA structure and the annealing process of the aptamers to the DNA gap structure. While the salt concentration of the final sensing buffer can be increased when using certain aptamers, this adjustment should be made only after the annealing process. This ensures highly stable DNA–aptamer complexes that can withstand the final salt concentration in the sensing buffer.
9. Free translocation measurements: Proper grounding and shielding are essential for achieving high signal-to-noise ratios,

particularly when detecting subtle events like single-molecule translocations. Phones, screens, and any vibrations should be placed far away from the setup. Care should also be taken to ensure that electrical devices are not plugged into the same electric outlet as the measurement device.

10. Visualization of bound DNA carriers: Before imaging the chamber surface, rinse the buffer to reduce background noise. If the concentration of tethered molecules is insufficient, first try improving the protocol by increasing the concentration of BSA–Bt in Step 2. After optimizing Step 2, adjust the DNA carrier concentration to get sufficient event and capture rates with the capillary.
11. Hybridization of ssDNA/RNA to DNA carrier: Because of the strong biotin–streptavidin interaction, it is possible to hybridize ssDNA strands to the DNA carrier even after the sample has been attached to the surface. It is also possible to remove part of the hybridized DNA by applying mild heating.
12. Tips to reduce unwanted pore–analyte interactions: When the translocation speed is too low, highly adhesive analytes (proteins) can irreversibly attach and contaminate the pore. Applying a reversal in voltage might kick the protein off, but failure to return to the original baseline usually means a stop in the experiment. Some mitigation strategies include coating the pore with a salinizing agent or applying a TWEEN 20 (P9416, Sigma-Aldrich) to the buffer.

Acknowledgments

A.R. and H.M. acknowledge support from the European Research Council under grant no. 101020445—2D-LIQUID and National Center of Competence in Research (NCCR) Bio-Inspired Materials. W.Y. would like to thank funding from Ambizione grant 223401.

References

1. Xue L et al. (2020) Solid-state nanopore sensors. Nat Rev Mater 5:931–951
2. Miles BN et al. (2012) Single molecule sensing with solid-state nanopores: novel materials, methods, and applications. Chem Soc Rev 42:15–28
3. Howorka S, Siwy Z (2009) Nanopore analytics: sensing of single molecules. Chem Soc Rev 38:2360–2384
4. Plesa C et al. (2013) Fast translocation of proteins through solid state nanopores. Nano Lett 13:658–663
5. Fragasso A, Schmid S, Dekker C (2020) Comparing current noise in biological and solid-state nanopores. ACS Nano 14:1338–1349
6. Wanunu M, Morrison W, Rabin Y, Grosberg AY, Meller A (2010) Electrostatic focusing of unlabelled DNA into nanoscale pores using a salt gradient. Nature Nanotech 5:160–165
7. Miljkovic H et al. (2025) Controlled sensing of user-defined aptamer-based targets using scanning ionic conductance spectroscopy. ACS Nano 19:13139–13148

8. Leitao SM et al. (2023) Spatially multiplexed single-molecule translocations through a nanopore at controlled speeds. Nat Nanotechnol 18:1078–1084
9. Bell NAW, Keyser UF (2015) Specific protein detection using designed DNA carriers and nanopores. J Am Chem Soc 137:2035–2041
10. Sze JYY, Ivanov AP, Cass AEG, Edel JB (2017) Single molecule multiplexed nanopore protein screening in human serum using aptamer modified DNA carriers. Nat Commun 8: 1552
11. Bošković F et al. (2023) Simultaneous identification of viruses and viral variants with programmable DNA nanobait. Nat Nanotechnol 18:290–298
12. Chen K et al. (2021) Dynamics of driven polymer transport through a nanopore. Nat Phys 17:1043–1049
13. Houghtaling J et al. (2019) Estimation of shape, volume, and dipole moment of individual proteins freely transiting a synthetic nanopore. ACS Nano 13:5231–5242
14. Steinbock LJ, Steinbock JF, Radenovic A (2013) Controllable shrinking and shaping of glass nanocapillaries under electron irradiation. Nano Lett 13:1717–1723
15. Navikas V et al. (2020) High-throughput nanocapillary filling enabled by microwave radiation for scanning ion conductance microscopy imaging. ACS Appl Nano Mater 3:7829–7834
16. Bell NAW, Keyser UF (2016) Digitally encoded DNA nanostructures for multiplexed, single-molecule protein sensing with nanopores. Nature Nanotech 11:645–651
17. Smeets RMM et al. (2006) Salt dependence of ion transport and DNA translocation through solid-state nanopores. Nano Lett 6:89–95
18. Plesa C, Van Loo N, Ketterer P, Dietz H, Dekker C (2015) Velocity of DNA during translocation through a solid-state nanopore. Nano Lett 15:732–737
19. Chen K et al. (2023) Super-resolution detection of DNA nanostructures using a nanopore. Adv Mater 35:2207434
20. Cho KR et al. (2022) Investigation of the AgCl formation mechanism on the Ag wire surface for the fabrication of a marine low-frequency-electric-field-detection Ag/AgCl sensor electrode. ACS Omega 7:25110–25121
21. Plesa C, Dekker C (2015) Data analysis methods for solid-state nanopores. Nanotechnology 26:084003
22. Leitao SM et al. (2021) Time-resolved scanning ion conductance microscopy for three-dimensional tracking of nanoscale cell surface dynamics. ACS Nano 15:17613–17622
23. Navikas V (2021) Biophysical applications of correlative scanning probe and super-resolution microscopy. EPFL, No. 9021, (vol 4, page 83). https://doi.org/10.5075/epfl-thesis-9021

Chapter 13

Polymer Translocation Through a Nanopore

Kuo Chen and Murugappan Muthukumar

Abstract

Polymers with distinct topologies—such as linear, star-shaped, ring, and bottlebrush architectures—exhibit diverse translocation behaviors when driven through solid-state nanopores due to differences in flexibility, hydrodynamic size, and steric interactions. These properties enable topology-sensitive single-molecule analysis. In this protocol, we describe detailed procedures for the electrophoretic translocation of PEG-based linear and star polymers through individual nanopores. This method enables precise characterization of polymer conformation and transport kinetics, with differences in dwell time, capture rate, and current blockage amplitude reflecting molecular architecture. The approach offers a practical platform for investigating polymer dynamics under nanoconfinement and supports topology-based analysis and separation.

Keywords Nanopore translocation, Polymer topology, Single-molecule analysis, PEG-based polymer, Molecular conformation, Electrophoretic transport, Topology-sensitive detection

1 Introduction

Polymer translocation through solid-state nanopores–with their tunable pore size and structurally well-defined, symmetric geometry–enables single-molecule analysis of polymer size, structure, and dynamics under confinement [1–5]. While initially developed for nucleic acid sensing [6–9], this technique has been extended to synthetic polymers with diverse architectures. Topological polymers, such as linear, star, bottlebrush, and ring structures, display distinct translocation behaviors due to differences in flexibility, hydrodynamic drag, and steric effects [10–12].

Topological polymers such as star-shaped and bottlebrush structures are commonly synthesized via graft-from, graft-onto, or core-first strategies [13–16]. However, structural heterogeneity is often unavoidable: star polymers may have varying arm numbers, while bottlebrush polymers frequently show differences in grafting density or backbone length [15]. These imperfections complicate the purification of defect-free topological polymers.

Fanjun Li (ed.), *Nanopore Technology*, Methods in Molecular Biology, vol. 3072,
https://doi.org/10.1007/978-1-0716-5527-6_13,

Conventional methods like size exclusion chromatography (SEC) and dialysis offer limited resolution and are time-consuming [17, 18]. In contrast, nanopore sensing provides a rapid, label-free, and topology-sensitive platform for analyzing such polymers at the single-molecule level [19–22]

Poly(ethylene glycol) (PEG)-based polymers exhibit polycation-like behavior in salt solutions [8,23–25], making them well-suited for studying topology-dependent electrophoretic transport. Their biocompatibility, commercial availability, and access to well-defined topologies—including linear, star, and ring architectures—make them ideal model systems for nanopore-based single-molecule analysis [26–29], as shown in Fig. 1. These structurally different polymers enable mechanistic insights into how molecular architecture influences capture and translocation dynamics under nanoconfinement [21, 22]. Such understanding is essential not only for polymer purification and gene delivery applications but also for elucidating general principles of macromolecular transport across free energy barriers present in biological and synthetic nanoscopic systems.

This chapter describes a protocol for the preparation, measurement, and analysis of PEG-based linear and star polymers translocating through solid-state nanopores under an electric field. Detailed methods are provided for sample handling, nanopore device setup, electrophoretic measurement, and current signal interpretation.

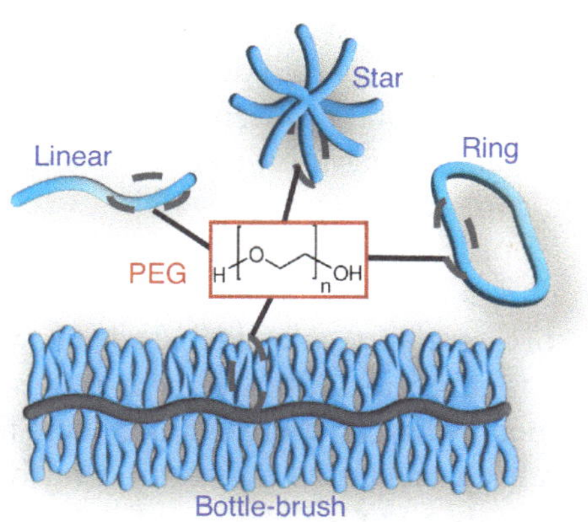

Fig. 1 Schematic illustration of PEG-based polymers with distinct topologies: linear, star, ring, and bottle-brush. Instead of PEG, any polymers made of charged monomers with their diverse architectures are suitable candidates for this technique

2 Materials

2.1 Platform, Membrane Chips, and Electrodes

Figure 2 shows the design of the solid-state nanopore device. The experimental platform consists of a custom-fabricated Teflon flow cell composed of two symmetric compartments that are separated by a freestanding silicon nitride (SiN_x) membrane chip. The SiN_x membrane functions as the only physical barrier between the two reservoirs, thereby defining the cis (donor) and trans (acceptor) sides of the nanopore system. Silicone rubber O-rings are placed on both sides of the membrane chip to ensure a watertight seal (*see* **Note 1**). The two halves of the flow cell are aligned and tightly clamped together using stainless steel screws.

Both compartments are filled with electrolyte buffer (e.g., 1 M KCl, 10 mM HEPES, pH 8.0) and serve as the cis and trans chambers for polymer translocation. Each chamber contains a port for inserting a silver/silver chloride (Ag/AgCl) electrode, enabling electrical measurements across the membrane (see Fig. 3). Ag/AgCl electrodes were prepared from 1 mm diameter silver wires chloridized in commercial bleach (see Sect. 3.2 for details). The SiN_x membrane, commonly employed in solid-state nanopore devices, is typically fabricated as a low-stress, freestanding film with a thickness of 10–20 nm, and is usually supported on a silicon substrate (*see* **Note 2**). A single nanopore is fabricated in the membrane via controlled dielectric breakdown, as described in Sect. 3.

2.2 Chemicals and Reagents

Analytical-grade potassium chloride (KCl), lithium chloride (LiCl), 4-(2-hydroxyethyl)-1-piperazineethanesulfonic acid (HEPES), and sodium hydroxide (NaOH) are used in all experiments. All reagents are obtained from Sigma-Aldrich or equivalent commercial suppliers unless otherwise specified.

Linear polyethylene glycol (PEG, $M_w = 10$ kDa) and star-shaped PEGs with four to eight arms (each arm $M_w = 5$ kDa) are used as model polymers. All PEGs are purchased from commercial vendors such as JenKem Technology.

Polymer stock solutions are prepared at a concentration of 1 g/L in nanopore testing buffer (e.g., 4 M KCl, 10 mM HEPES, pH 6.0). After thorough vortexing, the solutions are passed through a 0.22 μm polyvinylidene difluoride (PVDF) membrane filter (Millipore), which is routinely used to remove particulates prior to nanopore experiments, and subsequently stored at room temperature.

2.3 Buffer Solutions

Three buffer formulations are used for nanopore fabrication, nanopore conditioning, and polymer translocation experiments:

1. Fabrication buffer: 1 M KCl, 10 mM HEPES, adjusted to pH 8.0 using 1 N sodium hydroxide (NaOH). This buffer is used during controlled dielectric breakdown to create nanopores in the SiN_x membrane.

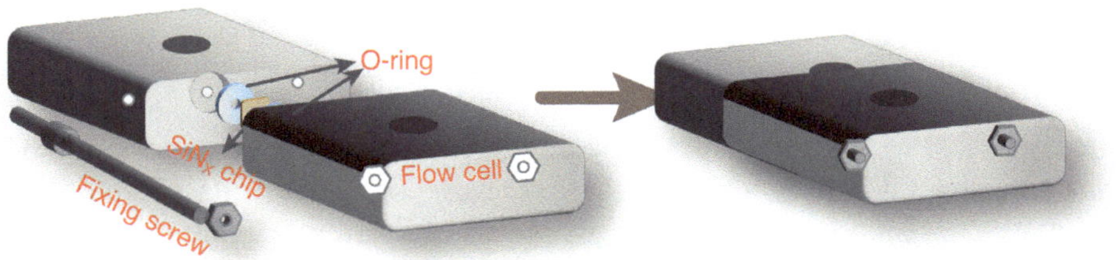

Fig. 2 Components of a solid-state nanopore device: the flow cell, fixing screw, membrane chip, and sealing O-rings, as well as the fully assembled integrated device

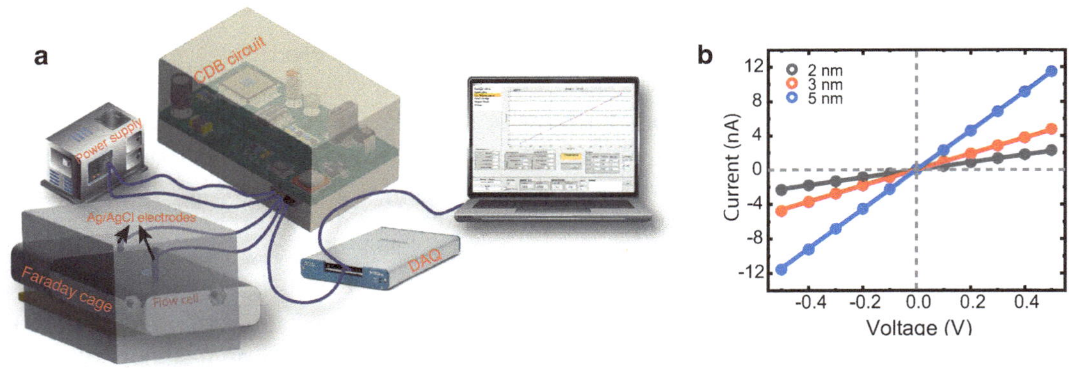

Fig. 3 (**a**) Schematic of the setup for solid-state nanopore fabrication via controlled dielectric breakdown, including the integrated device inside a Faraday cage, power supply, control circuit, data acquisition (DAQ) system, and control interface. (**b**) Representative *I–V* curves of nanopores with different diameters after fabrication. Solid lines represent linear fits to the data

2. Conditioning buffer: 3.6 M LiCl, 10 mM HEPES, pH 8.0. This high-conductivity buffer is used to gradually expand the nanopore diameter after initial formation by controlled dielectric breakdown.
3. Testing buffer: 4 M KCl, 10 mM HEPES, pH 6.0. This is used during translocation experiments to provide high ionic strength for improved signal-to-noise ratio and event resolution.

All buffer solutions are prepared using ultrapure water (Milli-Q, 18.2 MΩ·cm resistivity) and filtered through a 0.22 μm polyvinylidene difluoride (PVDF) membrane and stored at room temperature for short-term use (*see* **Note 3**).

2.4 Nanopore Fabrication Setup

The nanopore fabrication system is based on an automated controlled dielectric breakdown setup as described previously [30] (*see* **Note 4**). It includes a DC power supply, a National Instruments (NI) data acquisition (DAQ) device, a custom-built control circuit, and a solid-state nanopore flow cell. A grounded Faraday cage encloses the flow cell and electronics to reduce electromagnetic interference during current measurements.

2.5 Signal Acquisition Equipment

Ionic current recordings are performed using a patch-clamp amplifier (e.g., Axopatch 200B, Molecular Devices) connected to a Digidata 1550A digitizer and controlled by Clampex software (pClamp suite). The entire setup is enclosed in a grounded Faraday cage to minimize environmental noise.

3 Methods

3.1 Membrane Chip Cleaning

Warning

Piranha solution is highly corrosive and reacts violently with organic materials. Always prepare it fresh in a fume hood, wear appropriate personal protective equipment (PPE), and never store unused solution. Slowly add hydrogen peroxide to sulfuric acid–never the reverse (*see* **Note 5**).

1. Prepare fresh piranha solution immediately before use by slowly adding one part of 30% hydrogen peroxide (H_2O_2) to three parts of concentrated sulfuric acid (H_2SO_4) in a clean glass beaker under gentle magnetic stirring to ensure homogeneous mixing (*see* **Note 6**.)
2. Preheat the solution to 90°C on a hot plate under gentle magnetic stirring before introducing any materials. Monitor the temperature closely to prevent overheating or uncontrolled reactions.
3. Using polytetrafluoroethylene (PTFE) tweezers, place the SiN_x membrane chips on a PTFE chip rack and carefully lower the rack into the preheated piranha solution.
4. Incubate the chips for 30 min at 90°C. Bubble formation during this step indicates removal of organic surface contaminants.
5. Carefully remove the rack and rinse the chips thoroughly with deionized (DI) water, repeating the rinse step at least three times to ensure complete removal of residual chemicals.
6. Transfer the chips to a clean beaker containing fresh DI water and store them submerged until use. Chips should be used within 10 h for optimal surface quality.
7. Optionally, dry the chips with filtered nitrogen gas immediately before assembling the solid-state nanopore device.

3.2 Nanopore Device Preparation

1. Clean the Teflon flow cell with deionized (DI) water and 70% ethanol or acetone, then dry with nitrogen.
2. Place a silicone O-ring into the groove on one side of the flow cell. Before use, the O-ring should be sonicated in 70% ethanol or acetone, rinsed with DI water, and dried with nitrogen.

3. Carefully position the cleaned silicon nitride (SiN_x) membrane chip on top of the O-ring, ensuring that the membrane window aligns with the aperture.
4. Place the second silicone O-ring on top of the chip, aligning it with the aperture of the flow cell.
5. Tighten the flow cell screws evenly to secure the assembly. Avoid overtightening to prevent membrane fracture.
6. Prepare Ag/AgCl electrodes by immersing silver wires (1 mm diameter, >99.9% purity) in freshly prepared commercial bleach (5%–6% sodium hypochlorite) for at least 2 h to form a uniform silver chloride coating. Rinse the chlorinated wires thoroughly with DI water and allow them to dry completely before use. Insert the prepared Ag/AgCl electrodes into the designated ports of the cis and trans wells, ensuring direct and stable contact with the electrolyte solution for reliable current measurement.

3.3 Membrane Chip Wetting by Pretreatment

1. After assembling the flow cell, fill both the cis and trans chambers with 100% methanol (*see* **Note 7**).
2. Let the methanol soak the membrane for 1–2 min to ensure complete wetting of the SiN_x surface.
3. Remove the methanol by gently aspirating with a pipette, avoiding direct contact with the membrane.
4. Rinse both chambers three times with fabrication buffer to eliminate residual methanol (*see* **Note 8**).
5. Refill both chambers with fabrication buffer (1 M KCl, 10 mM HEPES, pH 8.0) and allow the system to equilibrate for 1–2 min.

3.4 Automated Nanopore Fabrication

1. Connect nanopore device to a custom-built control circuit interfaced with a National Instruments (NI) data acquisition (DAQ) device and a DC power supply. The electrodes and flow cell should be enclosed within a grounded Faraday cage to minimize electrical noise (see Fig. 3a).
2. Apply a 100 mV square-wave voltage to confirm membrane wetting by observing the capacitive current response.
3. Launch the automated controlled dielectric breakdown routine. The program begins by performing a voltage sweep from –0.5 V to +0.5 V in 100 mV increments (or as defined by the user) to acquire the *I–V* curve and assess membrane integrity. A linear *I–V* profile with sub-nanoampere current confirms that the membrane is intact and free of preexisting pores or leakage paths.
4. Once the membrane passes the integrity check, the system initiates a linear voltage ramp (typically 0.05–0.2 V/s) across the membrane to induce dielectric breakdown.

5. During the voltage ramp, the ionic current is continuously monitored in real time by the DAQ system. Upon detecting a sudden increase in current, the program identifies this as successful nanopore formation and automatically halts the voltage ramp to prevent overgrowth of the pore.
6. After halting the ramp, the system performs another voltage sweep to record the new *I–V* curve. The nanopore conductance, *G*, is extracted from the linear region of the curve and used to estimate the pore diameter, *d*, using the following empirical formula [31, 32]:

$$d = \frac{G}{2\sigma}\left(1 + \sqrt{1 + \frac{16\sigma l}{\pi G}}\right), \tag{1}$$

where G is the measured conductance (in Siemens (S)), σ is the buffer conductivity (in S/m), and l is the membrane thickness (in meters). Typical values for σ and l should be measured or verified beforehand (*see* **Note 9**).
7. If the estimated pore diameter is smaller than the desired range, the chip proceeds to the conditioning step (Sect. 3.5), where low-voltage conditioning in a high-conductivity buffer (e.g., 3.6 M LiCl) is used to gradually enlarge the pore. If the pore exceeds the target size, the chip is discarded and a new membrane is used.
8. The complete fabrication process—including integrity check, ramping, breakdown detection, and postfabrication characterization—typically takes 10–30 min, depending on the membrane thickness, buffer composition, applied voltage range, and the breakdown threshold of the individual chip.

3.5 Pore Size Conditioning

1. After nanopore formation, flush both cis and trans chambers three times with conditioning buffer (e.g., 3.6 M LiCl, 10 mM HEPES, pH 8.0) to replace the fabrication buffer. This high-conductivity solution stabilizes the pore and allows controlled enlargement under sub-threshold electrical stress.
2. Perform an initial current–voltage (*I–V*) sweep from −0.5 V to +0.5 V using the DAQ system to determine the open-pore conductance.
3. Estimate the nanopore diameter using the empirical formula given in Sect. 3.4 (*see* **Note 10**). If the diameter exceeds the target range, the pore is considered overgrown and should be discarded. If it is smaller than desired, proceed with electrochemical conditioning by applying a symmetric square-wave voltage pulse (e.g., ±2.5 V amplitude, 5 s period, 20 s total duration) to gradually expand the pore.

4. After each pulse train, repeat the *I–V* sweep and re-calculate the conductance and pore diameter. Depending on the updated size, adjust the voltage amplitude or duration of the square pulses in the next conditioning cycle (*see* **Note 11**). Careful control minimizes overshoot and avoids irregular pore geometries.
5. Continue the conditioning process iteratively until the pore diameter falls within the desired range. Figure 3b shows representative *I–V* curves corresponding to different pore sizes, with linear (ohmic) behavior confirming symmetrical pore formation. This iterative tuning process offers high precision and reproducibility while reducing the risk of catastrophic pore overgrowth.

3.6 Pore Stability Assessment

1. After completing the pore conditioning step, thoroughly flush both cis and trans chambers with the final measurement buffer (e.g., 4 M KCl, 10 mM HEPES, pH 6.0 for PEG translocation experiments). Ensure complete replacement of the conditioning buffer to prevent ionic strength gradients.
2. Let the system equilibrate at room temperature for 10–30 min without applying any voltage. This pause allows stabilization of the solid-state membrane and the electrode–electrolyte interface, which reduces thermal and chemical drift during measurements.
3. Apply a constant voltage bias across the membrane (e.g., + 100 mV) and continuously monitor the open-pore current over a 3–5 min period.
4. Assess the baseline current stability by calculating the drift rate, defined as the change in open-pore current (nA/min) normalized to the mean open-pore current. A drift rate of less than 1% per minute is generally acceptable and indicates that the pore is suitable for high-fidelity translocation measurements.
5. If the current trace exhibits significant drift (e.g., >1% per minute), frequent current spikes, or persistent partial blockages, the pore should be considered unstable. In such cases, discard the membrane chip and repeat the nanopore fabrication procedure as described in Sect. 3.4.

3.7 Sample Loading

1. Replace the buffer in both the cis and trans chambers with the designated testing buffer (e.g., 4 M KCl, 10 mM HEPES, pH 6.0). Carefully flush each chamber at least three times to ensure complete exchange. This step may be omitted if the same buffer has already been used during the pore stability assessment (Sect. 3.6).
2. Prepare a concentrated stock solution of polyethylene glycol (PEG) in the same testing buffer to ensure ionic compatibility. Vortex and briefly sonicate the solution to fully dissolve the polymer.

3. Immediately before the translocation experiment, pipette an appropriate volume of the PEG stock into the cis chamber to achieve a final concentration between 0.01 and 0.1 g/L (*see* **Note 12**). Gently mix without generating bubbles to ensure homogeneous distribution near the pore.
4. Leave the trans chamber filled only with the testing buffer, ensuring no polymer is introduced on the trans side.

3.8 Data Acquisition

1. Connect the flow cell to the headstage of a patch-clamp amplifier (Axopatch 200B), ensuring proper grounding and shielding. The complete data acquisition setup is illustrated in Fig. 4.
2. Set the digitizer to a sampling rate of 250 kHz and enable a 100 kHz low-pass Bessel filter on the amplifier. These settings are optimized to capture transient translocation signals of polyethylene glycol (PEG) molecules with high temporal resolution (*see* **Note 13**).
3. Apply a constant bias voltage across the nanopore, with the cis side grounded. This field drives the electrophoretic transport of PEG molecules through the nanopore toward the trans side (*see* **Note 14**).
4. Allow the system to stabilize for 2–3 min. Monitor the open-pore current to ensure that the baseline is steady and free of drift or transient spikes before recording begins (*see* **Note 15**).

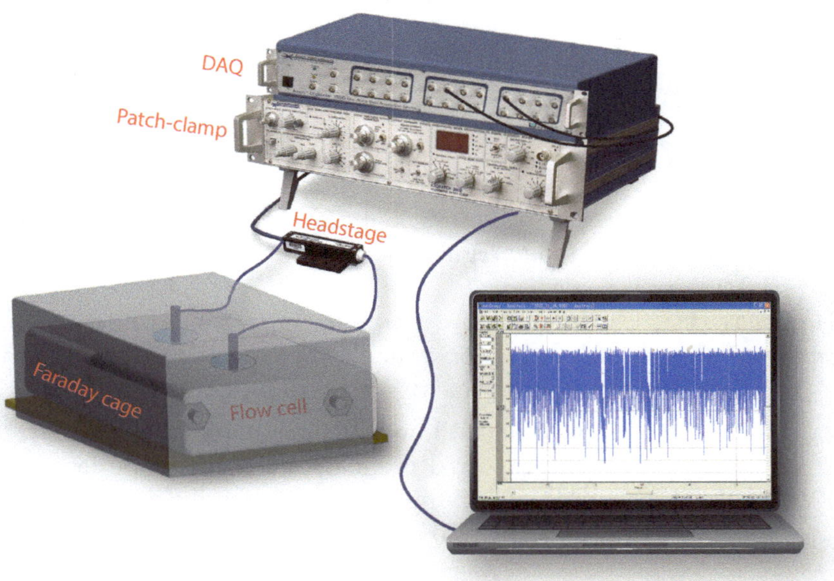

Fig. 4 Illustration of the nanopore measurement setup, including the nanopore device, patch-clamp amplifier, data acquisition system (DAQ), and recording interface

5. Launch the pClamp software to record ionic current traces. For robust statistical evaluation, acquire no fewer than 1000 individual PEG translocation events under each condition. A representative recording interface and ionic current trace are shown in Fig. 4 (*see* **Note 16**).
6. Save the raw data files in Axon Binary Format (ABF) for subsequent offline processing. Confirm that each trace contains well-defined and analyzable blockade signatures indicative of PEG translocation.

3.9 Cleanup

1. At the end of the experiment, turn off all applied voltages via the amplifier to prevent accidental current surges. Safely disconnect the Ag/AgCl electrodes from the flow cell.
2. Carefully disassemble the flow cell and remove the silicon nitride chip. If the pore is damaged, blocked, or oversized, discard the chip according to laboratory solid waste procedures.
3. Rinse all flow cell components—including the Teflon housing, silicone O-rings, and electrode ports—with copious amounts of deionized water to remove residual salts, PEG, or other analytes.
4. Follow with a rinse using 70% ethanol to disinfect and eliminate organic contaminants. This step is especially important if biological samples or high-molecular-weight PEGs were used.
5. Dry all parts thoroughly using filtered nitrogen gas. Ensure that narrow channels and ports are completely free of moisture to prevent microbial growth during storage.
6. Store all cleaned components in a clean, dust-free, and dry environment such as a sealed polypropylene container or desiccator.
7. Dispose of used buffers, PEG solutions, and waste liquids in accordance with institutional chemical safety protocols. Label and segregate waste if required by local environmental health and safety regulations.

3.10 Event Detection and Data Analysis

1. Import the raw ionic current traces (ABF format) into a data analysis platform such as MATLAB or custom Python scripts designed for nanopore signal processing (*see* **Note 17**).
2. Apply a digital low-pass filter (typically 5–10 kHz, depending on the expected event duration and noise level) to suppress high-frequency noise while preserving event fidelity. A representative filtered trace using a 10 kHz Bessel filter is shown in Fig. 5a.
3. Perform event detection using a threshold-based algorithm. Define translocation events as downward deviations in current exceeding 5%–10% of the baseline, sustained for a minimum

dwell time (e.g., >20 μs for PEG). Alternatively, define the threshold as five times the standard deviation (5×SD) of the baseline noise to enhance sensitivity.

4. For each detected event, extract key features including:
   - Blockade depth (ΔI), defined as the difference between the baseline and minimum current during the event;

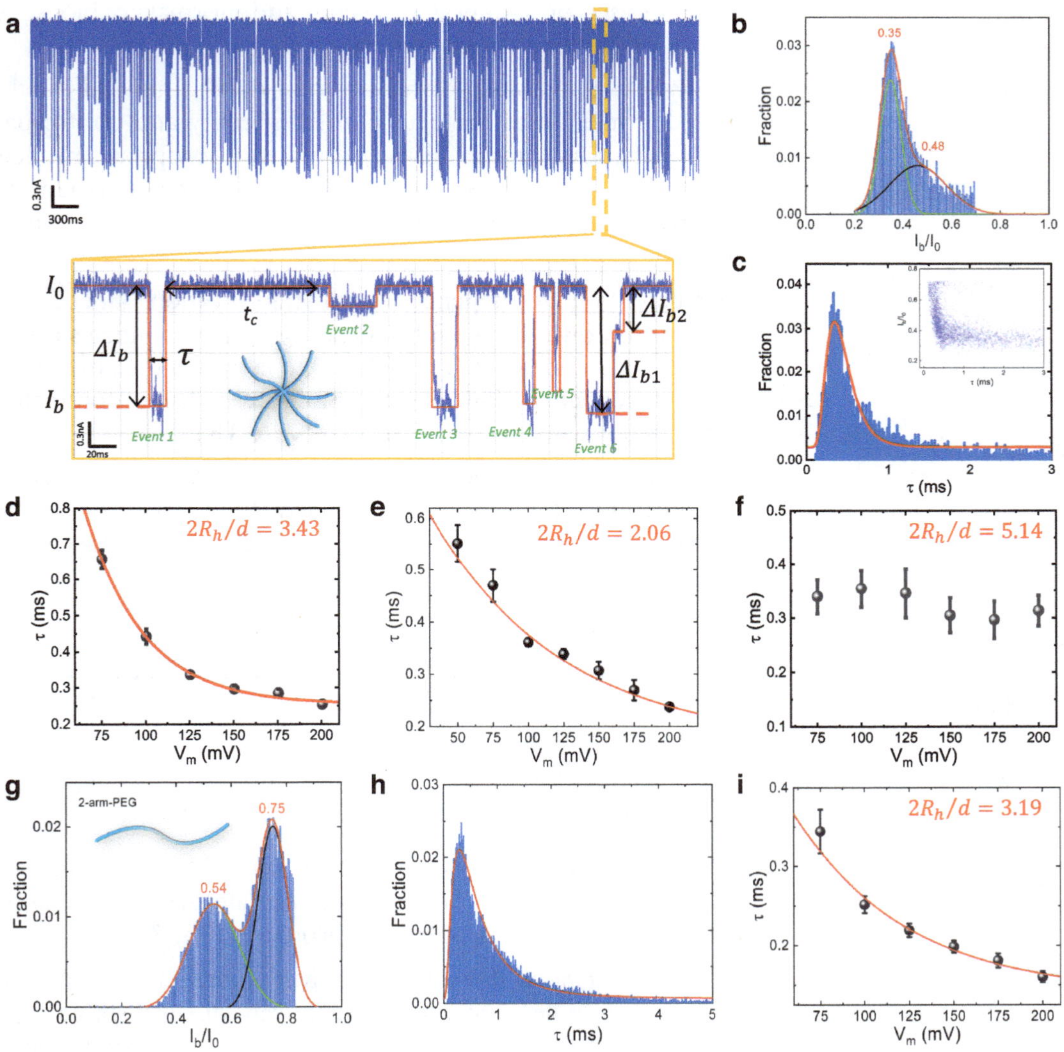

Fig. 5 (**a**) Representative ionic current trace of 8-arm-PEG translocation through a 3 nm nanopore at 100 mV. Inset: zoom-in showing six translocation events. (**b**) Histogram of I_b/I_0 for over 5000 8-arm-PEG translocation events under the same conditions. Solid line: Gaussian fit; peak indicated. (**c**) Dwell time (τ) distribution with log-normal fitting. Inset: scatter plot of individual events. (**d–f**) Voltage dependence of τ for 8-arm-PEG in nanopores with varying $2R_h/d$ ratios, as labeled. Solid lines: exponential fits. (**g**) Distribution of I_b/I_0 for 2-arm-PEG translocation through a 2 nm pore at 100 mV. (**h**) τ distribution for the same condition as in (**g**). (**i**) Voltage dependence of τ with exponential fitting. Figure adapted from our previous work [21]

- Dwell time (τ), calculated as the duration between event onset and return to baseline;
- Inter-event interval (t_c), measuring the time between consecutive translocation events.

5. These features are illustrated in the zoomed-in trace in Fig. 5a. Save the extracted values for downstream statistical analysis.
6. Remove atypical or ambiguous events based on user-defined quality control criteria–such as irregular baseline recovery, spike noise, or compound events–and normalize blockade depth by baseline current if necessary.
7. Analyze the statistical distribution of dwell time and blockade depth. Due to molecular polydispersity and conformation variation, fit the distributions to appropriate models (e.g., log-normal [33], exponential [4, 8], or Gaussian [4, 7]) to extract kinetic parameters. Example fits are shown in Fig. 5b, c.
8. Evaluate voltage dependence of translocation kinetics. A decrease in dwell time with increasing voltage confirms active translocation behavior (Fig. 5d, e). In contrast, the absence of such dependence (Fig. 5f) may indicate collision-dominated interactions with the nanopore (*see* **Note 18**).
9. Compare the translocation signatures of different polymer architectures (e.g., linear vs. branched PEGs) to elucidate the impact of molecular topology. Figure 5g–i shows representative results for linear PEG.

4 Notes

1. The flow cell and sealing O-ring are also commercially available from Norcada Inc.
2. Membrane chips are commercially available from vendors such as Norcada and SiMPore and can alternatively be fabricated in-house using published protocols [34, 35].
3. The buffer should be thoroughly degassed before use to prevent bubble formation, which can disrupt stable current recordings or block the nanopore during experiments. Filtration through a PVDF membrane is effective during initial buffer preparation. For later use, especially after storage or handling, brief sonication is recommended to eliminate newly formed microbubbles.

4. The solid-state nanopore fabrication setup based on controlled dielectric breakdown is also commercially available from Northern Nanopore Instruments.
5. Both oxygen plasma and piranha solution treatments improve membrane wettability. Piranha solution is generally more effective due to its aqueous environment.
6. Always prepare fresh piranha solution. Slowly add H_2O_2 to H_2SO_4 (never the reverse). Use only glass containers and discard waste into a neutralizing solution. Handle with extreme caution under a fume hood.
7. This methanol-based wetting step is optional, as prior cleaning with piranha solution or oxygen plasma usually provides sufficient surface wettability.
8. Ensure complete removal of organic solvents (e.g., methanol) before buffer exchange. Residual solvents may alter ionic conductivity and interfere with nanopore formation or sensing accuracy.
9. More advanced *I–V* fitting functions may incorporate access resistance and pore geometry effects for more accurate pore size estimation [36, 37].
10. The pore size may slightly increase during buffer exchange due to mechanical disturbances or electrochemical effects. It is therefore advisable to reassess pore size after each buffer replacement.
11. If the pore diameter is substantially smaller than desired, apply higher voltage and longer pulse durations during conditioning. As the pore approaches the target size, gradually reduce both parameters to avoid overgrowth.
12. The optimal polymer concentration depends on molecular weight and applied voltage. Higher concentrations increase event frequency but raise the risk of clogging, while lower concentrations may lead to longer acquisition times and increased pore drift.
13. Sampling rate and filtering settings should be optimized based on the analyte's molecular weight, translocation kinetics, and expected dwell time to ensure accurate event resolution.
14. Avoid strong light exposure, vibrations, or air currents during current acquisition. Minor environmental disturbances can cause picoampere-level artifacts.
15. If excessive noise, drift, or unstable baselines are observed, pause the experiment and inspect potential causes, such as contaminated buffers, displaced electrodes, or compromised membrane integrity. Replace faulty components as necessary before resuming.

16. If no events are detected after polymer loading, verify polymer concentration, pore size, and voltage polarity.
17. Nanolyzer (Northern Nanopore Instruments) provides an intuitive GUI for event detection and analysis. Additionally, open-source tools and custom scripts are available for batch processing and model fitting [38–40].
18. In some cases, certain polymers exhibit unusual voltage-dependent translocation behaviors (e.g., increased dwell time with voltage), possibly due to strong pore–polymer interactions or voltage-induced trapping. For PEGs, however, successful translocation events consistently show a monotonic decrease in dwell time with increasing voltage.

Acknowledgments

We thank Professor Jason Dwyer (University of Rhode Island), Brian Sheetz (University of Rhode Island), and Kyle Briggs (Northern Nanopore Instruments Inc.) for helpful discussions. Acknowledgment is made to the National Science Foundation (grant no. DMR 2309539) and AFOSR (grant no. FA9550-23-1-0584).

References

1. Li J, Stein D, McMullan C, Branton D, Aziz MJ, Golovchenko JA (2001) Ion-beam sculpting at nanometre length scales. Nature 412:166–169
2. Storm AJ, Storm C, Chen J, Zandbergen H, Joanny JF, Dekker C (2005) Fast DNA translocation through a solid-state nanopore. Nano Lett 5:1193–1197
3. Dekker C (2007) Solid-state nanopores. Nat Nanotechnol 2:209–215
4. Wanunu M, Sutin J, McNally B, Chow A, Meller A (2008) DNA translocation governed by interactions with solid-state nanopores. Biophys J 95:4716–4725
5. Muthukumar M (2011) Polymer translocation. CRC Press, Boca Raton
6. Kasianowicz JJ, Brandin E, Branton D, Deamer DW (1996) Characterization of individual polynucleotide molecules using a membrane channel. Proc Natl Acad Sci USA 93:13770–13773
7. Chen P, Gu J, Brandin E, Kim YR, Wang Q, Branton D (2004) Probing single DNA molecule transport using fabricated nanopores. Nano Lett 4:2293–2298
8. Krasilnikov OV, Rodrigues CG, Bezrukov SM (2006) Single polymer molecules in a protein nanopore in the limit of a strong polymer–pore attraction. Phys Rev Lett 97:018301
9. Xue L, Yamazaki H, Ren R, Wanunu M, Ivanov AP, Edel JB (2020) Solid–state nanopore sensors. Nat Rev Mater 5:931–951
10. De Gennes PG (1974) Coil–stretch transition of dilute flexible polymers under ultrahigh velocity gradients. J Chem Phys 60:5030–5042
11. Ge H, Pispas S, Wu C (2011) How does a star chain (nanooctopus) crawl through a nanopore Polym Chem 2:1071–1076
12. Li L, He C, He W, Wu C (2012) How does a hyperbranched chain pass through a nanopore? Macromolecules 45:7583–7599
13. Voit BI, Lederer A (2009) Hyperbranched and highly branched polymer architectures: synthetic strategies and major characterization aspects. Chem Rev 109:5924–5973
14. Johnson JA, Lu YY, Burts AO, Lim YH, Finn MG, Koberstein JT, Turro NJ, Tirrell DA, Grubbs RH (2011) Core-clickable PEG-branch-azide bivalent bottle-brush polymers by ROMP: grafting-through and clicking-to. J Am Chem Soc 133:559–566
15. Ren JM, McKenzie TG, Fu Q, Wong EHH, Xu J, An Z, Shanmugam S, Davis TP, Boyer C, Qiao GG (2016) Star polymers. Chem Rev 116:6743–6836

16. Xiao L, Chen Y, Zhang K (2016) Efficient metal-free "grafting onto" method for bottlebrush polymers by combining RAFT and triazolinedione–diene click reaction. Macromolecules 49:4452–4461
17. Hanton SD, Liu XM (2000) GPC separation of polymer samples for MALDI analysis. Anal Chem 72:4550–4554
18. Wang WJ, Kharchenko S, Migler K, Zhu S (2004) Triple-detector GPC characterization and processing behavior of long-chain-branched polyethylene prepared by solution polymerization with constrained geometry catalyst. Polymer 45:6495–6505
19. He L, Karau P, Tabard-Cossa V (2019) Fast capture and multiplexed detection of short multi-arm DNA stars in solid-state nanopores. Nanoscale 11:16342–16350
20. He J, Zhou J, Yang J, Zhu M, Li L (2022) Exploring the feasibility of utilizing nanopore-based ultrafiltration for the purification of graft-onto polymerization products. Macromolecules 55:10210–10219
21. Chen K, Li M, Muthukumar M (2024) Electrophoretic translocation of star-shaped polymers through single solid-state nanopores. Macromolecules 57:6693–6704
22. Ren M, Liu D, Qin F, Chen X, Ma W, Tian R, Ting W, Wang D, Astruc D, Liang L (2025) Single-molecule resolution of macromolecules with nanopore devices. Adv Colloid Interface Sci 338:103417
23. Bezrukov SM, Vodyanoy I (1993) Probing alamethicin channels with water-soluble polymers: effect on conductance of channel states. Biophys J 64:16–25
24. Robertson JWF, Rodrigues CG, Stanford VM, Rubinson KA, Krasilnikov OV, Kasianowicz JJ (2007) Single-molecule mass spectrometry in solution using a solitary nanopore. Proc Natl Acad Sci USA 104:8207–8211
25. Zhou C, Ji C, Nie Y, Yang J, Zhao J (2022) Poly(ethylene oxide) is positively charged in aqueous solutions. Gels 8:213
26. Yan ZG, Yang Z, Price C, Booth C (1993) Cyclisation of poly(ethylene glycol)s and related block copolymers. Makromol Chem Rapid Commun 14:725–732
27. Lapienis G (2009) Star-shaped polymers having PEO arms. Prog Polym Sci 34:852–892
28. Leitgeb A, Wappel J, Slugovc C (2010) The ROMP toolbox upgraded. Polymer 51:2927–2946
29. Li Z, Tang M, Liang S, Zhang M, Biesold GM, He Y, Hao SM, Choi W, Liu Y, Peng J et al (2021) Bottlebrush polymers: from controlled synthesis, self-assembly, properties to applications. Prog Polym Sci 116:101387
30. Kwok H, Briggs K, Tabard–Cossa V (2014) Nanopore fabrication by controlled dielectric breakdown. PLoS One 9:e92880
31. Hall JE (1975) Access resistance of a small circular pore. J Gen Physiol 66:531–532
32. Kowalczyk SW, Grosberg AY, Rabin Y, Dekker C (2011) Modeling the conductance and DNA blockade of solid-state nanopores. Nanotechnology 22:315101
33. Jeon BJ, Muthukumar M (2016) Electrostatic control of polymer translocation speed through α–hemolysin protein pore. Macromolecules 49:9132–9138
34. Dominic A, Parambath MS, Nasa S, Varma M (2023) Practical guide for in-house solid-state nanopore fabrication and characterization. J Vac Sci Technol, B 41:043204
35. Dutt S, Karawdeniya BI, Bandara YMNDY, Afrin N, Kluth P (2023) Ultrathin, high-lifetime silicon nitride membranes for nanopore sensing. Anal Chem 95:5754–5763
36. Frament CM, Dwyer JR (2012) Conductance–based determination of solid-state nanopore size and shape: an exploration of performance limits. J Phys Chem C 116:23315–23321
37. Wen C, Zhang Z, Zhang S-L (2017) Physical model for rapid and accurate determination of nanopore size via conductance measurement. ACS Sens 2:1523–1530
38. Raillon C, Granjon P, Graf M, Steinbock LJ, Radenovic A (2012) Fast and automatic processing of multi-level events in nanopore translocation experiments. Nanoscale 4:4916–4924
39. Plesa C, Dekker C (2015) Data analysis methods for solid-state nanopores. Nanotechnology 26:084003
40. Bandara YMNDY, Saharia J, Karawdeniya BI, Kluth P, Kim MJ (2021) Nanopore data analysis: baseline construction and abrupt change-based multilevel fitting. Anal Chem 93:11710–11718

Check for updates

Chapter 14

Threading of Single Protein Molecules Through a Solid-State Nanopore

Km Mamata Patel, Saurabh Chaubey and Prabhat Tripathi

Abstract

The building blocks and machinery of cells that maintain homeostasis and functional biological architecture are proteins. Protein analysis is a crucial aspect of modern biology, biophysics, biochemistry, and medicine, with numerous applications ranging from disease diagnostics to drug development. In recent years, nanopore technology has emerged as a promising approach for probing and characterizing proteins at the single-molecule level. A tiny hole in a biological or solid-state membrane is known as a nanopore, which can be used to detect the physical and chemical behavior of single protein molecules as they pass through the analyzer nanopore. This instructional book chapter helps users with a manual and a point of reference for employing nanopores for single-molecule measurements and analysis of protein translocation through a nanopore. Nanopore technology requires simple and easy protein sample preparation and can detect ultra-low-abundance of water-soluble proteins. One of its most important advantages is the capacity to deliver real-time analysis of label-free protein's structure and dynamics with sub-microseconds time resolution. As proteins pass through the nanopore, they can disrupt the ionic current flowing through the pore, which can give information about the protein's size, shape, sequences, conformation changes (folding and unfolding), short-lived intermediate states, transition states, and metastable states.

Keywords Nanopore technology, Solid-state nanopore, Proteins, Translocation

1 Introduction

A nanopore technology refers to a revolutionary approach that utilizes incredibly tiny nanoscale pores embedded in a thin film, which helps to analyze single biomolecules DNA, RNA, protein, as well as other molecules [1–5]. The nanopore is positioned in an electrically resistant membranes, such as SiN membrane or lipid bilayer membrane, when a voltage is placed across the membrane, an ionic current flows through the pore [3, 6–10]. Whenever the nanopore and the target protein molecule come into contact, the ionic current instantaneously drops to a current level (or levels, Fig. 1) which gives dynamic information about the protein molecule blocking the nanopore [2, 11]. Label-free biomolecules can be analyzed using fluctuations

Fanjun Li (ed.), *Nanopore Technology*, Methods in Molecular Biology, vol. 3072,
https://doi.org/10.1007/978-1-0716-5527-6_14,

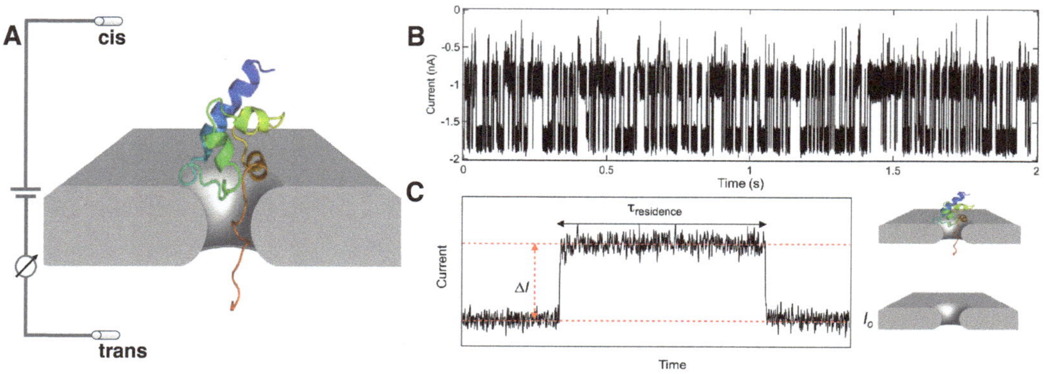

Fig. 1 Schematics of the threading of a single protein molecule through a solid-state nanopore. (**A**) The trajectory of a single molecule as it threads through a nanoscale pore (typically ~1–3 nm) on a SiN membrane can be directly monitored in real time by measuring ion current flowing through the nanopore. (**B**) An example of ion current traces of lambda repressor mutant (6–85 fragment) in 1 M KCl, 10 mM HEPES, pH 7.5. (**C**) Expanded view of an ionic current disruption event and description of the parameters. When a protein passes through the pore, the ionic current disrupts instantaneously by an amount ΔI from the open pore current I_0 for the duration of $\tau_{residence}$ (which represents the total time a protein spends in the pore before the complete translocation). In the typical threading experiments for a given protein, the distributions of $\Delta I/I_0$ and $\tau_{residence}$ are measured at different transmembrane voltages and pore diameters

in measured ionic current during the interaction of a single molecule of interest with the nanopore. The target molecule can be identified by monitoring the distinctive changes in the ionic current that certain compounds exhibit [1, 2, 12–14]. Because of these advantages, nanopore technology is under the development as a potent instrument for single-molecule research [6]. For over 30 years, nanopore sensing has been emerged to be powerful tool to analyze the complex biological molecules [13, 15]. Its uses include label-free detection of single molecule with a broad range of applications in biotechnology, medicine, forensic investigations, and DNA/protein sequencing [1, 16–19]. The primary driving force behind the creation and implementation of these nanopores was DNA sequencing [20], drug screening, illness diagnostics [21], and virus detection [22–24]. One of the earliest instances of modern nanopore sensors constructed on a micron-sized aperture was the Coulter counter [25, 26]. The method is called stochastic sensing, and it is also frequently referred to as the resistive pulse sensor (RPS) [20]. The idea is that current drops are caused by volume exclusion and aperture blockage. The aperture, which includes Ag/AgCl electrodes to apply potential, is positioned between two reservoirs that are filled with an electrolytic solution. A constant current is created when the particles go through the nanopore, and the dissolved ions in the electrolytic solution move to the counter-electrode (Fig. 1). Target species in the mixture may be quantified, and the physiochemical characteristics of individual molecules can be identified through the interpretation of ion current variations. Single-stranded DNA (ssDNA) was utilized to demonstrate the translocation through α-hemolysin nanopore using the counter-coulter concept. In 1996, Kasinowicz et al. used α-hemolysin pore to characterize the single-stranded polynucleotide translocation for the first time [27]. Recently, biological nanopore was used to observe the nanopore ionic current signatures of peptide sequences [28]. Although biological nanopores [19, 29] offer several benefits, including high signal-to-noise ratio (S/N), biocompatibility, structural stability, reproducibility of ionic current signals, solid-state nanopores had to be developed because of the lipid bilayer's limited pore dimensions and instability against pH and applied potential [30, 31]. In contrast to biological nanopores on a lipid-bilayer platform, solid-state nanopores have provided better control over pore size and were more resilient toward voltage breakdown [6, 32]. For better biomolecule sensing, researchers are also interested in hybrid nanopore generation [33, 34]. Numerous methods, including controlled dielectric breakdown [35] (CBD), laser irradiation (IR), focused ion beam (FIB), focused electron beam (though TEM) (FEB), helium ion beam (HIB), electrical pulse (EP), photothermal sculpting (PTS), nanoparticle etching, and electrochemical etching, can be used to fabricate nanopores on the solid-state membranes. Since its inception, nanopore sensing has allowed researchers to examine how different molecules, such as BSA [36], cytochrome c, lambda repressor, and functionally important proteins, move through these microscopic channels [37–39].

1.1 Fundamentals of Protein Translocation Through the Solid State Nanopore

There are various fundamental processes involved in the translocation of protein through the solid-state nanopore.

1.1.1 Electrokinetic Driving Forces

Electrophoresis (EP): The movement of charged protein molecules under the influence of an electric field is known as electrophoresis [36, 40]. Negatively charged molecules move electrophoretically to the positive side of the nanopore when a voltage is applied across it. According to the Helmholtz–Smoluchowski limit for the high ionic strength, when Debye screening length < protein diameter, the electrophoretic velocity of the protein (charged particle) $\nu_{\text{eph}} = {}^{\varepsilon_0 \varepsilon_r \zeta_p E(z)}/_{\eta}$, in which $E(z)$ is the electric field in the z direction, ε_0, ε_r are vacuum and relative permittivity of water, respectively. ζ_p is the zeta potential of the protein, and η is the viscosity of the medium [40]. The electric field's intensity, the molecule's net charge, and the characteristics of the medium all affect how strong the electrophoretic force acting on the molecule. The electrophoretic mobilities of the ions, the size, and shape of the nanopore result in the open pore current (I_0), given by Eq. 1 [41].

$$I_0 = (\mu_K + \mu_{Cl}) n_{KCl} e \left(\frac{4h_{eff}}{\pi d_p^2} + \frac{1}{d_p} \right)^{-1} V, \quad (1)$$

here, μ_K and μ_{Cl} are the electrophoretic mobility of the K^+ and Cl^-, n_{KCl} is the number density of the KCl, and e is the elementary charge, d_p is the diameter of the pore, h_{eff} is the effective thickness of the membrane of the nanopore. As the protein passes through the nanopore, it reduces the ionic current flux of the open pore current (I_0). Since h_{eff} is inversely proportional to the open pore current, so decreasing the thickness of the membrane of the pore results in better I_0 and thus gives significant current difference (ΔI) during the interaction of the molecule with a nanopore [41].

Electro-osmosis flow: When an electric field is applied, the formation of the electrical double layer (EDL) close to the nanopore walls occurs due to the fluid motion of the bulk solution. Counterions are drawn to the surface, forming an electrical double layer. Si_3N_4, solid-state nanopores, have walls that are negatively charged; the surface charge on the wall of the nanopore depends on both the pH of the electrolyte solution and the concentration of the bulk solution (electrolyte solution) (KCl) [42]. In the EDL, positive counterions are drawn by an electric field and drag the bulk fluid along with them. This in turn produces what is known as an electro-osmotic flow (EOF), which, depending on the analyte's charge, can either promote or impede its electrophoretic movement [12, 40].

Several important variables that affect the electrokinetic driving force

(a) Molecular charge [43].
(b) Electric field intensity.
(c) Viscosity and ionic mobility strength [40].
(d) pH of the solvent, which affects the electrophoretic behavior by changing the net charge on the protein or by changing the charge density of pore wall [36, 40, 44].
(e) The surface charge of the solid-state nanopore [40, 42, 44].

The net driving motion of an analyte at the nanopore is the result of both electrophoresis, electroosmosis, and diffusion [40], and the dipole moment of analyte also plays a crucial role in the capture and translocation of the protein through the solid-state nanopore [39]. In DNA translocation, electrophoresis is the dominant force, while in protein or peptide translocation, electroosmosis plays a major role [45]. When the analyte is added to the trans side of the nanopore, the electroosmotic force (F_{eof}) helps in the capture of the protein inside the pore, and if the protein molecule carries charge, then additional electrophoretic force (F_{eph}) acts on the protein molecule; in this way, random Brownian motion and directional driving motion balance each other. The driving force takes over after the analyte enters in a critical hemisphere region, increasing the probability of analyte capture over its diffusion out of the nanopore. The driving force and the directional velocity are given below in Eqs. 2 and 3.

$$F_{\mathrm{dir}} = F_{\mathrm{eof}} + F_{\mathrm{eph}} = \alpha q E(z) + 2\pi r_p \eta \overline{L} \frac{\nu_{\mathrm{eof}}}{\lambda_D}, \tag{2}$$

$$\nu_{\mathrm{dir}}(z) = \nu_{\mathrm{eof}}(z) + \nu_{\mathrm{eph}}(z) = \frac{d_p^2}{8z^2}\left(\nu_{eof,p} + \frac{\alpha q E_0}{6\pi\eta r_a}\right), \tag{3}$$

where q is the charge of the analyte protein, α is a screening factor (value lie between 0 and 1), r_a is the radius of the protein, $E(z)$ is the electric field in z direction, d_p is the diameter of the pore, $\overline{L}$ is the average length of the nanopore, η the viscosity of the electrolyte solution (water), ν_{eof}, velocity of the bulk solution due to electroosmotic flow inside the nanopore, and λ_D is the Debye Length [46]. Using a buffer with a pH around the proteins' isoelectric point (pI), where their net charge diminishes, is one method to drastically lower the electrophoretic component [12]. Therefore, it is anticipated that the electrophoretic mobility (μe) of the proteins will decrease at a pH close to the pI, allowing for higher temporal resolution and a longer residence period in the nanopore. On the other hand, depending on the analyte's charge, electroosmotic flow can either speed up or slow down its movement.

1.1.2 Pore Dimensions vs Protein Size

The relationship between protein size and pore diameters is essential for establishing resolution of experiments in detecting single molecules and signal quality in solid-state nanopore sensing. Signal-to-noise ratio reduction occurs when the pore is very large in comparison to the protein because of excessive ionic current leakage, and the protein passes through the pore too quickly to get significant signals. The protein's passage through the pore results in a notable and quantifiable alteration in the ionic current flowing through the pore when the nanopore dimensions are similar or smaller than the cross-sectional area of the protein. This enables improved protein characterization and differentiation in terms of size, charge, structure, and sequences discriminations [14, 39, 47].

Challenge with a large protein and flexible domains: The analysis of large proteins with flexible domains presents specific challenges in nanopore sensing.

(a) Protein unfolding: In order for large proteins to fit through the pore, they may need to unfold entirely or partially. This unfolding may result in substantial conformational changes, which makes it more difficult to understand signals.

(b) Heterogeneity: The various conformations of flexible domains in proteins result in a variety of intricate nanopore signals that are challenging to decipher and evaluate.

(c) Slow translocation: According to Larkin et al., larger proteins take longer to acquire signals because they move through the pore more slowly [48]. As a result, the number of events that may be recorded is similarly limited, and the protein may unfold or interact with the pore walls in an unintended way.

1.1.3 Viscous Drag and Friction

The kinetics of protein translocation via solid-state nanopores are significantly influenced by viscous drag and frictional forces. Proteins encounter resistance from the pore walls and the surrounding fluid when they pass through a nanopore. Protein translocation rates and signal properties are strongly influenced by the viscosity of the fluid. The protein and the confined electrolytic solution interact to produce the viscous drag forces, which are significantly impacted by the viscosity of the solution. Longer dwell durations and improved temporal resolution result from slower translocation caused by higher viscosity, although stalling or blockage are the drawbacks as well. Protein orientation, speed, possible unfolding events, and protein conformations can all be altered by physical interactions between the protein surfaces and nanopore material, including electrostatic, van der Waals, and hydrophobic interactions [2, 49–51]. Depending on the surface chemistry of the nanopore and the protein's charge distribution, these surface pressures can either promote or inhibit translocation.

1.1.4 Protein–Nanopore Surface Interaction

Among the various components of protein–nanopore interactions, electrostatic interactions, Van der Waals interactions, and hydrodynamic interactions are the most commonly employed [52]. The linearized Poisson–Boltzmann equation for particles with constant electrical potential of the surfaces (constant ζ potentials) is used to determine the electrostatic interaction [46]. Both the screening of static fluctuations and the retardation of electromagnetic fields are components of the Van der Waals interaction. Protein and nanopore wall hydrodynamic interactions are important factors that affect the translocation dynamics of proteins and other biomolecules. These interactions may be controlled by varying buffer parameters such as pH and ionic strength [53].

1.2 Solid-State Nanopore Fabrication and Characterization

Various types of materials are used in the fabrication of the solid-state nanopore [54]. Silica-based nanopore—silicon dioxide (SiO_2), silicon nitride (SiN) membrane, alumina-based—Al_2O_3, and nanopores based on 2D materials and Graphene [26, 55, 56]. Other 2D materials, such as WS_2 [57], Molybdenum disulfide (MoS_2) [58–60], HfO_2 [48], and boron nitride (BN) [61, 62], are also being extensively used. Due to their thinness, the 2D nanopores are expected to deliver better resolution for solid-state nanopore-based sensing and sequencing. For the fabrication of a solid-state nanopore, several methods have been applied depending upon the material of the solid-state membrane. For example, Focused Ion Beam (FIB) is used for SiN, SiO_2, SiC, and other 2D materials. FIB drilling provides precise control over pore dimensions >10 nm. Currently, FIB drilling is the preferred method for creating nanopores with dimensions >10 nm on 2D membranes like graphene [56, 63–65], Boron Nitride (BN), and MoS_2. Carefully removing material with a concentrated electron beam, a process known as FEB drilling, which is frequently carried out using TEM (transmission electron microscopy) systems, creates nanopores in thin films. TEM-based fabrication allows better pore fabrication and control down to 1 nm resolution and is being preferred for making pores for small protein (<15 kD) analysis. Controlled Dielectric Breakdown is another way to form the pores that utilizes the electrical breakdown of the SiN membrane by applying high voltage across the membrane. There are two types of approaches to nanopore formation: "top down" and "bottom up." The bottom-up strategy creates the nanopore during the production of the 2D material, whereas the top-down method creates it inside the material after it has been synthesized. The top-down approach allows for greater control over the nanopore's shape and manipulation [54].

1.2.1 Nanopore Fabrication in Freestanding SiN Membrane (Drilling) Using TEM

Using a TEM with a field emission gun (FEG), the nanopore can be created on a free-standing SiN membrane. There is a ~5 μm × 5 μm square-shaped membrane present (Fig. 2), which is supported by the 3 mm×3 mm in the silicon chip. Nanopore chips are usually fabricated (or can be purchased from Norcada Inc.) from a 4-inch silicon-based wafer, in which there are five layers: SiN of thickness of ~50 nm, a 2 μm SiO_2, 300 μm of silicon, a 2 μm SiO_2, and ~50 nm of SiN. Reactive ion etching (REI) is used after the photolithography and to expose the SiO_2 in the photolithography patterned region. For the exposé of silicon layers, a buffered oxide etch (BOE) is used. After 6 h of anisotropic KOH (40% w/w, 80°C) etching and exposure in 1:10 BOE once more to eliminate the remaining silicon oxide layer, a freestanding 25–50 nm silicon nitride membrane can be created [41, 66]. (Fig. 2) Before being used, all of the chips are usually washed using a hot piranha solution (1:3, H_2O_2/H_2SO_4) to maintain them as hydrophilic and clean [12, 67, 68].

Characterization of the nanopore: Various methods are used to characterize the nanopores, such as Transmission Electron Microscopy (TEM), Scanning Tunneling Electron Microscopy, Electrical Characterization (*I–V* Characterization), and Atomic Force

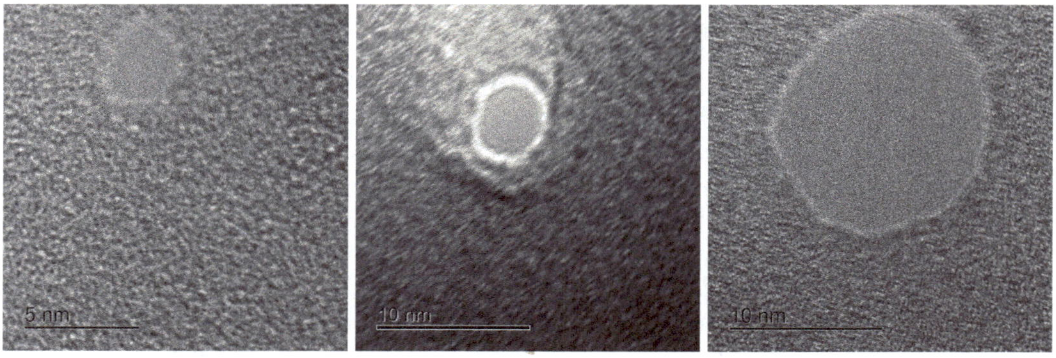

Fig. 2 TEM images of fabricated nanopores on the freestanding SiN silicon nitride membranes, which are drilled by the TEM (Transmission Electron Microscopy)

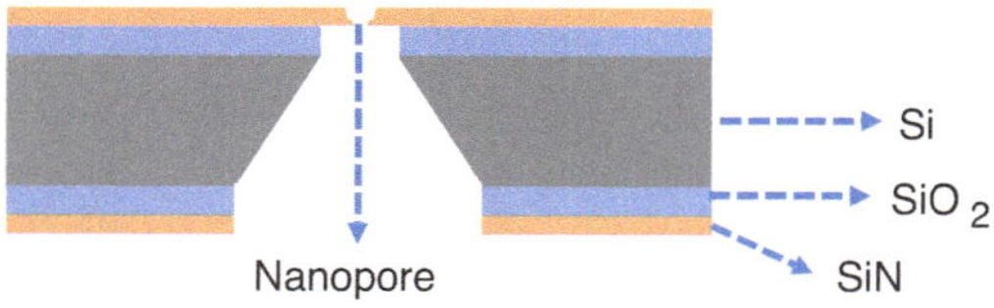

Fig. 3 Schematic of a freestanding SiN membrane chip

Microscopy (AFM). Measurements of a double stranded translocation can be also used to characterize the effective pore length and pore diameters. The TEM images of some nanopores fabricated on SiN membrane are presented in Fig. 3.

1.3 Strategies for the Controlled Protein Threading

1.3.1 Optimizing Driving Forces and Buffers

Voltage or electric field can optimize the translocation speed of the protein throughout the nanopore, pH, and ionic strength of protein and electrolyte solution can affect the translocation of the protein. The viscosity also plays a major role in the translocation of the protein through the nanopore and helps to get a better signal resolution.

(1) **Voltage control:** Optimizing translocation speed due to voltage regulation, we can stabilize the unfolded protein states and thus can speed up the protein translocation through the solid-state nanopore (see Section 3.11).

(2) **pH and ionic strength:** In modulating protein charge and ionic current, pH of the electrolyte plays a crucial role. The pH of the solution significantly influences the protein translocation through the solid-state nanopore by affecting the surface charge of both the protein and the nanopore, which results in the altered balance of the electrophoretic and electroosmotic forces governing the protein movement and translocation in

the nanopore. Since proteins are ampholytes, they have no net electrical charge at a particular pH at the isoelectric point (pI). Protein mobility in the electric field is affected by pH variations that are higher or lower than the pI because they provide a net negative or positive charge to the protein. Additionally, the pH alters the surface charge of the nanopore. The direction and strength of electroosmotic flow (EOF), which can help or impede the protein's electrophoretic passage through the pore, are determined by the surface charges of both proteins and nanopores [44, 46].

(3) **Viscosity modifiers:** The viscosity of the solution majorly affects the translocation of the protein through the solid-state nanopore because it mainly influences the diffusion coefficient (D) and, therefore, the mobility of the protein inside the nanopore. Higher viscosity results in greater resistance to the translocation speed and dwell time of the protein. On increasing viscosity, the reduction in the protein drift velocity happens due to the viscous drag opposition of the electric-field-induced motion and affects the EOF. A high-viscosity buffer also helps to reduce the very fast translocation kinetics of the protein, enabling better resolution and characterization during translocation through the nanopore.

1.3.2 Surface Engineering of Nanopores

Surface engineering of a nanopore focuses on modifying the internal pore surfaces to optimize the properties, such as surface charge, hydrophobicity, hydrophilicity, chemical reactivity, and biomolecule immobilization, which in turn influence sensing performance, molecule interactions, and transport behavior [69]. The atomic layer deposition of materials like HfO_2 or Al_2O_3 conformally covers the nanopore, provides atomic layer thickness control, and makes the pore surfaces more stable under various conditions of the translocation through the nanopore. In silicon nitride membranes, Chen et al. showed that controlled deposition of Al_2O_3 by ALD is a viable method to lower 1/f noise, regulate the diameter, and neutralize the surface charge of nanopores created by ion beam sculpting [70]. The lipid-coated solid-state nanopore facilitated more sensing and sequencing of the nanopore [14].

1.3.3 Denaturants-Induced Unfolding

Denaturants like Urea and guanidinium chloride (GdmHCl) can induce controlled unfolding for linear threading through the solid-state nanopore [39, 71]. Denaturants disrupt the noncovalent interaction in protein causing them to unfold into extended or partially extended conformations. Researchers can control how much of the protein unfolds before or during the translocation by varying the denaturant concentration. Proteins can be further stretched and unfolded by voltage or force, exposing structural features and variations in stability [50].

2 Materials

(1) **Free-standing SiN membrane:** Silicon nitride (SiN) membrane supported by Si/SiO_2 chip, TEM (transmission electron microscopy), for drilling and characterization (size and shape) of the nanopore in the SiN membrane.

(2) **Electrolyte buffer solution:** KCl and HEPES Buffer solution, at pH = 6.5–8.5, Ultra-pure Milli-Q water with resistivity of 18 Ωcm, protein denaturing agents (urea, Gdm-HCl).

(3) **Proteins:** Cytochrome c, λ repressor family λ_{6-85} at very low concentration in μM-nM,

(4) **Measurement Device:** A high-bandwidth instrument such as Axopatch 200B [37–39], Custom Voltage sensor (Chimera VC 100) [38], 10 MHz nanopore reader (Elements Srl) for measuring the current blockade with a high single-to-noise ratio.

(5) Solid-state nanopore flow cell with cis and trans compartments connected with electrode-like Ag/AgCl electrode.

(6) Analysis software: Pyth-Ion software for data extraction and filtration, software for characterization of the signals (AxoGraph, Igor Pro [72, 73]).

3 Method

Select a clean SiN chip with no cracks and leakage to fabricate nanopore of desired pore size that is relevant to the protein molecule of interest, then follow the steps below.

3.1 Fabrication of Nanopore (Drilling) Through TEM

1. Using a TEM with a field emission gun (FEG), the nanopore of desired diameter (usually in the range of ~1–3 nm for protein threading experiments) can be created on a SiN membrane(*see* **Notes 1 and 2**).
2. Before being used, all of the chips should be washed using a hot piranha solution (1:2–5, H_2O_2/H_2SO_4) to make the pore wet, hydrophilic, and clean devoid of any impurities.

3.2 Cleaning of the Nanopore Chip

For cleaning purposes, hot piranha solution (2:1, H_2SO_4/H_2O_2) is used for 20 min, and hot deionized water is used after the piranha wash. After Piranha treatment, the chip must be washed with hot deionized water and dried for further use (*see* **Note 3**).

3.3 Assembling the Nanopore Chip

After cleaning, the nanopore chip should be assembled properly in the flow cell. Silicon elastomers are used for securely sealing the space between the chip and the flow cell, and this also helps to reduce the capacitive noise(*see* **Notes 4 and 5**).

3.4 Electrolyte Solution Preparation

Generally, KCl salt is used as a electrolyte solution. Making 0.5–2 M KCl, 10 mM HEPES buffer solution, ultrapure Milli-Q water is used, for regulating the pH of the buffer solution to a pH of 6.5–8.5; KOH and HCl is added to the solution (*see* **Note 6**).

3.5 Measurement of the Length of the Nanopore

After electrolyte solution addition, apply different voltages to measure the currents and obtain the slope of the current vs voltage (I vs V) data; this will allow us to evaluate the conductance of the nanopore (G). The length of the nanopore can be calculated by the formula $G = \sigma\left[{}^{4l}/_{\pi d^2} + {}^{1}/_{d}\right]^{-1}$. Where σ is the specific conductivity of the electrolyte solution (for 1 M KCl solution, $\sigma = 11.13nS/\text{nm}$), l (nm) is the effective length of the nanopore, and d (nm) is the diameter of the nanopore (*see* **Note** 7).

3.6 Electrolyte Addition and protein addition

Add electrolyte solution, generally 0.5–2 M KCl, with 10 mM HEPES buffer solution for regulating the pH of the medium for protein. Sometimes, a solution with a different pH is used, making the neutral protein into ionic form for enhancing the electrophoresis through the solid-state nanopore. For better protein translocation measurements, use a very low concentration of the translocating protein, such as in the nM–μM range, and various denaturing agents like guanidinium hydrochloride (Gdm-HCl).

3.7 Measurement Procedure of Threading of Protein

An ionic solution, for example, KCl buffer solution, fills both sides of the membrane. As seen in Fig. 1, a pair of electrodes (usually Ag/AgCl) is used to apply an electrical bias voltage across the pore with the appropriate polarity, creating a steady baseline ionic current flow through the pore. Significant large electric fields draw biomolecules on the cis side of the pore to the trans side. A brief rise (or reduction) in the magnitude of ionic current is caused by the interaction of proteins with the nanopore, and then we collect hundreds to thousands of such events at different voltages (Fig. 4) to construct and analyze the distributions of event parameters (Fig. 1) to classify that the current blockade events are actually due to the translocation. Depending on salt content, chemical environment, and voltage polarity, the transient current modulation may cause the ionic current to rise or decrease. Regardless of the direction of the current shift, these electric signals will be referred to as blockades. The translocation signals contain information on the physical characteristics of a biomolecule (*see* **Notes 8–10**).

3.7.1 Extraction, Isolation, and Analysis of the Nanopore Current Signals

For the analysis of the signals, various combinations of software solutions are currently in use, such as Pyth-Ion (https://github.com/rhenley/Pyth-Ion/), AxoGraph, MATLAB, and Igor Pro. Igor Pro

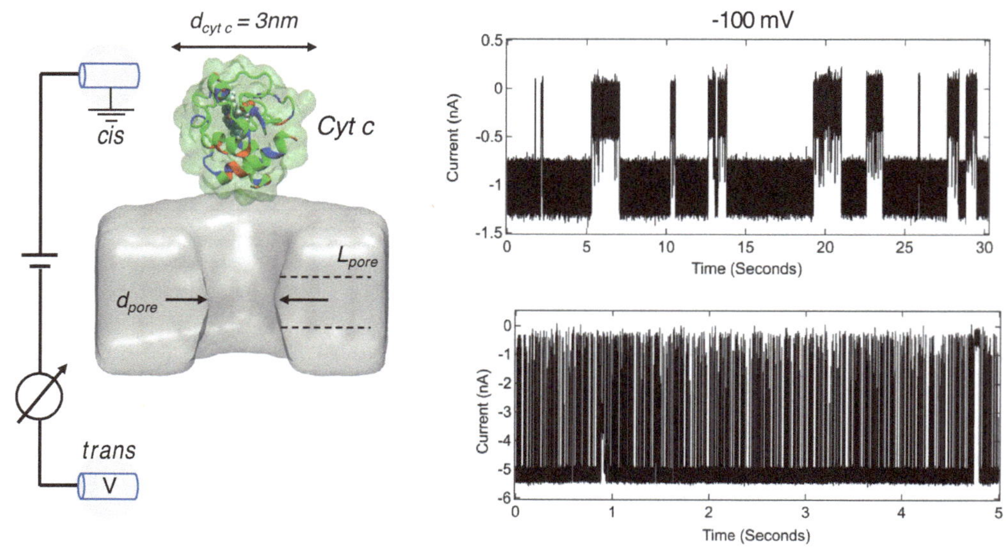

Fig. 4 Example ionic current traces for the translocation of cytochrome c with a solid-state nanopore. Reproduced from [39] with permission from author

gives robust data analysis, data visualization, and publication-quality graphics [74]. The extracted data of ionic current trace of protein interaction and translocation can be illustrated and analyzed with the Igor Pro software. Here are some key steps involved in the analysis of data.

Extraction of data—choose that fraction of data which is continuous and contains both open pore and current blockade events, values, and extract the data with Pyth-Ion software in a definite time domain at a given electric field.

Filtering of data—filter the data in Pyth-Ion software by applying a low-pass Bessel filter at the required bandwidth.

Analysis of data—by using different software, data can be analyzed. With the help of parameters such as baseline and threshold current level, data are analyzed in Igor Pro software or MATLAB.

3.7.2 Characteristics of a Protein

When a protein molecule enters and passes through the nanopore, it temporarily obstructs the flow of ions through the open nanopore (Figs. 1 and 4). These blockades cause a measurable drop in the ionic current. The distributions of current blockade amplitude ($\Delta I/I_O$) and residence time ($\tau_{residence}$) provide information about the translocating protein molecules.

(a) **Amplitude of the current drop**: The magnitude of the current drop is proportional to the size, shape, and charge of the protein. A single interaction of cyt c with the pore is shown in Fig. 4. When protein translocates through the nanopore, the distribution of $\Delta I/I_O$ provides information about the protein stretching, squeezing, and folding–unfolding.

(b) **Residence time:** The time it takes for the protein to pass through the nanopore. This is influenced by the size of the protein, its dipole moment, conformational state, and its interaction with the surface of the nanopore and the applied voltage. The $\tau_{\text{residence}}$ gives information regarding the rate at which a protein translocates through the nanopore and how it is coupled to energetics of conformational dynamics. When $d_{\text{pore}} < d_{\text{protein}}$, protein gets excited to higher energy levels, which are partially unfolded intermediate states of the protein en route to the unfolded state (U).

3.8 Electrical Unfolding of Cytochrome c

The electric field at the nanopore is a powerful tool to unfold the protein through the solid-state nanopore without any chemical denaturing agents. Using ultrathin SiN pores of varying diameters (1.5–5.5 nm) and controlled voltage (100–900 mV), the measured distribution of ionic current blockades ($\Delta I/I_0$) shifts from higher to lower value (Fig. 5). Distinct current blockade levels and residence times can reveal the multiple translocation pathways, i.e., intact squeezing, partial unfolding, and complete unfolding [37–39]). For cytochrome, the electric field threshold generally lies in the range of 30–100 MV m^{-1} to pull through the narrow pore (~2.5 nm). For a smaller pore (<2.5 nm), the cytochrome undergoes conformational changes to unfold at the mouth of the pore and translocate via threading in high-voltage regime [39].

3.9 Observation of Single-Protein Transition State Passage Using Solid-State Nanopore

By monitoring the ionic current jumps associated with the conformational transitions during the protein translocation, the transition state can be captured using a solid-state nanopore. The high-resolution signals of current blockade events corresponding to folding and unfolding transitions of proteins can be identified at the

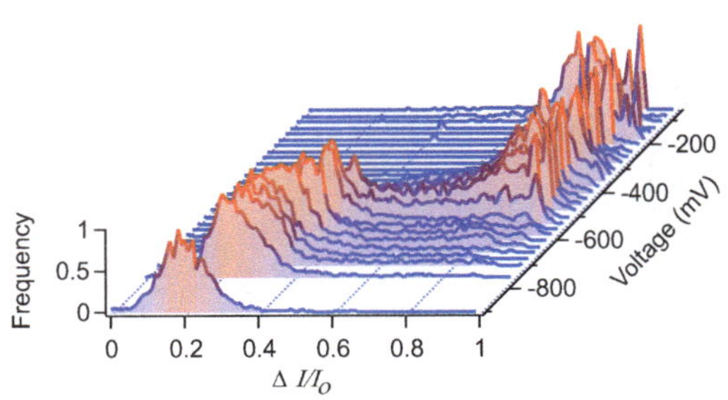

Fig. 5 The electrical unfolding of cytochrome c during the translocation through the pore. The diagram shows the single-molecule measurements of distributions of ionic current blockade ratio ($\Delta I/I_0$) as a function of the transmembrane voltage for the cytochrome c protein (reproduced from [39], with permission from the author). The observed shifts in the distribution as a function of voltages are due to thermodynamic stabilization of partial unfolded intermediates and unfolded protein states during the translocation process

single-molecule level (Fig. 6). The ionic current helps to resolve the current signature associated with native, unfolded, and transition state conformations and transitions between them (Figs. 6 and 7 and [38]). This can allow us to characterize the saddle points in the energy landscape of the protein where it transitions between different conformational states. In comparison to other techniques, this allows the transit time of a protein to be directly observed for

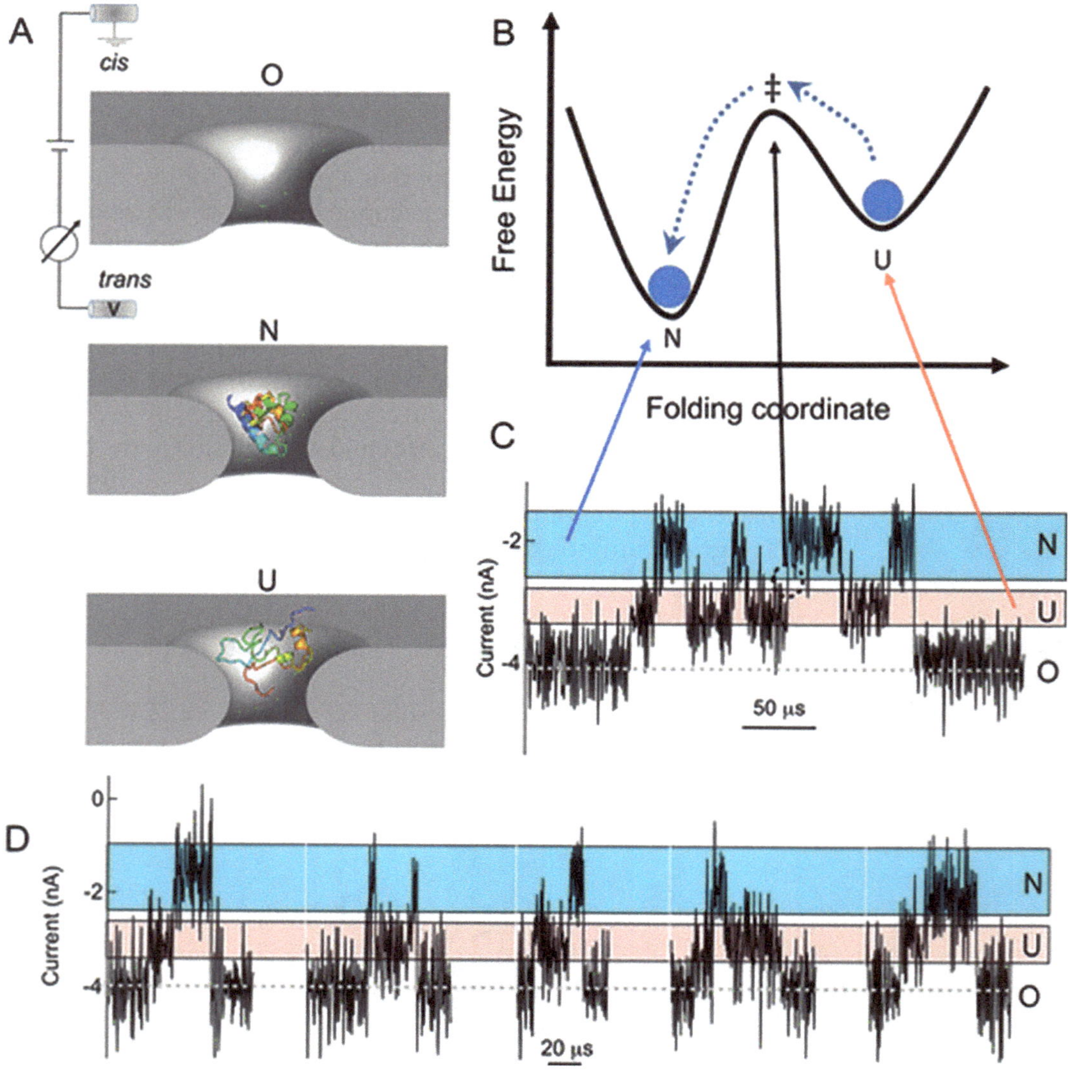

Fig. 6 Schematic of transition state between folding and unfolding transition of a two-state folder of λ-repressor protein family. (**A**) Application of voltage in trans chamber, pulls the protein toward cis to trans. When protein molecule translocates through the solid-state nanopore, there are changes in their conformation which are in dynamic equilibrium between native and unfolded state. (**B**) Potential energy surface for the folding reaction coordinate. (**C**) Example ionic current trace for the passage of the transition state, i.e., N→U and U→N transitions. (**D**) Other current trace shows the arrival of protein, conformational transitions, and translocation to restore an open pore (Reproduced from [38], with permission from Copyright Clearance Centre permission)

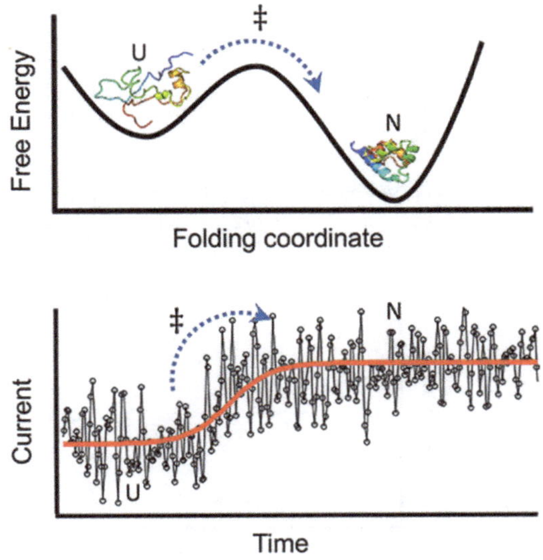

Fig. 7 Transition state passage of a single protein for the transition between unfolded state (*U*) and native folded state (*N*) and vice versa can be observed using nanopore ionic current jumps (reproduced from [38] from copyright clearance center permission)

a sub-microsecond duration. Usually, when the protein transitions between different states, the ionic current level jumps from one level to another (Figs. 6 and 7). High-bandwidth amplifier allows us to zoom into the transition events to characterize the transition state passage and obtain the distribution of transit time [38].

3.10 Comparison of the Energy Landscape of Protein Mutants

Interdependency of protein translocation and protein conformational dynamics lies at the heart of biological processes. Mutations in proteins can speedup folding/unfolding by optimizing the population of intermediate states, by stabilizing transition states, or by tuning the ruggedness of the energy landscape. One way to quantify the ruggedness of a protein's energy landscape is to count the total number of conformational states of mutants. It can be possible that the number of conformational states via mutation can lead to a speedup of the protein translocation process. Conventionally, the free energy of protein states has been manipulated by chemical denaturants, pressure, and temperature. In recent reports, it was demonstrated that electric-field-controlled single protein conformational dynamics of the λ-repressor fragment mutants (μs relaxation time) during nanopore threading can reveal the translocation of kinetically selected states, in analogy to what occurs at the inner bacterial membrane [37]. These single-molecule experiments show that the more rugged mutants translocate faster compared (Fig. 8) to the less rugged mutant, indicating that proteins retain some memory of

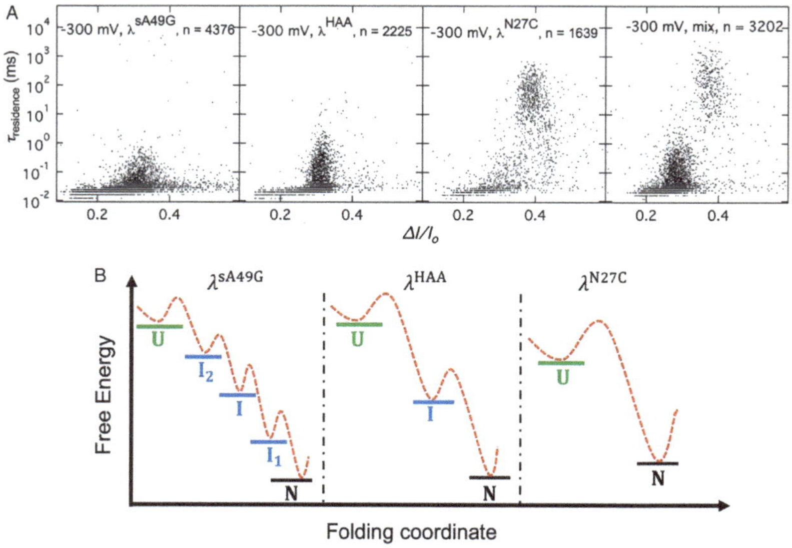

Fig. 8 Threading of single proteins through a solid-state nanopore can be used to compare their energy landscapes and identification of mutants in the solution mixture. In this example, the scatter plots of residence time and current blockade ratio for the three λ repressor mutants (λ^{N27C}, λ^{HAA}, and λ^{sA49G}) and their mixture solution are shown. (**A**) the scatter plot of τ residence vs $\Delta I/I_0$ for the individual mutants and a ternary mixture of all mutants' protein at voltage $V = -300$ mV). (**B**) Summary of differences in the energy landscape between the mutants (reproduced from [37] from the author's permission). Mutant sA49G has more intermediate states and thus translocates faster than the faster-folding three- and two-state mutants HAA and N27C, respectively. Such dependency of translocation rate on the energy landscape may allow to separate protein mutants in the solution mixture

conformational dynamics during translocation, thus paving the path for single-molecule-level protein mutant separation.

3.11 Energetics of Unfolding and Kinetics of Threading

If the time scale of the protein conformational transition (Fig. 9) is much faster than that of translocation time of the protein, then the observed shift (Fig. 5) in the average value of $\Delta I/I_o$ from the distribution can be described for a protein undergoing two-state $M \leftrightarrow U$ dynamic interconversion during the translocation (as [37] and [39])

$$\left\langle \frac{\Delta \mathrm{I}}{\mathrm{I}_o} \right\rangle_{MU} = \frac{\phi_M\, e^{\frac{\Delta\Delta G_{MU} - \Delta p_{MU} E_{app}}{k_B T_0}} + \phi_U}{1 + \exp\{(\Delta\Delta G_{MU} - \Delta p_{MU} E_{app})/k_B\, T_0\}}.$$

Here, ϕ_M is the average current blockade ratio ($\Delta I/I_o$) of pure M state and ϕ_U average current blockade ratio of pure U-state. Where $\Delta\Delta G_{MU}$ stands for the free energy difference between M and U states in absence of transmembrane voltage bias and $-\Delta p_{MU} E_{\mathrm{app}}$ (Δp_{MU} being difference in the dipole moments of U and M states, respectively) factor comes from the free energy correction due to the

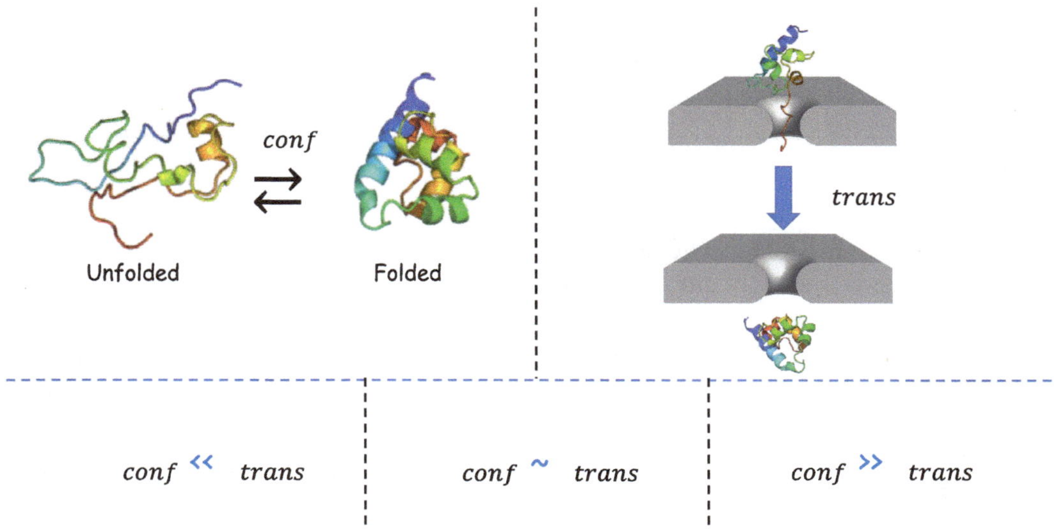

Fig. 9 Consequence of time scale separation. If the time scale of the translocation is much slower compared with the time scale of conformational dynamical interconversion, then the model predicts that equilibrium conformational fluctuation will govern the nonequilibrium translocation through the nanopore

application of electric field ($E_{\text{app}} = V/L_{\text{pore}}$). It comes from the idea that the more unfolded state U has a larger electrical dipole moment and thus a more extended conformation will have higher free energy. The application of electric field will stabilize the more polar and elongated unfolded conformations. Since more unfolded and elongated conformations allow more current to pass through nanopore the current blockade ϕ_U is lower in comparison to ϕ_M and thus application of transmembrane voltage causes a shift in $\Delta I/I_o$ distributions from a higher value to lower value as in Fig. 5 [37–39]. The quantitative analysis of the $\Delta I/I_o$ distributions with above thermodynamic equation allow us to get the fundamental physical quantities $\Delta\Delta G_{MU}$ and Δp_{MU} for a given protein molecule.

Similarly, one can further analyze the $\tau_{\text{residence}}$ distribution to understand how translocation of protein is corelated with its structural fluctuation. For a two-state protein, the observed translocation rate constant can be derived to be proportional to thermodynamic probability of unfolded protein states [37–39].

$$k_{tr(M\leftrightarrow U)} = \frac{k_{oU}\, e^{-\frac{\{\Delta\Delta G_{MU} - \beta_U^{\star} E_{app}\}}{k_B T_0}}}{1 + e^{-\frac{\{\Delta\Delta G_{MU} - \Delta p_{MU} E_{app}\}}{k_B T_0}}}.$$

Here, k_{oU} is the zero-field translocation rate of unfolded protein state and $\beta_U^{\star} = \beta_U + \Delta p_{MU}$, β_U being the amount of translocation energy barrier reduced per unit application of electric field. This analysis allows us to determine the energy barrier for the translocation. More details on the energetics/kinetics data and analysis can be found in references [37–39].

4 Notes

1. The electron beam intensity (e nm^{-2}) and exposure time (s) and energy of the electrons (keV) are the major factors to control the size of the nanopore.
2. During the nanopore fabrication contraction and expansion happen, which cause size deviation from desired size of the nanopore.
3. During preparation of the Piranha solution, we must wear a lab coat, safety gloves, and goggles and prepare the solution in the fume-hood because during heating, Piranha solution can be vigorously in boiling and fuming condition.
4. Chip should be properly assembled in the flow cell; otherwise, it caused leakage and interrupt the signal.
5. Silicon elastomer should be applied carefully on the chip so that it does not seal the membrane and nanopore area.
6. Regulating the pH, ensure that the concentration of acid and base (HCl, KOH) should be high concentration so that concentration of KCl and HEPES remains constant.
7. The slope of the *I–V* curve helps to determine the effective length of the pore, which is further essential for protein translocation rate evaluation.
8. For nanopore threading, the size of the nanopore should be smaller or similar to the size of the protein.
9. Achieving reproducible linear threading for precise analysis; proteins typically enter into, with the random orientations, which results in the complicating readouts signal. For resolving these complications, we can use some strategies like voltage tuning, chemical denaturation, or coupling to processive enzymes for consistent linearized unfolding and threading for repeatable signatures.
10. Faster acquisition rates for rapid events since the dynamic interconversion of different conformational states of the protein is in microseconds, we must have such type of electronics having high bandwidth (in MHz) with high signal-to-noise ratio, integration of ultra-low-noise amplifiers, and a high-speed digitizer is critically important.

Acknowledgments

We thank ANRF and DST for the funding support (ANRF/SRG/2023/000234).

References

1. Wanunu M (2012) Nanopores: a journey towards DNA sequencing. Phys Life Rev 9:125–158
2. Ahmad M et al (2023) A generalizable nanopore sensor for highly specific protein detection at single-molecule precision. Nat Commun 14:1374
3. Gershow M, Golovchenko JA (2007) Recapturing and trapping single molecules with a solid-state nanopore. Nat Nanotechnol 2:775–779
4. Fyta M, Melchionna S, Succi S (2011) Translocation of biomolecules through solid-state nanopores: theory meets experiments. J Polym Sci Part B Polym Phys 49:985–1011
5. Schmid S, Dekker C (2021) Nanopores: a versatile tool to study protein dynamics. Essays Biochem 65:93–107
6. Dekker C (2007) Solid-state nanopores. Nat Nanotechnol 2:209–215
7. Ananth AN et al (2018) Spatial structure of disordered proteins dictates conductance and selectivity in nuclear pore complex mimics. eLife 7:e31510
8. Rahman M, Sampad MJN, Hawkins A, Schmidt H (2021) Recent advances in integrated solid-state nanopore sensors. Lab Chip 21:3030–3052
9. Yang H, Qing G (2021) Solid-state nanopores and nanochannels for the detection of biomolecules. Chem Phys Rev 2:021306:1-19
10. Hall AR (2012) Solid-State nanopores: from fabrication to application. Microsc Today 20:24–29
11. Dutt S et al (2023) High accuracy protein identification: fusion of solid-state nanopore sensing and machine learning. Small Methods 7:2300676
12. Han A et al (2008) Label-free detection of single protein molecules and protein–protein interactions using synthetic nanopores. Anal Chem 80:4651–4658
13. Kasianowicz JJ, Robertson JWF, Chan ER, Reiner JE, Stanford VM (2008) Nanoscopic porous sensors. Annu Rev Anal Chem 1:737–766
14. Yusko EC et al (2017) Real-time shape approximation and fingerprinting of single proteins using a nanopore. Nat Nanotechnol 12:360–367
15. Garoli D, Yamazaki H, Maccaferri N, Wanunu M (2019) Plasmonic nanopores for single-molecule detection and manipulation: toward sequencing applications. Nano Lett 19:7553–7562
16. Schneider GF, Dekker C (2012) DNA sequencing with nanopores. Nat Biotechnol 30:326–328
17. Li J, Gershow M, Stein D, Brandin E, Golovchenko JA (2003) DNA molecules and configurations in a solid-state nanopore microscope. Nat Mater 2:611–615
18. Storm AJ et al (2005) Fast DNA translocation through a solid-state nanopore. Nano Lett 5:1193–1197
19. Dorey A, Howorka S (2024) Nanopore DNA sequencing technologies and their applications towards single-molecule proteomics. Nat Chem 16:314–334
20. Alfaro JA et al (2021) The emerging landscape of single-molecule protein sequencing technologies. Nat Methods 18:604–617
21. Lin Y, Ying Y-L, Shi X, Liu S-C, Long Y-T (2017) Direct sensing of cancer biomarkers in clinical samples with a designed nanopore. Chem Commun 53:11564–11567
22. Taniguchi M et al (2019) High-precision single-molecule identification based on single-molecule information within a noisy matrix. J Phys Chem C 123:15867–15873
23. Alam I et al (2024) Single-molecule analysis of SARS-CoV-2 double-stranded polynucleotides using solid-state nanopore with AI-assisted detection and classification: implications for understanding disease severity. ACS Appl Bio Mater 7:1017–1027
24. Arima A et al (2018) Identifying single viruses using biorecognition solid-state nanopores. J Am Chem Soc 140:16834–16841
25. Zhe J, Jagtiani A, Dutta P, Hu J, Carletta J (2007) A micromachined high throughput Coulter counter for bioparticle detection and counting. J Micromech Microeng 17:304
26. Graham M (2022) Don. The Coulter principle: a history. Cytometry A 101:8–11
27. Kasianowicz JJ, Brandin E, Branton D, Deamer DW (1996) Characterization of individual polynucleotide molecules using a membrane channel. Proc Natl Acad Sci 93:13770–13773

28. Brinkerhoff H, Kang ASW, Liu J, Aksimentiev A, Dekker C (2021) Multiple rereads of single proteins at single–amino acid resolution using nanopores. Science 374:1509–1513
29. Crnković A, Srnko M, Anderluh G (2021) Biological nanopores: engineering on demand. Life 11:27
30. Fragasso A, Schmid S, Dekker C (2020) Comparing current noise in biological and solid-state nanopores. ACS Nano 14:1338–1349
31. Horne RI, Sandler SE, Vendruscolo M, Keyser UF (2025) Detection of protein oligomers with nanopores. Nat Rev Chem 9:224–240
32. Kasianowicz JJ (2002) Nanometer-scale pores: potential applications for analyte detection and DNA characterization. Dis Markers 18:185–191
33. Mojtabavi M, Greive SJ, Antson AA, Wanunu M (2022) High-voltage biomolecular sensing using a bacteriophage portal protein covalently immobilized within a solid-state nanopore. J Am Chem Soc 144:22540–22548
34. Hall A et al (2011) Hybrid biological/solid-state nanopores. Biophys J 100:168a
35. Goto Y, Matsui K, Yanagi I, Takeda K (2019) Silicon nitride nanopore created by dielectric breakdown with a divalent cation: deceleration of translocation speed and identification of single nucleotides. Nanoscale 11:14426–14433
36. Fologea D, Ledden B, McNabb DS, Li J (2007) Electrical characterization of protein molecules by a solid-state nanopore. Appl Phys Lett 91:053901
37. Tripathi P, Firouzbakht A, Gruebele M, Wanunu M (2022) Threading single proteins through pores to compare their energy landscapes. Proc Natl Acad Sci 119:e2202779119
38. Tripathi P, Firouzbakht A, Gruebele M, Wanunu M (2022) Direct observation of single-protein transition state passage by nanopore ionic current jumps. J Phys Chem Lett 13:5918–5924
39. Tripathi P et al (2021) Electrical unfolding of cytochrome C during translocation through a nanopore constriction. Biophys J 120:99a
40. Firnkes M, Pedone D, Knezevic J, Döblinger M, Rant U (2010) Electrically facilitated translocations of proteins through silicon nitride nanopores: conjoint and competitive action of diffusion, electrophoresis, and electroosmosis. Nano Lett 10:2162–2167
41. Wanunu M et al (2010) Rapid electronic detection of probe-specific microRNAs using thin nanopore sensors. Nat Nanotechnol 5:807–814
42. Lin K et al (2021) Surface charge density inside a silicon nitride nanopore. Langmuir 37:10521–10528
43. Arjmandi N, Van Roy W, Lagae L, Borghs G (2012) Measuring the electric charge and zeta potential of nanometer-sized objects using pyramidal-shaped nanopores. Anal Chem 84:8490–8496
44. Saharia J et al (2021) Modulation of electrophoresis, electroosmosis and diffusion for electrical transport of proteins through a solid-state nanopore. RSC Adv 11:24398–24409
45. Westermeier R (2016) Electrophoresis in practice: a guide to methods and applications of DNA and protein separations. John Wiley & Sons
46. Wen C, Schmid S, Dekker C (2024) Understanding electrophoresis and electroosmosis in nanopore sensing with the help of the nanopore electro-osmotic trap. ACS Nano 18:20449–20458
47. Tripathi P et al (2023) A Marcus-type inverted region in the translocation kinetics of a knotted protein. J Phys Chem Lett 14:10719–10726
48. Larkin J, Henley RY, Muthukumar M, Rosenstein JK, Wanunu M (2014) High-bandwidth protein analysis using solid-state nanopores. Biophys J 106:696–704
49. Zhang Y, Ding M (2025) Probing nanopores: molecular dynamics insights into the mechanisms of DNA and protein translocation through solid-state and biological nanopores. Soft Matter 21:2385–2399
50. Freedman KJ, Haq SR, Edel JB, Jemth P, Kim MJ (2013) Single molecule unfolding and stretching of protein domains inside a solid-state nanopore by electric field. Sci Rep 3:1638
51. Ritmejeris J, Chen X, Dekker C (2024) Single-molecule protein sequencing with nanopores. Nat Rev Bioeng 3:303–316
52. Hu G et al (2023) Nanopore sensors for single molecular protein detection: research progress based on computer simulations. IET Nanobiotechnol 17:257–268
53. Freedman KJ et al (2014) Nonequilibrium capture rates induce protein accumulation and enhanced adsorption to solid-state nanopores. ACS Nano 8:12238–12249
54. Kudr J et al (2015) Fabrication of solid-state nanopores and its perspectives. Electrophoresis 36:2367–2379
55. Robinson JP, Coulter WH (2013) Decades of invention and discovery. Cytometry A 83:424–438
56. Merchant CA et al (2010) DNA translocation through graphene nanopores. Nano Lett 10:2915–2921
57. Mojtabavi M, VahidMohammadi A, Liang W, Beidaghi M, Wanunu M (2019) Single-molecule sensing using nanopores in two-dimensional transition metal carbide (MXene) membranes. ACS Nano 13:3042–3053

58. Wang F et al (2023) MoS_2 nanopore identifies single amino acids with sub-1 Dalton resolution. Nat Commun 14:2895
59. Gu C et al (2019) Experimental study of protein translocation through MoS_2 nanopores. Appl Phys Lett 115:223702
60. Si W et al (2022) Navigated delivery of peptide to the nanopore using in-plane heterostructures of MoS_2 and SnS_2 for protein sequencing. J Phys Chem Lett 13:3863–3872
61. Zhou Z et al (2013) DNA translocation through hydrophilic nanopore in hexagonal boron nitride. Sci Rep 3:3287
62. Muthukumar M, Plesa C, Dekker C (2015) Single-molecule sensing with nanopores. Phys Today 68:40–46
63. Traversi F et al (2013) Detecting the translocation of DNA through a nanopore using graphene nanoribbons. Nat Nanotechnol 8:939–945
64. Geim AK (2009) Graphene: status and prospects. Science 324:1530–1534
65. Wells DB, Belkin M, Comer J, Aksimentiev A (2012) Assessing graphene nanopores for sequencing DNA. Nano Lett 12:4117–4123
66. Wanunu M, Sutin J, McNally B, Chow A, Meller A (2008) DNA translocation governed by interactions with solid-state nanopores. Biophys J 95:4716–4725
67. Wei G, Hu R, Lu W, Wang Z, Zhao Q (2024) Bidirectional peptide translocation through ultrasmall solid-state nanopores. Langmuir 40:20831–20839
68. Storm AJ, Chen JH, Ling XS, Zandbergen HW, Dekker C (2003) Fabrication of solid-state nanopores with single-nanometre precision. Nat Mater 2:537–540
69. Eggenberger OM, Ying C, Mayer M (2019) Surface coatings for solid-state nanopores. Nanoscale 11:19636–19657
70. Chen P et al (2004) Atomic layer deposition to fine-tune the surface properties and diameters of fabricated nanopores. Nano Lett 4:1333–1337
71. Street TO, Courtemanche N, Barrick D (2008) Protein folding and stability using denaturants. Methods Cell Biol 84:295–325
72. Zhang N et al (2011) Desktop software for patch-clamp raw binary data conversion and preprocessing. J Electr Comput Eng 251215
73. Igor Pro® | Igor Pro by WaveMetrics. https://www.wavemetrics.com/products/igorpro
74. Plesa C, Dekker C (2015) Data analysis methods for solid-state nanopores. Nanotechnology 26:084003

Chapter 15

Metrology of Nanoscale Objects Using Nanopores

Kun Li, Kittipitch Yooprasertchuti, Ishita Agrawal, Tihomir Marjanović and Slaven Garaj

Abstract

Solid-state nanopore serves as a versatile tool for single-particle analysis of nanoscale objects, enabling the measurement of their geometrical and physicochemical properties with molecular precision. When a bias drives an analyte to translocate through a nanopore, ionic current blockades encode information on the object's size, shape, charge, and dynamics. Compared with biological nanopores, solid-state nanopores offer superior mechanical stability, broader operational ranges, and easier integration with optical or mechanical characterization systems. This chapter introduces the concept of nanopore tomography (NT), which focuses on extracting quantitative geometrical parameters rather than genomic sequence information. We discuss the experimental considerations in nanopore characterization, including noise sources, sampling rate limitations, and filtering effects that distort short translocation events. The event charge deficit (ECD) emerges as a robust parameter preserved under signal filtering, allowing reliable particle fingerprinting. Furthermore, multiphysics coupled finite element modeling is presented to simulate electrokinetic processes and translocation dynamics, revealing the nonlinear dependence of current blockade on voltage and the contributions of electrophoretic and electroosmotic forces. Together, these insights advance nanopore-based metrology toward high-throughput, label-free, and nondestructive characterization of biomolecules, viruses, and nanoparticles.

Keywords Nanopore tomography, Nanometrology of single-entity, Temporal resolution, Signal distortion, Current blockade modeling

1 Introduction

Solid-state nanopore sensors can measure the shapes and other geometrical attributes of nanoscale objects. They can probe individual particles, also collecting sufficiently large statistics to explore ensemble-average properties, heterogeneity, and rare configurations. For example, this technology could be used to detect viruses and fingerprint viruses [1, 2] and proteins [3–5], and to study DNA–protein complexes [6–8], DNA knots [9–11], and nanoparticles [12, 13].

The nanopore sensor consists of two chambers filled with aqueous salt solution, separated by a thin membrane. A single nanoscale

Fanjun Li (ed.), *Nanopore Technology*, Methods in Molecular Biology, vol. 3072,
https://doi.org/10.1007/978-1-0716-5527-6_15,

pore (diameter D_p in the range of 1–300 nm) perforates the membrane, creating the only pathway connecting the two chambers (see Fig. 1). When a voltage is applied between two Ag/AgCl electrodes immersed in the opposing chambers, the ionic current through the pore can be measured. The current is highly sensitive to the geometry of the pore and its surface charge. When a nanoscale object is electrophoretically (or electroosmotically) driven into and across the nanopore, it briefly occludes part of the pore, reducing the cross-sectional area available for ion transport and producing an instantaneous current drop—i.e., a current blockade. The instantaneous value of the current blockade $I_B(t)$ depends on the object's obstructing size at a given time and may also reflect its physicochemical properties such as surface charge. The overall shape of $I_B(t)$ can therefore convey information about both the geometrical and physicochemical attributes of the nanoscale object.

Three distinctive types of nanopore sensors have been developed: biological pores [14] (biopores), solid-state nanopores [15] (ss-pores), and two-dimensional nanopores [16–18] (2D-pores). The most widely used application of nanopore sensing—DNA sequencing commercialized by Oxford Nanopore Technologies—relies on biopores. Biological nanopores are transmembrane protein channels embedded in a lipid bilayer. In 1996, Kasianowitz, Deamer, and Branton [14] first demonstrated nanopore sensing by characterizing individual polynucleotides using the α-hemolysin (α-HL) pore, suggesting the feasibility of nanopore-based DNA sequencing. Crucial advances toward sequencing were later achieved with controlled translocation [19] and nucleobase recognition [20]. Biopores are protein channels with highly reproducible structures down to near-atomic precision, but their usable pore diameters ($D_p < 2$ nm) and chemical robustness are limited by protein biochemistry, constraining their applicability beyond sequencing.

In comparison to biological nanopores, solid-state nanopores are less affected by ambient conditions, exhibiting better stability across wider ranges of pH and temperature, as well as longer operational

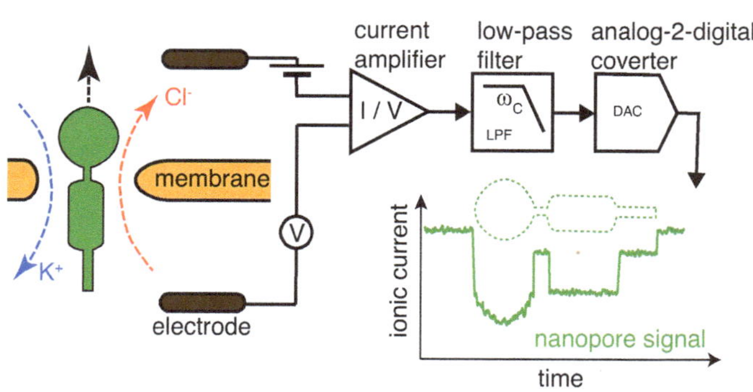

Fig. 1 Principle of operation of the nanopore detector

lifetimes. Moreover, the geometric shape and surface chemistry of solid-state nanopores can be tailored to match the properties of specific analytes, expanding the range of applications and improving detection performance. Solid-state nanopores can also be fabricated in large numbers and readily integrated with mechanical or optical detection techniques [21–25]. This versatility allows for the characterization of a wide range of analytes, including DNA, proteins, nanoparticles, and viruses (Fig. 2).

To avoid confusion in interdisciplinary discussions, we distinguish between nanopore DNA sequencing—common in biological research—and the extraction of geometrical parameters of nanoscale objects. We refer to the latter as *nanopore tomography* (*NT*). For example, one may detect viruses using metagenomic (nanopore) sequencing, or one may detect individual viral particles as they traverse a nanopore.

In this chapter, we explore the parameters used in nanopore tomography, experimental limitations, and different modes of operation, with an emphasis on practical implementation. We ask what quantifiers can reliably be extracted from NT signals and what the main sources of signal distortion are.

2 Nanopore Measurements and Characterization

Nanopore sensors enable high-resolution electrical detection of single molecules and nanoscale particles by monitoring the ionic current that flows through a nanometer-sized aperture in a thin membrane. The performance of such sensors relies critically on: (1) particular design of a nanopore chip, (2) the microfluidic architecture in which the pore is embedded, (3) the stability and bandwidth of the electronic measurement system, and (4) implementation of extensive noise-reduction setup. Together, these components form

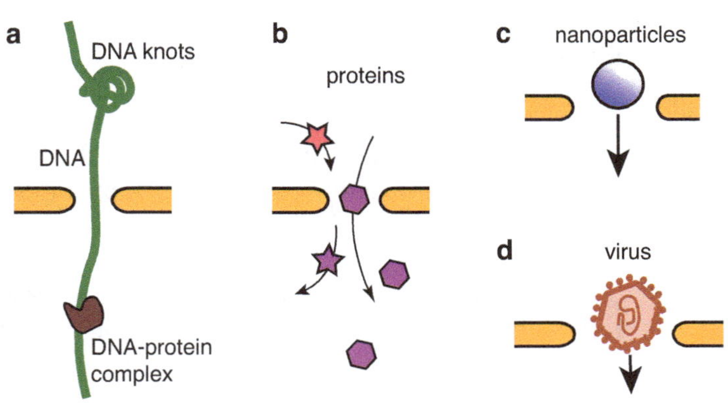

Fig. 2 Solid-state nanopore sensor has been used to detect (**a**) DNA, DNA knots, and DNA–protein complexes, (**b**) proteins, (**c**) nanoparticles, and (**d**) viruses

an integrated platform that allows precise control over the ionic environment and reliable acquisition of transient current blockades associated with molecular translocation.

Nanopore chip: A solid-state nanopore chip lies at the core of this platform. The chip contains a single pore drilled in a SiN_x, SiO_2, graphene, or other membrane, and the pore's diameter and surface chemistry largely determine the sensitivity, selectivity, and temporal resolution of the sensor.

Microfluidics: The chip is mounted between two electrolyte reservoirs, commonly referred to as the cis and trans chambers, which together form a microfluidic cell. The two chambers are filled with buffered ionic solution—typically 1 M KCl—to ensure stable ionic conduction. Polydimethylsiloxane (PDMS) gaskets provide a leak-tight seal around the membrane, eliminate parasitic current paths, and contribute to reducing electrical noise. Dedicated fluidic access ports allow independent buffer exchange and analyte introduction, enabling analytes to be delivered to the cis reservoir and collected in the trans reservoir after translocation through the nanopore.

Electronics: To drive ionic current through the pore, Ag/AgCl electrodes are immersed in each chamber and connected to a low-noise patch-clamp amplifier. The amplifier establishes a stable potential difference across the pore and converts the resulting ionic flux into a measurable current signal. Modern nanopore instruments typically operate at bandwidths between 100 kHz and several megahertz, allowing detection of sub-microsecond events with current noise levels below 10 pA.

Noise reduction setup: The measurement system is usually housed in a Faraday cage and mounted on a vibration-isolated platform to suppress electromagnetic interference and mechanical disturbances. Such shielding is essential when high bandwidth and high sensitivity are required, particularly for resolving fast translocation events from proteins, small nucleic acids, or virus particles.

Because the sensing volume of a nanopore is extremely small, maintaining the quality and cleanliness of the pore is essential for reliable operation. Nanometer-scale contaminants—such as hydrocarbons, fabrication residues, or adsorbed biomolecules—can alter the surface charge, impede pore wetting, or even partially block the pore, resulting in unstable conductance or suppressed translocation rates. Therefore, rigorous surface cleaning and thorough quality control are performed prior to measurements. A stable and reproducible open-pore conductance, combined with linear current–voltage characteristics, is a strong indicator that the pore is well wetted, free of surface contamination, and electrically neutral.

In a nanopore experiment, we are pushing a single molecule or nanoparticle, with sizes of a few nanometers to tens of nanometers in diameter, through a single pore of a comparable size. Such nanoscale system is very sensitive to perturbation. Any small contamination of the nanopore, such as a nonspecific coating of nanopore by hydrocarbons or impurity molecules, will affect the experiments. It might prevent the translocation of the molecules, leading to pulse variations in the current, or block of the pore completely. Despite the utmost care required in experiment preparation, nanopore cleaning, and sample handling, robust quality control remains essential for our experiments.

Figure 3a shows TEM images of a 20-nm SiN_x nanopore and a 13-nm graphene nanopore fabricated by electron-beam drilling at 200 keV. Figure 3b presents the corresponding open-pore current–voltage curves measured in 1 M KCl. Both nanopores exhibit linear, ohmic behavior over the applied voltage range, demonstrating that the pores are clean and that ion transport is dominated by bulk conduction rather than surface trapping or rectification. The frequency content of the current noise provides an additional diagnostic for pore quality. Figure 3c displays the power spectral density (PSD) of current noise recorded at 100 mV. The PSD reveals the characteristic noise regimes of solid-state nanopores, including low-frequency $1/f$ ("flicker") noise, frequency-independent thermal (white) noise, dielectric noise arising from polarization of the membrane, and high-frequency capacitive noise associated with the chip and wiring.

The $1/f$ noise component is particularly important because it dominates the low-frequency regime ($f < 100$ Hz) and often dictates

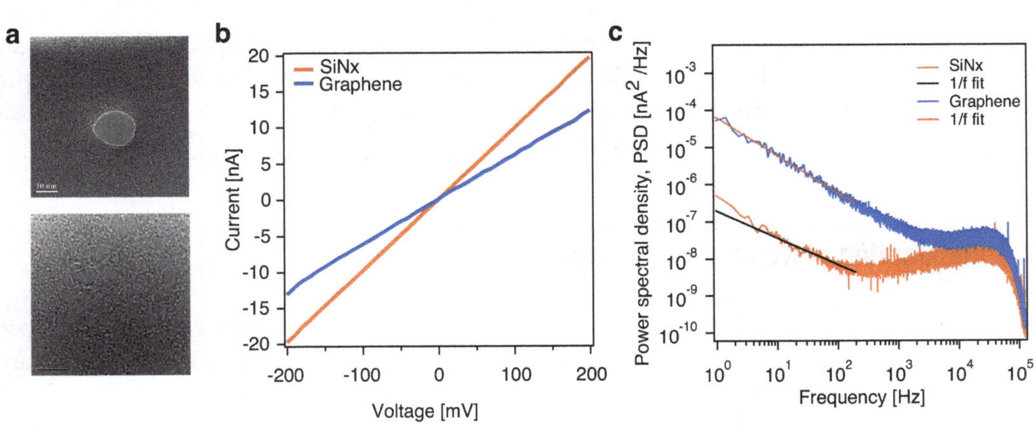

Fig. 3 Structural and electrical characterization of SiN_x and graphene nanopores. (**a**) TEM images of a SiN_x nanopore with 20 nm diameter and a 13 nm wide graphene nanopore. (**b**) Open-pore current-voltage characteristics of SiN_x and graphene nanopore in 1 M KCl. Both SiN_x and graphene nanopore displayed a linear *I–V* curve, indicating ohmic behavior. (**c**) Power spectral density of $SiNx$ and graphene nanopore at 100 mV bias voltage. The graphene pore exhibited two order of magnitude higher $1/f$ noise (<100 Hz, fitted black line) when compared to $1/f$ noise in 20 nm thick SiN_x pore

the ultimate signal-to-noise ratio of nanopore measurements. This flicker noise is commonly described by Hooge's empirical relation [26]

$$\frac{S_1}{I^2} = \frac{A}{f} = \frac{\alpha}{N_c f'}$$

where S_I is the current noise spectral density, i is the open-pore current, N_c is the number of charge carriers, and α is Hooge's parameter. Experiments on SiNx nanopores show that varying the salt concentration [27]—which changes the number of charge carriers—modulates the $\frac{1}{f}$ noise amplitude in accordance with Hooge's relation. However, the effective Hooge parameter also depends strongly on surface chemistry and pore preparation. Aggressive cleaning methods, such as piranha treatment [28] and high-voltage conditioning pulses [29], both reduce the noise amplitude, presumably by removing adsorbates and improving hydrophilicity.

In two-dimensional (2D) material nanopores, such as graphene, the $\frac{1}{f}$ noise amplitude is substantially higher—typically by two orders of magnitude compared with SiNx pores of similar diameter. This enhanced flicker noise could arise from mechanical fluctuations of the ultrathin free-standing membrane, as well as from dynamic interactions between ions and the hydrophobic carbon surface. In these systems, reducing the unsupported membrane area is an effective strategy for lowering the noise [30] and improving the fidelity of translocation measurements.

Together, these electrical and mechanical characteristics define the functional quality of a nanopore sensor and determine its suitability for high-resolution nanopore experiments. A clean, stable, low-noise pore is essential for accurate interpretation of current blockades reliable detection of molecular substructures, and reproducible quantitative measurements in nanopore-based biophysics and materials science.

3 Translocation of Nanoscale Objects

When an electrical bias voltage is applied across a nanopore embedded in a thin insulating membrane separating two electrically isolated fluid reservoirs filled with an electrolyte solution (e.g., KCl, NaCl, LiCl), ions migrate through the pore, producing a stable baseline ionic current (I_0) [31, 32]. In a high ionic strength environment, such as 1 M KCl, even a relatively low applied voltage ($\approx$100 mV) can exert sufficient force (~20 pN) on a long, charged biopolymer like λ-DNA entering the nanopore, pulling and stretching it through the pore. This process results in an instantaneous reduction—termed a current blockade—in the baseline signal [33, 34].

Figure 4a displays a typical DNA translocation characteristic event from the translocation event involving a 48.5 kbp long λ-DNA molecule in a 20 nm SiN_x pore in 1 M KCl solution. Figure 4b shows the knot event that the unknotted parts are color-coded with blue, while parts assigned to knots are presented in purple [10]. Figure 4c shows the translocation of a long, rigid rod-like fd virus [35].

The instantaneous amplitude of the current blockade during translocation $I_B(t)$ determined by the volume and geometry of the portion of the biomolecule occupying the active region of the nanopore at a given time, as well as by the physicochemical properties of the biomolecule, the nanopore walls, and the surrounding ionic electrolyte [36, 37]. By analyzing key features of a current blockade event—such as its shape, amplitude, duration, and frequency—it is possible to infer critical physical and structural characteristics of the translocating biomolecule, including size, length, shape, conformation, surface charge, and dynamic behavior in the ionic environment [38, 39]. An example is the translocation of a partially folded λ-DNA molecule shown in Fig. 4a, where the first step corresponds to the passage of a folded segment through the nanopore, followed by the translocation of the remaining single-file portion. The initial blockade level exhibits approximately twice the amplitude of the second step because the folded segment contains two DNA strands simultaneously occupying the nanopore.

Since current blockade signals can exhibit a wide range of shapes that are not easily correlated with the molecular structure—and because molecules may adopt multiple conformations and orientations—see Fig. 4a, b—it is often insufficient to rely on a single translocation trace. However, nanopore sensing is inherently high-throughput, and statistical analysis enables extraction of meaningful information. To achieve this, each translocation event is

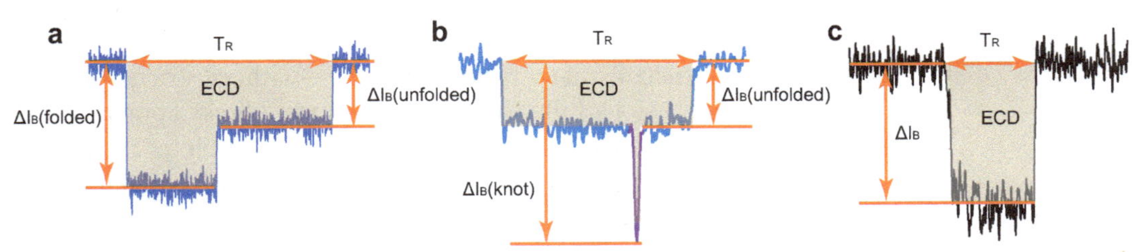

Fig. 4 Representative translocation signals of different biological entities through nanopores. (**a**) Typical current trace showing a translocation event of a 48.5 kbp dsDNA (λ DNA) translocating through a 20 nm diameter SiN_x nanopore in 1 M KCl environment under an applied bias voltage of +250 mV. Two transient current blockade levels, where the second blockade level is exactly double that of the first blockade, indicate that DNA is translocating in a partially folded conformation. I_0 represents open-pore current, ΔI_B is transient current blockade resulting from DNA translocation, and ΔTR denotes translocation time, i.e., time spent by the molecule inside the nanopore. (**b**) Typical current trace showing a translocation event of DNA prime-L knot translocating through nanopore [10]. (**c**) Typical current trace showing a translocation event of a long, rigid rod-like fd virus [35]

reduced to a set of global parameters: translocation time T_R, average current blockade $\langle \Delta I_B \rangle$, and event charge deficit ECD.

The **translocation time,** T_R, is defined as the interval between the moment when the ionic current first deviates from the open-pore baseline and the moment when it returns to that baseline. This quantity reflects the dynamics of the biomolecule inside the nanopore and depends on factors such as the applied bias voltage and the molecule's surface charge [4, 40].

The **average current blockade** is defined as

$$\langle \Delta I_B \rangle = -\frac{1}{T_R} \int_0^{T_R} \Delta I_B(t) \cdot dt,$$

where $\Delta I_B(t) = I_B(t) - I_0$ In our nomenclature, ΔI_B is positive when the current decreases during translocation.

Another useful quantity is the **event charge deficit (ECD),** defined as:

$$\mathrm{ECD} = \Delta I_{avg} \cdot T_R,$$

which represents the total ionic charge prevented from passing through the nanopore during the event. The robustness of ECD makes it a valuable metric for particle fingerprinting, as we will demonstrate shortly.

In Fig. 5a, each DNA translocation event is represented on the $\langle \Delta I_B \rangle$–$T_R$ plane, with the color scale indicating the density of events within each bin. Signals corresponding to linear, folded, and partially folded DNA conformations all appear in this representation. Remarkably, regardless of conformation, all events cluster along the same hyperbolic curve defined by a constant event charge deficit (ECD). This reflects the fundamental fact that for molecules of identical contour length, the total ionic charge displaced during translocation is conserved.

For a molecule that translocates freely—without stochastic stick–slip interactions—a fully folded DNA strand blocks approximately twice the current of a single-file molecule but completes the passage in roughly half the time. Its ECD, therefore, remains equal to that of the single-file configuration. Partially folded DNA produces traces with two sequential blockade levels and an intermediate translocation time [41], yet the integrated charge deficit again matches that of the other conformations.

Taken together, these observations demonstrate that ECD is a robust, conformation-insensitive descriptor of a molecule's effective volume. This makes ECD exceptionally powerful for distinguishing different molecular species within complex or heterogeneous mixtures.

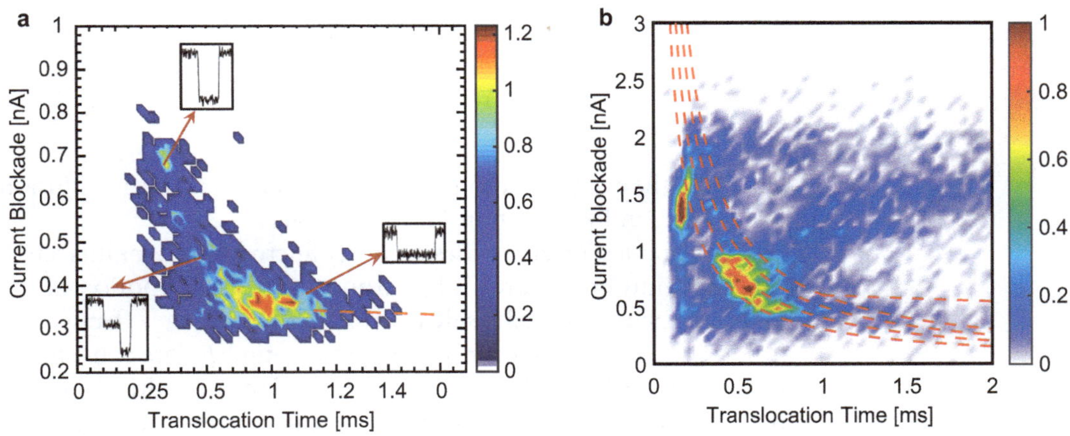

Fig. 5 ECD distributions of λ DNA (48.5 kbp) translocation through a nanopore. (**a**) ECD plot between average event current blockade and translocation time of double-stranded λ DNA in 1 M KCl at pH 8 under +200 mV bias voltage. Red-dashed line is hyperbolae of constant event charge deficit for DNA molecules translocating in various conformations shown in insets (I-unfolded molecule, II-partially folded Molecule, and III-fully folded molecule). (**b**) ECD plot of the stick-slip translocation λ DNA through a nonspecific adsorption (due to insufficient piranha cleaning) in 1 M KCl at pH 8 under +200 mV bias voltage

ECD is also valuable for identifying undesired nonspecific interactions between DNA and the nanopore walls. If the nanopore is not properly cleaned or processed, residual surface contaminants can promote stochastic stick–slip interactions as the molecule translocates. These interactions are random in time, lack biochemical specificity, and cannot be controlled or used for meaningful structural analysis. Under such conditions, translocation events deviate strongly from the constant-ECD hyperbola: instead of collapsing onto a single curve, the events become broadly dispersed in the $\langle \Delta I_B \rangle$–$T_R$ plane, as illustrated in Fig. 5b. This spread directly reflects the irregular pauses and intermittent binding events that alter both the instantaneous current blockade and the effective translocation time.

Even in the regime free from surface interactions—where the ECD remains conserved—individual events do not collapse onto a single, infinitesimally thin hyperbola. Instead, the data exhibit a measurable scatter *perpendicular* to the constant-ECD curve, producing a visibly "fat" band rather than a single line. This broadening originates from intrinsic fluctuations in the translocation velocity of each molecule. Schrödinger's first-passage-time distribution, which assumes Gaussian velocity fluctuations, captures the full distribution of translocation times extremely well and allows extraction of two key dynamical parameters: the drift velocity of DNA through the nanopore and its effective diffusion coefficient inside the pore [42]:

$$F(t) = \frac{L}{\sqrt{4\pi D t^3}} \exp\left(-\frac{(L - vt)^2}{4Dt}\right),$$

where t is the event duration, L is the pore length, D is the diffusion coefficient, and v is the translocation velocity.

The physical origin of these velocity fluctuations is multifaceted. The most intuitive source is longitudinal Brownian motion of the DNA segment confined inside the pore. However, the effective diffusion constants extracted from first-passage-time fits are consistently larger than the bulk diffusion coefficient and scale quadratically with the applied electric field [40]—behavior characteristic of Taylor–Aris dispersion arising from the coupling of axial drift with transverse diffusion in the nonuniform electro-osmotic flow field inside the pore. This mechanism [43] has been identified both in nanopore DNA experiments and in theoretical treatments of electrokinetic transport.

Electrostatic and hydrodynamic effects also modulate the fluctuations. The electrokinetic pulling force depends strongly on the lateral position of the DNA within the pore [43]: a molecule centered in the pore experiences a larger electrophoretic force than one located closer to the pore wall, resulting in event-to-event variability in drift velocity. For long flexible polymers, configurational fluctuations of the cis-side coil introduce additional stochasticity in the entry dynamics, producing contour-length-dependent dispersion [44]. For stiff, rod-like filaments (e.g., filamentous viruses), the spread in velocities increases with applied voltage [45], partly due to buckling of the trans-side segment [46].

For compact or spherical particles, velocity broadening arises predominantly from **geometric confinement** and **particle–wall interactions**, which suppress diffusivity relative to bulk and remain largely voltage-independent [47]. In all these cases, a finite perpendicular spread in the $\langle \Delta I_B \rangle$–$T_R$ plot, reflecting the diverse sources of stochastic velocity modulation inherent to nanoscale transport. Importantly, the velocity perturbations arising from specific short-range interactions can be exploited to identify or classify individual viral particles in nanopore measurements [48].

4 Sampling

Free translocation of biomolecules and nanoparticles through a nanopore is usually a very fast process. The translocation of 16 μm long λ-DNA (48.5 kbp) usually takes [11] $T_R \sim 500$ μs – 1 ms, whereas the free translocation of a short DNA/RNA snippet, viruses or proteins could be anywhere between 1 and 20 μs, sometimes beyond the detection limit [5, 49, 50]. Intentionally or unintentionally induced interaction between the biomolecules and the pore walls could increase the translocation time significantly, so the short molecules could be detected [5, 49, 50], but with significantly reduced diagnostic capabilities due to the stochastic nature of the interactions.

The instruments limit the capacity to detect short signals and their accuracy in multiple ways. The noise floor of the current-to-voltage amplifier is the most critical limiting factor, primarily determined by the feedback circuit in the headstage, intrinsic input noise, and capacitance noise of the nanopore liquid setup. This noise floor sets the minimum resolvable ionic current signal in the selected bandwidth. The shape of the amplifier's transfer function can distort the transient events, while the sampling rate of the digitizer constrains the temporal resolution of the recorded translocation signal. Those effects could prevent the detection of the short signals and small molecules.

Nanoampere-scale ionic currents through nanopores are typically sampled at rate $f_s = 100 - 500$ kHz using standard amplifier systems available on the market[1].This results in the temporal resolution between two subsequent digitized datapoints in the range $t_S = 2 - 10$ μs. Generally, the sample rate should be five to ten times larger than the signal bandwidth in order to be able to adequately approximate the shape of the signal without interpolation.

Figure 6 demonstrates the effect of the sampling frequency on the perceived shape of the signal. The red curve in Fig. 6a–c is the unfiltered simulated current blockade for a spherical particle passing through a nanopore with a high aspect ratio (membrane thickness is much smaller than the pore diameter). The signal duration is $T_R = 20$ μs and the maximum amplitude is $\Delta I_{\max} = 5$ nA. The gradual drop in current blockade signal is due to the gradual increase in blocked cross-section of the pore, and a spherical particle progressively jams into the pore.

At the sampling rate of $f_S = 10$ MHz, the measured digitized signal faithfully represents the current blockade, since the recorded time resolution is much finer than the event duration $T_R = 200 \times t_S$ (Fig. 6a). At the lower sampling rate of $f_S = 500$ kHz, i.e, $T_R = 10 \times t_S$, the intrinsic signal (red line) is still well represented by the digitized data points (Fig. 6b). But at $f_S = 250$ kHz ($T_R = 5 \times t_S$), the recorded trace does not represent the true signal shape (Fig. 6c). In this case, the measured blockade amplitude $\Delta I_{\max}^{(meas)} = 3.5$ nAis severely underestimated compared to the intrinsic value $\Delta I_{\max}^{(real)} = 5$ nA. Therefore, the under-sampled measurements introduce significant errors in particle size estimation based on the magnitude of the current blockade. Furthermore, even a small timing shift of the signal relative to the sampling points can produce noticeably different digitized waveforms with different blockade depths, increasing the spread in the reconstructed particle sizes.

[1]Axon Axopatch 200B (Molecular Devices) or Heka EPC 10 USB (HEKA Electronics)

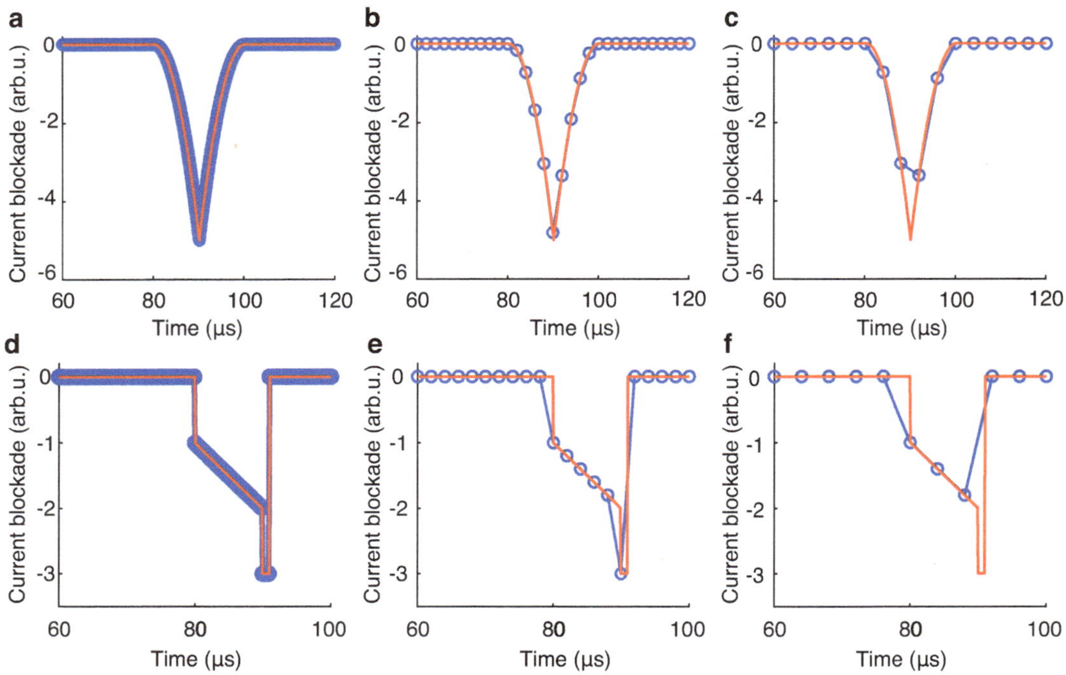

Fig. 6 Effect on the sampling frequency on the signal shape. (**a–c**) Red line represents the simulated shape of the signal of a circular nanoparticle passing through a thin nanopore, with event duration $T_R = 20$ μs. The blue line represents data samples at 0.1 μs, 2 μs, and 4 μs time resolution (10 MHz, 500 kHz, and 250 kHz sampling) (**d–f**). A translocation signal with pronounced fine features (red line) and the sampled data points for different sampling resolutions

In a more striking example, we observe a more complicated current blockade signal with a detailed substructure consisting of a gradual, sloping drop in current, followed by a sharp spiky drop of width ~1 μs (Fig. 6d, e). Such a signal could easily arise if we have a decorated DNA molecule [6, 8], or we aim to detect a small characteristic feature within a protein or a virus. We can see that the spike feature is lost in the digitized data at the sampling frequency $f_S = 500$ kHz and especially at $f_S = 250$ kHz.

One approach to increase the longitudinal resolution of the sensor is to reduce the translocation speed of the molecules in a controllable fashion. Increasing nanopore–wall interactions via nonspecific interactions between the nanopore walls and biomolecules is not the solution, as it will result in random stick-slip movement of the molecules and large spread in translocation times [37]. This is not conducive to a precise analysis of the molecules as it will not preserve the information of the lateral positions on the features along the translocating molecules as a predictor for the nanopore tomography. Increasing the solution viscosity [51] or replacing the salt with 4 M LiCl solution [52] could increase the translocation time tenfold, while retaining the tight *ecd* profile indicative of smooth translocation that preserved analytical capacity of the nanopore translocation.

A more engaging process in precisely controlling the translocation of the molecules involves optical tweezers [52], magnetic tweezers [53], or AFM [54, 55]. It relies on external mechanical force to directly pull the molecules through the nanopores. The translocation speed slows down by more than three orders of magnitude, but these methods need to anchor the molecule on a controllable actuator, which not only makes the operation difficult but also greatly increases the cost of detection. On the one hand, it is also difficult to achieve high-throughput parallel detection, which is contrary to the advantages of nanopore detection technology.

On top of slowing down the translocating molecules, an alternative line of development has focused on optimizing the recording bandwidth and sampling resolution beyond traditional range of the Axopatch (~100 kHz bandwidth sampled at 500 kHz with 16-bit resolution). For example, the VC100 amplifier [54, 55] (Chimera Instruments) has demonstrated operation at 1 MHz bandwidth with sampling at 4 MHz in 14 bit resolution, while the Nanopore Reader [56] (Elements SRL) is operating up to 10 MHz bandwidth with sampling at 40 MHz. Designing such amplifiers is not trivial. It requires parallel improvements in both liquid cell and nanopore sensor design to minimize both amplifier's intrinsic noise and the total input capacitance of the chip and connections, which together define the noise floor and practically limit the useful bandwidth, which add to the complexity of such approach.

5 Filtering

The sampling rate is defined by the analog-to-digital converter, i.e., data acquisition card DAC (see Fig. 1), which converts analog signal from the current amplifier using digital signal with the sampling frequency f_S. However, any current amplifier will have intrinsic filtering of the output signal due to the design of its electronic circuitry, but it often includes a dedicated low-pass filter with a well-defined cutoff frequency f_C to add more control to the signal conditioning. Such filter suppresses all the high-frequency components of the signal above f_C according to its design, i.e., to its transfer function [57]. For current blockade signals, consisting of sharp transient drops in current, the most appropriate is Bessel filter, as it avoids ringing effects. The hardware filter could be replaced by a software filter in the signal postprocessing, but to prevent antialiasing distortions of the signal, the sampling frequency should be at least twice the filtering frequency, $f_S \geq 2f_C$.

Signal or a feature in the signal with characteristic timescale T_R comparable to the filtering timescales $1/f_c$, will be significantly distorted. However, if we assume a particular signal shape, the filter effect could be deconvoluted from the signal with the standard assumption that the filter is lossless, i.e., its transfer function preserves the signal's energy [57].

Figure 7 depicts the effects of different filters on short current blockade signals. We simulated the effect of the analog 4-pole Bessel filter with a range of cutoff frequencies. This configuration corresponds to filters that are employed in the Axon amplifiers, one of the most used filters in NT experiments. Figure 7a depicts the initial signal as a square well with duration of $T_{R0} = 20$ μs and depth of 1 (relative units), as well as the evolutions of the signal after passing through filters with a range of cutoff frequencies. We observe that

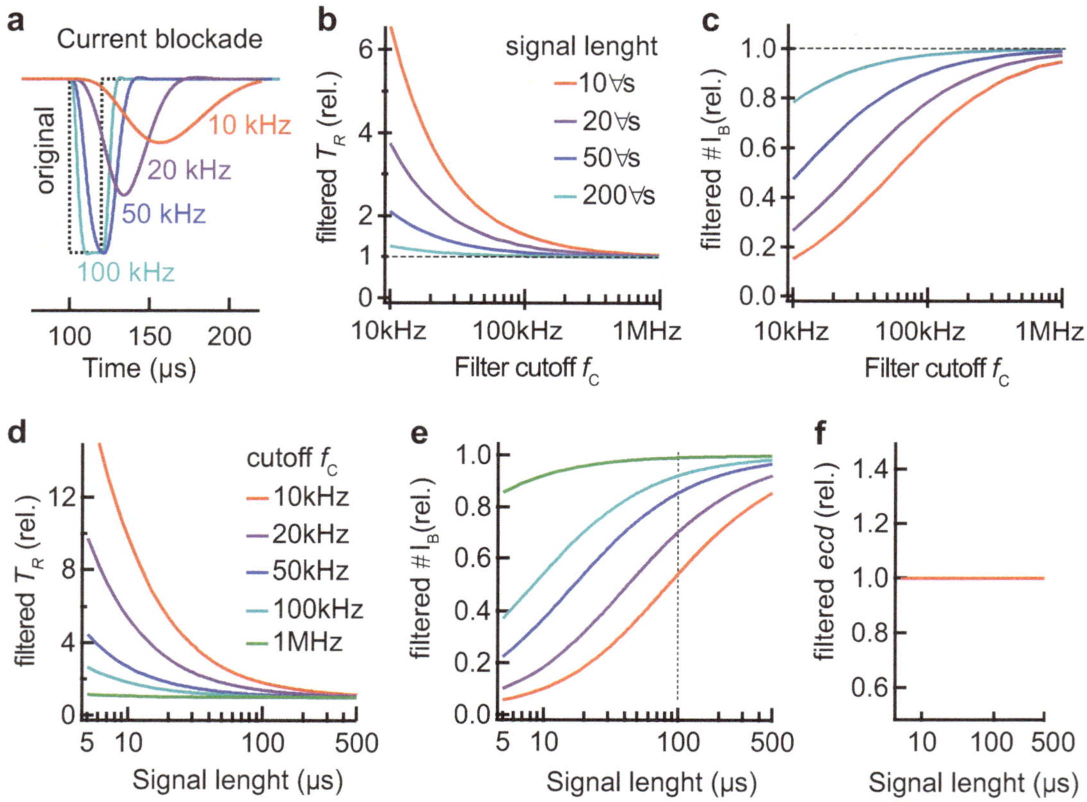

Fig. 7 Effect of filter on the nanopore signal. (**a**) The original signal defined by a square well of duration $T_R = 20$ μm (dotted black line). After passing through a 4-pole Bessel low-pass filter with, the signal will be distorted depending on the cutoff frequency of the filter ($f_C = 10$ kHz, 20 kHz, 50 kHz or 100 kHz). (**b**) Duration of the filtered signal relative to the original signal duration as a function of the filter cutoff frequency, calculated for different signal durations. (**c**) Average current blockade ΔI_B as a function the filter cutoff frequency, calculated for the same signal durations as in (**b**). (**d**) Relative signal length T_R after filtering as function of the original signal length, calculated for different cutoff frequencies. (**e**) Relative average current blockade ΔI_B as function of the original signal length, calculated for the same cutoff frequencies as in (d). (**f**) Relative *ecd* as a function of signal length is preserved, for all the filter frequencies, under the assumption that the filter is lossless

the signal changes shape, and the duration of the filtered signal $T_R^{(f)}$ progressively increases, and the filtered maximum current blockade $\Delta I_{\max}^{(f)}$ and filtered average current blockade $\Delta I_B^{(f)}$ decrease with the drop in f_C. Generally, for $f_C T_{R0} \geq 1$, the maximum current blockade does not change, i.e., $\Delta I_{\max}^{(f)}(f_C) = 1$. The overall evolution of the current blockade length $T_R^{(f)}(f_C)$ and average current blockade $\Delta I_B^{(f)}(f_C)$ for different filters is shown in Fig. 7b, c.

More informative, when evaluating the distributions of current blockade signals, is to look at $T_R(T_{R0})$ and $\Delta I_B(T_{R0})$ as a function of the intrinsic signal duration T_{R0}, for a fixed filter cutoff f_C (Fig. 7e, f). Since the T_R and ΔI_B values are used to deduce lateral and longitudinal dimensions of the translocating molecule, the filter effect could skew the conclusion. This is especially the case for the investigation of viruses, nanoparticles, and proteins, since they are small and translocate very fast. For example, recent report on nanopore characterization of adenoviruses [58] uses measurements setup with $f_C = 10$ kHz, and reports the distributions of current blockade signal with dominating weight in the range of $T_R^{(f)} \approx 20 - 100$ μs Following the simulations in Fig. 7d, e, those values led to overestimation of the signal length by two to six times, and underestimation of the current blockade by 50% – 80%, bringing in question the rest of the conclusions of the paper. In contrast, Taniguchi et al. [1] measured viral particles with $f_S = 1$ MHz and $f_C = 260$ kHz, which fully validates their conclusions.

Important point to notice is that, under a reasonable assumption of the lossless filtration, the electric charge deficit $ecd(f_C)$ is conserved, regardless of the signal length and the filter cutoff (Fig. 7f). Hence, *ecd* is a good measure of the overall geometrical properties of the particles, related to the overall volume, even in the case of over-filtering.

6 Characterization of Axopatch 200B Frequency Response

To understand the signal acquisition limitations of the Axopatch 200B patch-clamp amplifier, we characterized its transverse function. We used a function generator (Rigol DG4162) and a mixed-signal oscilloscope (Rigol MSO2072A), computer-controlled via a custom automation script. The generator was configured to produce a fixed-amplitude sine wave with the frequency logarithmically stepped across 500 points between 1 Hz and 1 MHz. The output voltage was simultaneously applied to the oscilloscope and to a custom-built wide-bandwidth voltage-to-current converter (conversion ratio 1V = 100 nA), whose output was directly fed into the head-stage of the amplifier under test. The oscilloscope measured both the amplitude and the phase shift between the converter input and the amplifier output. The stimulus amplitude was adjusted to yield an amplifier output of approximately 5 V at low frequencies under both tested conditions ($\beta = 1$ and $\beta = 0.1$). The α-gain setting does not influence the frequency response.

At high gain ($\beta = 1$) and the highest frequency setting (100 kHz), the frequency response of the Axopatch amplifier was modeled as a cascade of filters (Fig. 8): a simple RC low-pass filter at ~200 Hz, followed by a reconstruction filter with a low-frequency zero (~200 Hz) and a high-frequency pole (~130 kHz), and finally a second-order Bessel filter centered at ~80 kHz. Transfer function fitting and parameter estimation were performed in MATLAB. The time-domain response of the amplifier was reconstructed in two complementary ways. First, experimental data were interpolated onto the required frequency grid, followed by an inverse Fourier transform, thereby preserving the measured system behavior, including imperfections and noise. Second, the model-based time response was computed analytically by converting the fitted Axopatch transfer function into the time domain in MATLAB, yielding an idealized prediction without measurement noise or unmodeled nonlinearities.

The characterized magnitude response and the corresponding model fit are illustrated in Fig. 9a. From the characterized magnitude response, two baseline reference values of 4.9 and 4.4 nA were identified. Accordingly, the cutoff frequencies determined under the two definitions were 48.91 kHz and 57.57 kHz, respectively, with one of them consistent with simple estimations reported by other groups [59]. Furthermore, the transfer function of the Axopatch 200B was

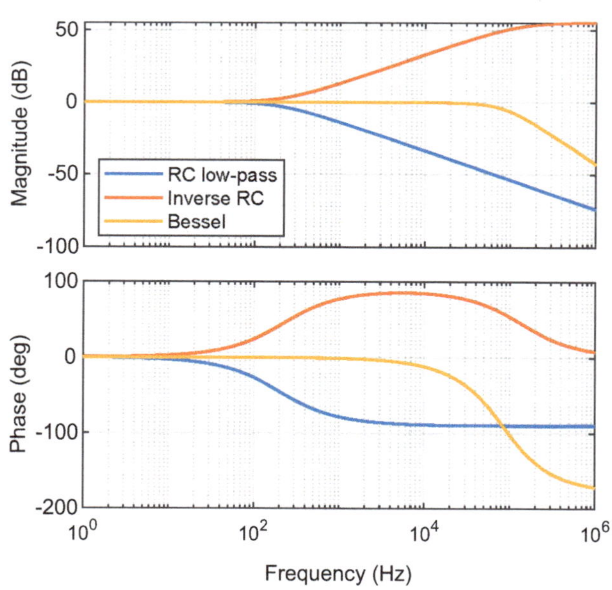

Fig. 8 The theoretical cascade of filters that was used to model the frequency response of the Axopatch 200B amplifier: (blue line) Simple RC low-pass filter (~200 Hz); (red line) Reconstruction filter with a low-frequency zero (~200 Hz) and a high-frequency pole (~130 kHz); (yellow line) Second-order Bessel filter with a center frequency of ~80 kHz

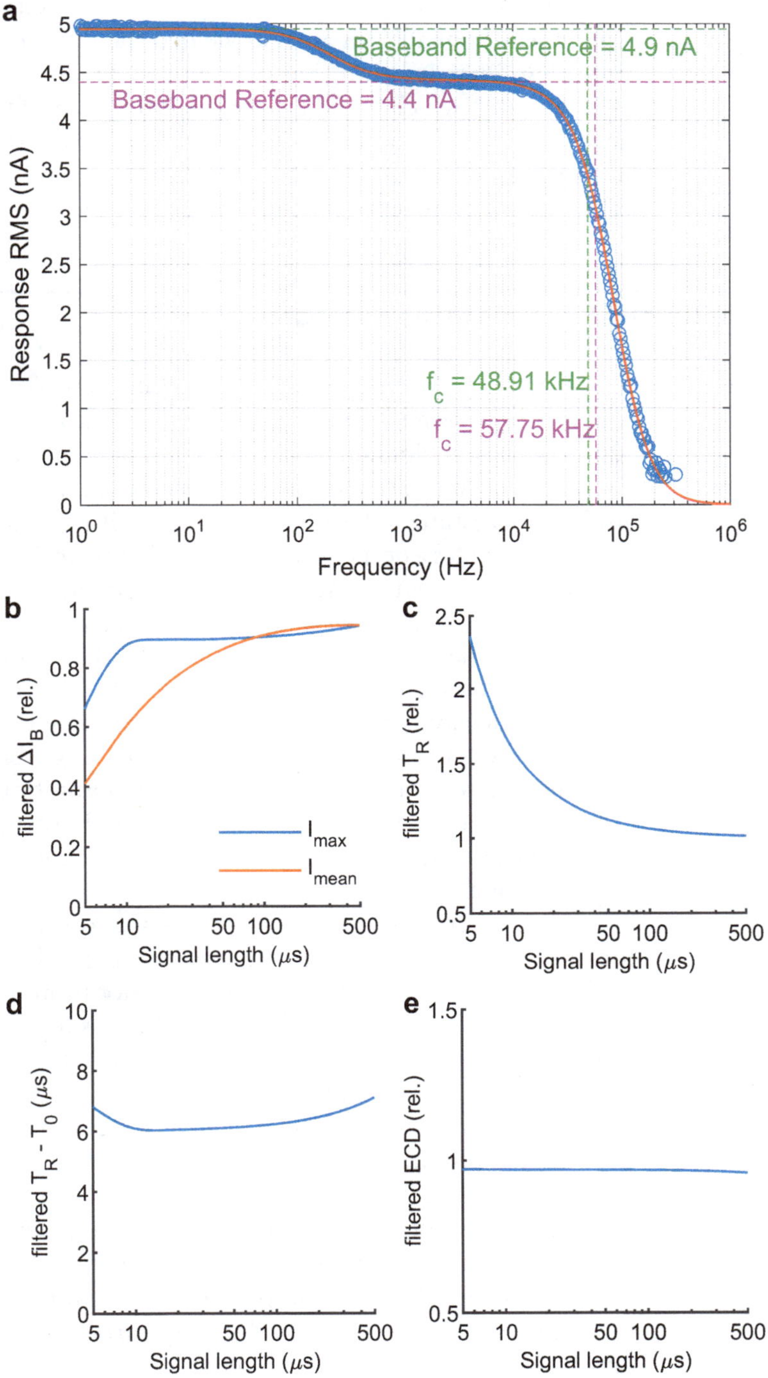

Fig. 9 Frequency response and signal metrics of the AXOPATCH 200B. (**a**) Magnitude response of the AXOPATCH 200B (Serial Number: 821481, with $\beta = 1$ and nominal cutoff frequency $f_c = 100\ \mathrm{kHz}$) and the transfer function fitted using the composite filter model. The fitted transfer function was then applied to a square-wave signal to evaluate changes in (**b**) current amplitude (blue: maximum, red: average), (**c**) duration ratio, (**d**) difference in event duration, and (**e**) ECD ratio before and after filtering

fitted using a composite filter model and subsequently applied to a square-wave signal. This analysis yielded the ratios of signal amplitude (Fig. 9b), duration (Fig. 9c), and ECD (Fig. 9e) before and after filtering, which could enable the estimation of the true signal amplitude and duration from measured data. Notably, the difference in signal duration between pre- and post-filtered waveforms remained relatively stable at approximately 6.5 μs (Fig. 9d).

7 Modeling the Current Blockade in the Nanopore

Understanding the complex physics of analyte translocation through a nanopore is essential for deducing their physical properties from the resulting current blockade signal. While a simple approximation based on the obstruction of the nanopore's cross-sectional area is often sufficient to estimate the analytes' sizes, a full analysis of all physical variables provides a deeper insight into each element's contribution to the observed current blockade. One clear example is the involvement of the electric double layer surrounding the charged surfaces, where localized nonneutral charge distribution could constitute extra ionic current and flow-current coupling. Another example is the motion of the translocating particle that drags along the surrounding fluid. All of these contribute to the two primary forces governing the process: the electrophoretic force and the electroosmotic drag force. To accurately capture the roles of these forces and other parameters, it is beneficial to use a fully coupled physics calculation for a complex geometry like a nanopore.

As for a simple example, we would like to investigate a translocation of an elongated particle or a charged polyelectrolyte and find its pseudo-steady velocity of the translocation. For simplification, a charged polymer can be reduced to a simple charged rigid body interacting with the surrounding electrolyte solution. This allows for the use of multiphysics software like COMSOL Multiphysics to implement the Finite Element Method (FEM) to solve the coupled physics equations, which include: the Low-Reynolds number Navier-Stokes equation for fluid flow, the Poisson equation for electrostatic interactions, and the Nernst–Planck equations for the conservation of cations and anions. The model consists of two chambers of fluid with an impermeable membrane partitioning the fluid chambers and the translocating particle being a long (also impermeable) rigid charged cylinder.

As illustrated in Fig. 10, we created an hourglass-shaped nanopore with a diameter at the bottleneck of 5 nm on a membrane with a thickness of 50 nm in COMSOL. The particles and the membrane are impenetrable for ions and electric fields, assuming no electric current passing through them. The two chambers are drawn with regions far enough from the nanopore to be in equipotential with

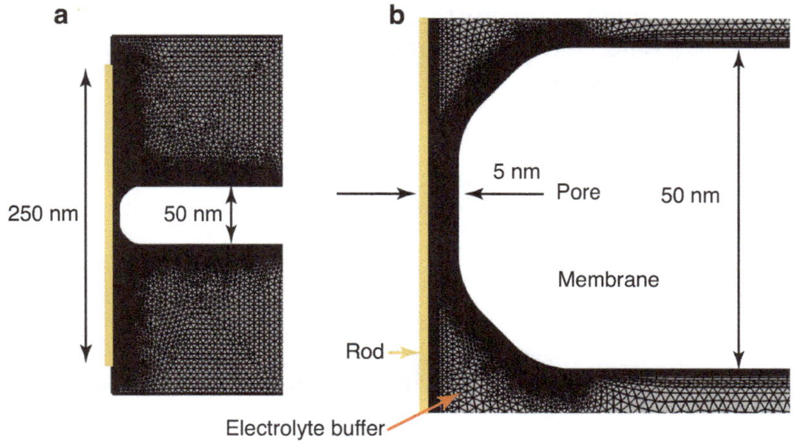

Fig. 10 Schematic depiction of the 2D axisymmetric COMSOL Model: (**a**) for the entirety of the model, and (**b**) a magnified view of the same model around the nanopore

their respective electrodes (constant electric potentials). The long polyelectrolyte is assumed to be vertically stretched during the translocation in which we will verify this assumption through the force distribution. Translocating cylindrical rod, with a length of 250 nm with its diameter being 1 nm, is placed perfectly at the center axially aligned. Such a configuration allows a 2D simplification by axisymmetric geometry. Beyond the advantages of lower computational resources, this simplification can be justified by its minimized electrostatic interaction energy [43]. Both surfaces of the particle and the pore membrane are set to have a constant surface charge boundary of $-0.136\,C/m^2$ and $-0.06\,C/m^2$. They are nonslip surfaces, with the particle being a moving boundary with its velocity directed along the vertical direction. The surrounding electrolyte solution is set to be a 1 M KCl solution.

Within the model, there exist regions that need extra precautions in mesh creations to facilitate high gradients. The presence of electric double layers surrounding the charged surfaces and a highly dense electric field near the nanopore requires a localized dense meshing near the surfaces and the region around the nanopore. Once the model is set and completed, after processes of numerical convergence within COMSOL, a time-independent solution of the physics variable of fluid pressure, velocity, electric potential, and ionic concentrations are simultaneously solved as coupled partial differential equations.

To achieve a steady motion translocation, mechanical equilibrium condition must be achieved—electrophoretic force balanced with electroosmotic drag. An optimization routine can be used to tune the translocating speed with its goal to find the conditions with the total force acting on the particle to be zero. The total force on the translocating particle is calculated by integrating electric stress tensor and fluid viscous stress tensor on the particle's surface.

$$\vec{F} = \oint_{Particle} (\boldsymbol{\sigma} + \boldsymbol{\tau}) \cdot \hat{n} \; dS,$$

where $\boldsymbol{\sigma}$ denotes the electric stress tensor with its element being

$$\sigma_{ij} = \epsilon \left(E_i E_j - \frac{1}{2} \delta_{ij} E^2 \right)$$

and $\boldsymbol{\tau}$ denotes the viscous stress tensor with its element being

$$\tau_{ij} = \left(-\delta_{ij} p + \mu \left(\nabla_i u_j + \nabla_j u_i \right) \right).$$

These field variables require FEM in order to solve including: E_i the electric field components, p the hydraulic pressure and u_i the fluid velocity components; as well as the material's property constant, including ϵ the electric permittivity and μ the fluid viscosity.

By inspecting the stress distribution as shown in Fig. 11a, the extension parts of the rods outside the nanopore contribute negligible driving forces in comparison to the parts in vicinity to the pore's cross-section. This also entails how long polyelectrolytes are stretched when approaching and recoil afterwards, returning to its random coil form. Furthermore, this allows us to reduce the model geometry without the need to include the full extension of polymer toward the exterior bulk, reducing the computation domain.

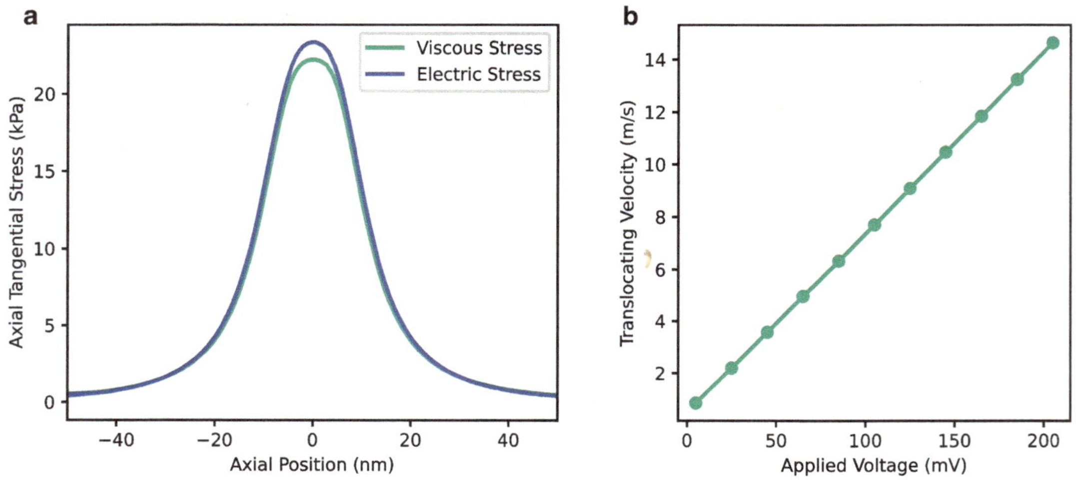

Fig. 11 Stress distribution and voltage-dependent translocation velocity during polymer translocation. (**a**) Plot of the distribution of axial tangential stress along the polymer strands with the most stress part being the region in vicinity to the pore's cross-section. (**b**) Plot depicting the linear proportionality between the translocation velocity and the applied voltage

By varying the voltage of the electric potential differences of the two electrodes and identifying the balanced force conditions, a linear relation for the translocating velocity on voltage dependence can be observed in Fig. 11b, which agrees with most of the models. The calculations of the current can be done by inspecting the current density within the pore's cross-section and integrate it over the nanopore's cross-sectional area. The current density consists of three contributions: convection, diffusion, and ion migration of the fluxes of individual chemical species, for this case, monovalent cation and anion.

$$j = F\sum_i z_i N_i$$

$$= \underbrace{Fu\sum_i z_i C_i}_{\text{Convection}} - \underbrace{F\sum_i D_i z_i \nabla C_i}_{\text{Diffusion}} - \underbrace{\frac{F^2}{RT}\nabla V \sum_i z_i^2 D_i C_i}_{\text{Ion Migration}},$$

where F is Faraday's constant, $\boldsymbol{u}$ is the velocity field, C_i, D_i, and z_i are the concentration, diffusion coefficient, and valence of species i , R is the ideal gas constant, and T is the temperature. A slight modification can be done by rewriting the diffusion current and the ion migration current to be electrochemical diffusion current as:

$$j = \underbrace{Fu\sum_i z_i C_i}_{\text{Convection}} - \underbrace{F\sum_i D_i z_i C_i \nabla \mu_i}_{\text{Electrochemical diffusion}}$$

with μ_i being the electrochemical potential for each ionic species defined as

$$\mu_i = \ln C_i + \frac{z_i F V}{RT}.$$

This helps with the display of the current within the nanopore without issue with local induced concentration gradient within the nanopore. The current distribution across the nannopore cross-section is displayed in Fig. 12a. The changes in the total current during translocation do not purely stem from the reduction in area conductance. A large portion of additional current exists near the charged polyelectrolytes' surface as a result of surface conductivity. This conductivity consisted of both the electrochemical drifting current and a convection current that is generated by the electroosmotic flow and the polyelectrolytes' motion. The magnitude of the speed of the translocation is not significant enough to provide strong drag to affect the convection current. This supports the assumption that many of the existing analytical models simplify the problem by fixing the polymer to be stationary [2]. However, most analytical

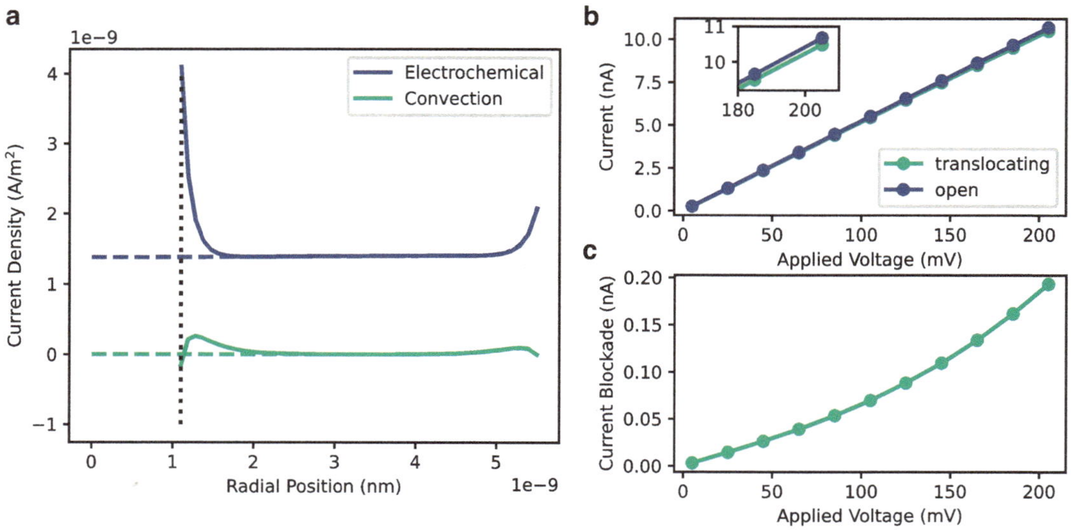

Fig. 12 Current components during polymer translocation. (**a**) The current distribution comparison constituted from electrochemical current (blue) and convection current (green) between during the polymer translocation (bold lines) and opened pore (dashed lines), and the model geometry can be seen in Fig. 10. **(b)** Current as a function of applied voltage for the translocating (green) and open (blue) states; inset shows detail near 200 mV. **(c)** Current blockade (difference between open-pore and translocating current) as a function of applied voltage

predictions of the current blockade did not anticipate nonlinear current rectifications, thus leaving most of current blockade to be a voltage-linear-dependent phenomenon. The current blockade result from the simulation depicts a super-linear trend, which occurs due to the differences in the nonlinear rectifications between the translocating current and the open pore. The rectification is significantly stronger during translocation due to the polyelectrolytes charged surfaces, resulting in higher blockade current at higher voltage. In summary, we can utilize COMSOL multiphysics to model and inspect the voltage nonlinearity of the current blockade. Using the model, we can systematically vary parameters such as polyelectrolyte sizes and surface charges to either predict the resulting current blockade magnitude or work backward (backtrack) to identify the optimal parameters that yield an experimentally observed blockade.

Additionally, it is assumed that the conductivity of the pore and object are assumed to be zero, or that their resistivity is infinite. Hence, no electric field penetrates through either of them, and the electric field lines are distorted as they pass through the pore. The membrane is modeled as insulating, solid-state, and embedded with a prescribed pore, the characteristic lengths of which is the diameter.

The utility of continuum models can be limited since the continuum approximation can break down for certain limiting cases. Given their limitations in sub-nm regimes, these models cannot be used to study phenomena at that scale, such as the effect of hydration radii, hydrogen bonds, etc. However, macroscopic quantities such as the ionic conductance, ion concentrations, or fluid velocities can still be determined. Continuum models have served to understand and interpret nanopore translocation experiments for more than a decade. For well-characterized pores, continuum models can predict the open pore conductance with high levels of accuracy. Furthermore, they have been used widely to complement experiments, understand the effects of wide parameter ranges, which are hard to do experimentally, and infer the molecule sizes based on the reported current blockades.

8 Conclusion

Nanopore signals contain a wealth of information about the molecules that pass through the pore, yet much of this information is not directly accessible without a clear understanding of the physical processes that shape the recorded current traces. The current blockade, its temporal structure, and the duration of translocation events all encode structural and dynamical properties of the analyte. However, these measured features arise from a combination of electrophoretic forces, configurational fluctuations, hydrodynamic drag, and position-dependent electrokinetic interactions that occur within the confined environment of the nanopore.

In addition to these intrinsic molecular effects, the recorded signals are influenced by limitations of the measurement apparatus. Electrical noise, amplifier bandwidth, digitization rate, and the choice of analog or digital filters can distort event shapes, suppress short-lived features, or broaden distributions in ways that mimic or obscure genuine molecular signatures. These instrumental artefacts need to be carefully considered when interpreting nanopore data and comparing measurements across different experimental setups.

Together, these factors highlight that reliable interpretation of nanopore measurements requires accounting for both the molecular dynamics inside the pore and the practical constraints of the measurement system. Understanding how these elements influence signal formation is essential for extracting meaningful structural and dynamical information from nanopore experiments and for improving the consistency and interpretability of nanopore-based analyses.

Acknowledgments

This work was supported by the funding from the Agency for Science, Technology and Research (A*STAR), Singapore, under the Advanced Manufacturing and Engineering (AME) Program (Award A18A9b0060) from Ministry of Health, Singapore

(PREPARE-OC-ETM-Dx-2023-005) and from the Ministry of Education, Singapore (Academic Research Fund Tier 1 FY2024 Grant No. 24-1563-A000).

References

1. Taniguchi M et al (2021) Combining machine learning and nanopore construction creates an artificial intelligence nanopore for coronavirus detection. Nat Commun 12: 3726
2. McMullen AJ, Tang JX, Stein D (2017) Nanopore measurements of filamentous viruses reveal a sub-nanometer-scale stagnant fluid layer. ACS Nano 11:11669–11677
3. Talaga DS, Li J (2009) Single-molecule protein unfolding in solid state nanopores. J Am Chem Soc. 131:9287–9297
4. Oukhaled A et al (2011) Dynamics of completely unfolded and native proteins through solid-state nanopores as a function of electric driving force. ACS Nano 5:3628–3638
5. Yusko EC et al (2011) Controlling protein translocation through nanopores with bio-inspired fluid walls. Nat Nanotechnol 6:253–260
6. Bell NAW, Keyser UF (2015) Specific protein detection using designed DNA carriers and nanopores. J Am Chem Soc 137:2035–2041
7. Bulushev RD et al (2016) Single molecule localization and discrimination of DNA-protein complexes by controlled translocation through nanocapillaries. Nano Lett 16:7882–7890
8. Yang W et al (2018) Detection of CRISPR-dCas9 on DNA with solid-state nanopores. Nano Lett 18:6469–6474
9. Plesa C et al (2016) Direct observation of DNA knots using a solid-state nanopore. Nat Nanotechnol 11:1093–1097
10. Kumar Sharma R, Agrawal I, Dai L, Doyle PS, Garaj S (2019) Complex DNA knots detected with a nanopore sensor. Nat Commun 10: 4473
11. Sharma RK, Agrawal I, Dai L, Doyle P, Garaj S (2021) DNA knot malleability in single-digit nanopores. Nano Lett 21:3772–3779
12. Goyal G, Freedman KJ, Kim MJ (2013) Gold nanoparticle translocation dynamics and electrical detection of single particle diffusion using solid-state nanopores. Anal Chem 85:8180–8187
13. Yazbeck R, Xu Y, Porter T, Duan C (2022) Nanoparticle-blockage-enabled rapid and reversible nanopore gating with tunable memory. Proc Natl Acad Sci U S A 119:e2200845119
14. Kasianowicz JJ, Brandin E, Branton D, Deamer DW (1996) Characterization of individual polynucleotide molecules using a membrane channel. Proc Natl Acad Sci U S A 93:13770–13773
15. Li J et al (2001) Ion-beam sculpting at nanometre length scales. Nature 412:166–169
16. Garaj S et al (2010) Graphene as a subnanometre trans-electrode membrane. Nature 467:190–193
17. Merchant CA et al (2010) DNA translocation through graphene nanopores. Nano Lett 10:2915–2921
18. Schneider GF et al (2010) DNA translocation through graphene nanopores. Nano Lett 10:3163–3167
19. Cherf GM et al (2012) Automated forward and reverse ratcheting of DNA in a nanopore at 5-Å precision. Nat Biotechnol 30:344–348
20. Manrao EA et al (2012) Reading DNA at single-nucleotide resolution with a mutant MspA nanopore and phi29 DNA polymerase. Nat Biotechnol 30:349–353
21. Ivankin A et al (2014) Label-free optical detection of biomolecular translocation through nanopore arrays. ACS Nano 8:10774–10781
22. Huang S, Romero-Ruiz M, Castell OK, Bayley H, Wallace MI (2015) High-throughput optical sensing of nucleic acids in a nanopore array. Nat Nanotechnol 10:986–991
23. Cai S, Sze JYY, Ivanov AP, Edel JB (2019) Small molecule electro-optical binding assay using nanopores. Nat Commun 10: 1797
24. Zrehen A, Huttner D, Meller A (2019) On-chip stretching, sorting, and electro-optical nanopore sensing of ultralong human genomic DNA. ACS Nano 13:14388–14398
25. Fried JP, Wu Y, Tilley RD, Gooding JJ (2022) Optical nanopore sensors for quantitative analysis. Nano Lett 22:869–880
26. Smeets RMM, Keyser UF, Dekker NH, Dekker C (2008) Noise in solid-state nanopores. Proc Natl Acad Sci U S A 105:417–421
27. Hoogerheide DP, Garaj S, Golovchenko JA (2009) Probing surface charge fluctuations with solid-state nanopores. Phys Rev Lett. 102:256804
28. Tabard-Cossa V, Trivedi D, Wiggin M, Jetha NN, Marziali A (2007) Noise analysis and reduction in solid-state nanopores. Nanotechnology 18:305505
29. Beamish E, Kwok H, Tabard-Cossa V, Godin M (2012) Precise control of the size and noise of solid-state nanopores using high electric fields. Nanotechnology 23:405301
30. Garaj S, Liu S, Golovchenko JA, Branton D (2013) Molecule-hugging graphene nanopores. Proc Natl Acad Sci U S A 110:12192–12196
31. Kowalczyk SW, Wells DB, Aksimentiev A, Dekker C (2012) Slowing down DNA

translocation through a nanopore in lithium chloride. Nano Lett 12:1038–1044
32. Ito S, Yamazaki H, Tsukahara M, Esashika K, Saiki T (2016) Salt dependence of DNA translocation dynamics through silicon nanopores detected by ultraviolet excitation. Appl Phys A: Mater Sci Process 122:342
33. Tabard-Cossa V et al (2009) Single-molecule bonds characterized by solid-state nanopore force spectroscopy. ACS Nano 3:3009–3014
34. Nelson EM, Li H, Timp G (2014) Direct, concurrent measurements of the forces and currents affecting DNA in a nanopore with comparable topography. ACS Nano 8:5484–5493
35. Li K et al (2023) Metrology of individual small viruses. Adv Mater Interfaces 10:2300385. https://doi.org/10.1002/admi.202300385
36. Cui S (2010) Current blockade in nanopores in the presence of double-stranded DNA and the microscopic mechanisms. J Phys Chem B 114:2015–2022
37. Wanunu M, Sutin J, McNally B, Chow A, Meller A (2008) DNA translocation governed by interactions with solid-state nanopores. Biophysi J 95:4716–4725. https://doi.org/10.1529/biophysj.108.140475
38. Yusko EC et al (2017) Real-time shape approximation and fingerprinting of single proteins using a nanopore. Nat Nanotechnol 12:360–367
39. Ren Z (2016) Molecular events during translocation and proofreading extracted from 200 static structures of DNA polymerase. Nucleic Acids Res 44:7457–7474
40. Liu Q et al (2012) Voltage-driven translocation of DNA through a high throughput conical solid-state nanopore. PLoS One 7:e46014
41. Fologea D et al (2005) Detecting single stranded DNA with a solid state nanopore. Nano Lett 5:1905–1909
42. Ling DY, Ling XS (2013) On the distribution of DNA translocation times in solid-state nanopores: an analysis using Schrödinger's first-passage-time theory. J Phys Condens Matter 25:375102
43. Lu B, Hoogerheide DP, Zhao Q, Yu D (2012) Effective driving force applied on DNA inside a solid-state nanopore. Phys Rev E Stat Nonlin Soft Matter Phys 86:011921
44. Lu B, Albertorio F, Hoogerheide DP, Golovchenko JA (2011) Origins and consequences of velocity fluctuations during DNA passage through a nanopore. Biophys J 101:70–79
45. McMullen A, de Haan HW, Tang JX, Stein D (2014) Stiff filamentous virus translocations through solid-state nanopores. Nat Commun 5: 4171
46. McMullen A, de Haan HW, Tang JX, Stein D (2018) Buckling causes nonlinear dynamics of filamentous viruses driven through nanopores. Phys Rev Lett 120:078101
47. Larkin J, Henley RY, Muthukumar M, Rosenstein JK, Wanunu M (2014) High-bandwidth protein analysis using solid-state nanopores. Biophys J 106:696–704
48. Arima A et al (2018) Identifying single viruses using biorecognition solid-state nanopores. J Am Chem Soc 140:16834–16841
49. Wanunu M et al (2010) Rapid electronic detection of probe-specific microRNAs using thin nanopore sensors. Nat Nanotechnol 5:807–814
50. Wei R, Gatterdam V, Wieneke R, Tampé R, Rant U (2012) Stochastic sensing of proteins with receptor-modified solid-state nanopores. Nat Nanotechnol 7:257–263
51. Fologea D, Uplinger J, Thomas B, McNabb DS, Li J (2005) Slowing DNA translocation in a solid-state nanopore. Nano Lett 5:1734–1737
52. van den Hout M, Vilfan ID, Hage S, Dekker NH (2010) Direct force measurements on double-stranded RNA in solid-state nanopores. Nano Lett 10:701–707
53. Peng H, Ling XS (2009) Reverse DNA translocation through a solid-state nanopore by magnetic tweezers. Nanotechnology 20:185101
54. Venta K et al (2013) Differentiation of short, single-stranded DNA homopolymers in solid-state nanopores. ACS Nano 7:4629–4636
55. Briggs K et al (2018) DNA translocations through nanopores under nanoscale preconfinement. Nano Lett 18:660–668
56. Chien C-C, Shekar S, Niedzwiecki DJ, Shepard KL, Drndić M (2019) Single-stranded DNA translocation recordings through solid-state nanopores on glass chips at 10 MHz measurement bandwidth. ACS Nano 13:10545–10554
57. Wanhammar L (2009) Analog Filters Using MATLAB. Springer US.
58. Karawdeniya BI et al (2020) Adeno-associated virus characterization for cargo discrimination through nanopore responsiveness. Nanoscale 12:23721–23731
59. Uram JD, Ke K, Mayer M (2008) Noise and bandwidth of current recordings from submicrometer pores and nanopores. ACS Nano 2:857–872

Part III

Droplet Interface Bilayer System

Check for updates

Chapter 16

Protein Transport Studied by a Model Asymmetric Membrane Army Arranged in a Dimple Chip

Jing Huang, Xin Li and Min Chen

Abstract

Reconstituted model membrane systems are powerful platforms to tackle interesting problems existing in membrane biology. One of the barriers to efficient drug delivery, as therapeutics to disease, is the physical membrane barrier of the cell. Small molecules can typically diffuse through the membrane; however, biomolecules such as proteins or nucleic acids cannot passively diffuse the bilayer and thus much research has been geared to engineering protein and/or nucleic acid delivery methods. One delivery method uses cell-penetrating peptides (CPPs). In this chapter, we introduce the model "membrane army" arranged in a dimple chip to study the delivery of β-galactosidase by a CPP known as Pep-1. This method uses droplet interface bilayer technology (DIB). It accelerates the speed to screen through the working conditions in CPP-assisted protein translocations because each chip provides dimples that can accommodate 36 pairs of droplets or 18 model bilayers. We will use one of the successful translocation conditions of β-galactosidase delivery as the example to illustrate how the model "membrane army" is built and utilized.

Keywords Model membrane, Droplet interface bilayers, Screening method, Cell-penetrating peptides, Protein delivery, Translocation, Asymmetric bilayer, Protein transport, Drug delivery

1 Introduction

Biological membranes are barriers of cells and their internal organelles. These barriers define the border of cells and control the cellular material exchange with the surrounding environment. The plasma membrane is usually a liquid crystalline lattice embedded with transmembrane proteins, glycoproteins, carbohydrates, cholesterol, as well as other molecules [1]. The structure of the plasma membrane is roughly identified by the fluid mosaic model, where the embedded components have a certain range of movement within the planar membrane, directing essential cellular activities such as molecular trafficking through the secretory and endocytic pathways [2]. However, the complexity of the biological membrane makes it difficult to use whole cells to study biological membrane processes, and it is quite challenging to isolate meaningful data considering

Fanjun Li (ed.), *Nanopore Technology*, Methods in Molecular Biology, vol. 3072,
https://doi.org/10.1007/978-1-0716-5527-6_16,

the interferences from several pathway networks. Model membrane systems are developed with simpler and more controllable parameters to make the experimental investigation more tractable.

Model membranes have various forms and are classified by the lipid composition and production method. Unilamellar vesicles, however, are widely used as the model membrane in drug delivery research [3, 4]. They are spherical and constituted of lipid bilayers. They can be made either as small unilamellar vesicles (SUVs, 20–100 nm), large unilamellar vesicles (LUVs, 50–200 nm), or giant unilamellar vesicles (GUVs, 1–100 μm) [5]. The high surface/volume ratio of SUVs and LUVs makes them compatible with many assays, such as binding assays and fluorescence spectrometric assays [6, 7]. The diameter of GUVs is above the diffraction limit, and so they are usually used in light microscope research. Another popular membrane system is the supported lipid bilayer (SLB). An example, the black lipid membrane, is the formation of a bilayer across an aperture of a supported Teflon film [8]. Nanodiscs, nanoscale phospholipid bilayers, are another example of a model membrane. They could provide flexibility in lipid bilayer content and stand out as advanced model membranes for membrane protein study [9, 10].

Another model membrane approach, named the droplet interface bilayer (DIB), is based on immersing two individual lipid-monolayer-coated aqueous droplets in an oil phase and then joining the two droplets to create lipid bilayer at the interface [11]. This DIB system was proved to stay stable for weeks. This model membrane system is not only stable in structure but also compatible with functional membrane protein insertion. For example, the pore-forming toxin, alpha hemolysin (αHL), was expressed within the droplet by an in vitro transcription/translation system and then detected to insert into the DIB by electrophysiology technique [12]. The size of the lipid bilayer interface can be precisely controlled by adjusting the distance between the embedded electrodes [13]. In addition, chemical reagents on both sides of the bilayer and the bilayer lipid composition can also be adjusted according to the experimental needs [14]. This DIB system is the prototype for the model “membrane army” discussed in this chapter.

The dimple chip-based DIB takes advantage of the high stability of DIBs and arranges pairs of droplets in the dimple chip filled with hexadecane, allowing this method to screen different conditions simultaneously. In earlier designs of DIB, both droplets are inserted with electrodes for tracing the current signal across the bilayer [15]. However, in this chapter, we are more interested in investigating the conditions that underline how a small 21-amino acid peptide, Pep-1, is able to deliver a ~500 kDa protein directly across the membrane rather than how the membrane is interfered by translocation [16]. We, therefore, removed the electrodes and used the fixed distance

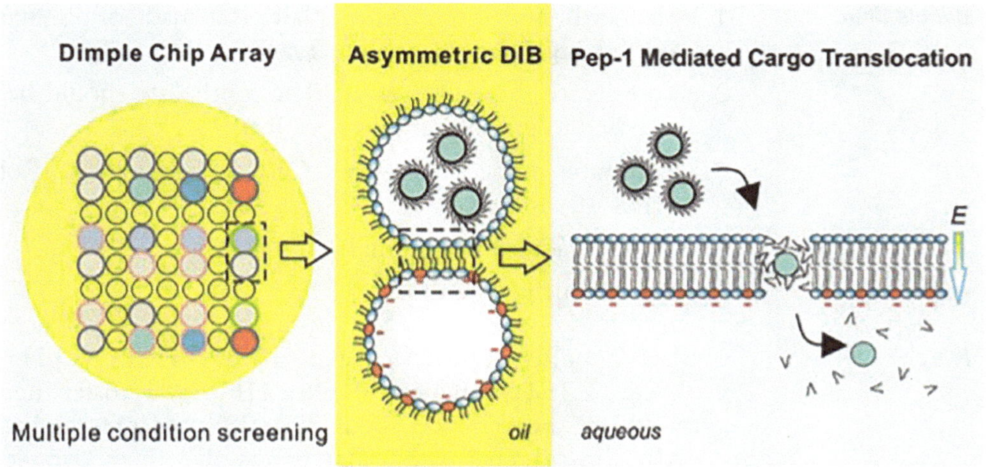

Fig. 1 Schematic illustrating the dimple chip membrane army methodology. Pep-1-mediated translocation of proteins is driven by a potential difference provided by the asymmetric droplet interface bilayer. [Reprinted (adapted) with permission from [17]. Copyright 2017 American Chemical Society]

between the adjacent wells to confine the bilayer size constant from well to well [17].

We can usually screen 36 droplets in six columns in one dimple chip and form 18 DIBs by pushing the adjacent droplets toward each other. In this chapter, we will describe the "membrane army" method applied to the Pep-1-assisted β-galactosidase (β-gal) translocation (Fig. 1). The components in each pair of droplets are different; the droplet with Pep-1–β-gal inside is called the source droplet, and the other droplet is named the capture droplet. The capture droplet contains the β-gal substrate, resorufin-β-D-galactopyranoside (RG), which could be turned into fluorescent resorufin by β-gal, so that the delivered β-gal can be quantitated by the fluorescence intensity turnover rate of the capture droplet. Recently, this direct analysis of transport across the DIB has been applied to further CPP investigations. Gehan et al. have used asymmetric DIBs to study the lipid compositions favoring the direct translocation of penetratin, a cationic CPP [18]. Faugeras et al. showed the capability to discriminate CPPs with tiny structural differences via the distinct permeability of diffusion through the DIB [19].

2 Materials

Prepare all solutions with ultrapure water and analytical grade reagents. All solutions are filtered through a sterile PES (polyethersulfone) membrane with 0.22 μm pore size to remove potential microbial and dust contaminants.

2.1 Dimple Chip

1. Standard 24-well cell culture plate: It is made of polystyrene and has high clarity (*see* **Note 1**).
2. Toothpaste and cotton swab: The toothpaste should be silica-based to polish the wells (*see* **Note 2**).
3. Computer numerical control (CNC) machine (e.g., Roland MDX-40).
4. 0.5 mm radius ball-end mill.
5. Microscope glass slides.

2.2 LUVs

1. LUVs buffer: 150 mM NaCl, 10 mM HEPES, pH 7.4. Weigh 2.192 g NaCl and 0.596 g HEPES. Transfer them to a 400 mL beaker. Add about 200 mL water into the beaker. Mix the reagents with water to dissolve them completely. Weigh 0.400 g NaOH into 10 mL water to make 1 M NaOH solution. Titrate the 1 M NaOH into the buffer to adjust the buffer pH to 7.4 (*see* **Note 3**). Add water to make the volume of solution up to 250 mL. Store the buffer at room temperature for instant usage (*see* **Note 4**).
2. PC solution: Dissolve 25 mg 1,2-diphytanoyl-sn-glycero-3-phosphocholine (DphPC, Avanti Polar lipids) powder in 2.5 mL pentane to make 10 mg/mL DphPC (*see* **Note 5**). Store in a vial at –20°C.
3. PG solution: Dissolve 25 mg 1,2-diphytanoyl-sn-glycero-3-phosphocholine (DphPG, Avanti Polar lipids) powder in 2.5 mL pentane to make 10 mg/mL DphPG (*see* **Note 5**). Another negatively charged lipid, PS (1,2-diphytanoyl-sn-gly-cero-3-phospho-L-serine), can be substituted here for PG.
4. Containers: 4 Dram glass vials with black caps, 1.5 mL Eppendorf tubes.
5. Speedvac vacuum concentrator.
6. Vortex mixer.
7. Water bath (37°C).
8. Liquid nitrogen.
9. Avanti 100 nm pore-size polycarbonate extrusion membranes.
10. Avanti Mini extruder set, which includes a mini extruder, heating block, two O-Rings, two gas-tight syringes, and 100 filter supports.
11. Cleaning reagents: Water, 70% ethanol, any commercial detergent.

2.3 Translocation

1. Pep-1: An amphipathic 21 amino acid peptide (Ac-KETWWETWWTEWSQPKKKRKV-Cya) where Ac is an acetyl moiety and Cya is a cysteamide. Add water to dissolve Pep-1

powder to make the final concentration 680 μM according to the manufacturer's protocol. Aliquot the stock Pep-1 solution into 20 μL in each Eppendorf tube. This process was operated on ice to conserve the Pep-1 structure/activity. Then flash freeze the aliquots in liquid nitrogen and store at −20°C.

2. β-gal: Add water to β-gal to make 20 μM β-gal stock. Aliquot the stock β-gal solution into 20 μL in each Eppendorf tube. This process is operated on ice to conserve the protein structure/activity. Then flash freeze all the aliquots in liquid nitrogen and store at −20°C.
3. RG: Add water to RG and make 50 mM RG stock solution. Keep RG solution in a brown vial and wrap it with aluminum foil to avoid exposure to light. Store at −20°C.
4. Fluorescence Microscope (e.g., Nikon Eclipse T3i microscope).

3 Methods

All procedures are operated at room temperature unless otherwise specified.

3.1 Fabrication of the Dimple Chip

1. Tape the individual well of the plastic plate to the workpiece table of the computer CNC machine.
2. Install a 0.5 mm radius ball-end mill into the spindle head.
3. Create the dimple chips by the programmed design. Each dimple chip contains 12 × 12 array of dimples. The dimple arrays are 0.7 mm apart from the center and the depth of each dimple is programmed to be approximately 250 μm.
4. Polish the dimple chip with silica-based toothpaste and cotton swab. This will minimize the defects created by the drill machine and will also remove the residual plastic debris. A smooth dimple surface enhances the stability of the DIB system. Figure 2 illustrates a fully fabricated dimple chip.
5. Wash the dimple chip with ample detergent and water until it is clean.

3.2 Large Unilamellar Vesicles of 100 nm in Size

1. Prepare DphPC lipids: Aliquot 169 μL of 10 mg/mL DphPC pentane solution as the lipids for the neutral leaflet of the model membrane. Add the solution to a 4 Dram glass vial.
2. Prepare DphPC/DphPG lipid mixture: Mix 118 μL of 10 mg/mL DphPC pentane solution and 51 μL of 10 mg/mL DphPG pentane solution. This lipid mixture is the composition of the negatively charged leaflet of the model membrane. Add the mixture in a 4 Dram glass vial.

Fig. 2 A CNC-generated dimple chip. A single 24-well with a 6 × 6 array of dimples etched onto the bottom. A glass slide attached to the bottom of the dimple chip for mounting to a microscope. For size comparison, a US one-cent coin is used. Each pair of dimples is placed one well away from each other

3. Place the above two lipid solutions in SpeedVac Concentrator with the vials open with no cap. Desiccate the lipids for 1 h until no pentane remains (*see* **Note 6**).
4. Add 1 mL LUVs buffer into the vials with the dried lipids. Resuspend the lipids in LUVs buffer thoroughly. The neutral lipid suspension contains 2 mM DphPC while the negatively charged lipid suspension mixture contains 1.4 mM DphPC and 0.6 mM DphPG. The suspensions appear translucent or milky.
5. Mix the suspensions by vortex mixer for 30 min, mixing every 3–5 min to fully hydrate the lipids. As the hydration proceeds, the liquid becomes less milky. Transfer the suspensions into 1.5 mL Eppendorf tubes.
6. Prepare a liquid nitrogen bath and a 37°C water bath. Place the suspensions in liquid nitrogen for 60 s or until frozen. Remove from liquid nitrogen and place in a 37°C water bath until completely thawed. Vortex suspension to mix. Repeat the alternating process between the liquid nitrogen bath and the 37°C bath for five rounds of freeze–thaw cycles in total. These freeze–thaw cycles help to hydrate the lipids into large multilamellar vesicles.
7. Fully assemble the mini-extruder set with a 100 nm pore-sized membrane (*see* **Note 7**).
8. Equilibrate the mini-extruder set: Fill the LUVs buffer into one of the gas-tight syringes and make sure the syringe plunger on the other end is set to zero. Push the plunger of the syringe filled with LUVs buffer inward so that the buffer automatically fills into the empty syringe through the membrane. Repeat the passing of LUVs buffer from one syringe to

the other for a total of two passes. Remove the syringe filled with LUVs buffer from the mini-extruder. Discard the buffer from the syringe (*see* **Note 8**).

9. Repeat Step 8 two more times.
10. Fill the hydrated DphPC lipid solution into the same syringe used to wash the mini-extruder with the LUVs buffer.
11. Gently push the plunger of the filled syringe until the suspension is transferred to the alternate syringe on the other end completely.
12. Gently push the plunger of the alternate syringe to transfer the suspension back to the original syringe.
13. Repeat Steps 11 and 12 until the suspension is clear. The clarity of the solution indicates the homogeneity of the vesicles; where the solution is clear, it contains a highly homogeneous mix of vesicles. In general, 21 passes through the membrane of the mini-extruder are recommended for a 2 mM lipid suspension. A lipid suspension with a higher concentration will probably need extra cycles of back-and-forth extrusion to reach an ideal level of homogeneity.
14. After 21 passes, the extruded lipid solution is filled in the alternate syringe. This is to avoid any possible contamination of vesicle aggregates.
15. Remove the syringe from the extrusion system and transfer the vesicles into a clean 1.5 mL Eppendorf tube. Store at 4°C. Vesicles are stable at 4°C for up to 2 weeks.
16. Clean the mini-extrusion set thoroughly in alternating water and ethanol washes. Disassemble the mini-extruder and the syringes. Rinse each piece with water and then with 70% ethanol twice. After the second 70% ethanol rinse, submerge all the pieces in a 95% ethanol bath and let sit for 15 min. Rinse the pieces with water and 70% ethanol again. Let the final rinse be with 95% ethanol. Allow the pieces to air-dry for 15–30 min.
17. Once dried, reassemble the mini-extruder to prepare the DphPC/DphPG vesicles. Repeat Steps 7–16. After cleaning and drying the mini-extruder pieces and syringes, store the pieces for future use.

3.3 Preparation of the Dimple Chip in an Oil Bath Over the Microscope

1. Fill the dimple chip with 400 μL hexadecane to create the oil bath for supporting the source and capturing droplets.
2. Take a clean microscope glass slide and position it right over the stage center of the microscope.
3. Drop about 30 μL hexadecane on the center of the glass slide and then put the oil bath on top of the wet surface. Make sure

the glass slide and the dimple chip are sealed well by the oil and no air bubbles in between.

3.4 Preparation of Source Droplets

1. Thaw an aliquot of the stock Pep-1 and β-gal on ice.
2. Dilute Pep-1 to 40 μM by adding 1 μL of 680 μM Pep-1 stock to 16 μL water.
3. Dilute β-gal to 500 nM by adding 1 μL of 20 μM β-gal stock to 39 μL water.
4. Incubate a mixture containing 20 μM Pep-1 with 250 nM β-gal (80:1 mole ratio) for 30 min at room temperature (25°C) to form the complexes.
5. Dilute 2 μL complexes with 18 μL of PC solution by pipetting the solution up and down to form complexes of 2 μM Pep-1 with 25 nM β-gal.
6. Pipette 0.3 μL of the Pep-1–β-gal complexes in PC solution in the hexadecane reservoir to create a droplet.
7. Allow the droplet to sit in the oil reservoir for 30 min for monolayer formation around the droplet. Alternatively, use a 20 μL pipettor to pipette the droplet back and forth in the oil reservoir to accelerate the monolayer formed around the droplet (*see* **Note 9**).
8. Transfer the relaxed source droplet to the oil bath (*see* **Note 10**).
9. Repeat Steps 4–8 to create more source droplets.

3.5 Preparation of Capture Droplets

1. Dilute the stock 50 mM RG with water to 500 μM. For example, mix 1 μL 50 mM RG stock with 99 μL water.
2. Resuspend 2 μL diluted RG with 18 μL PC/PG solution.
3. Pipette 0.3 μL of the RG/PC/PG solution in the hexadecane reservoir and pipette the droplet back and forth to create a complete PC/PG monolayer as described in Step 7 in Sect. 3.4.
4. Transfer the relaxed capture droplet to the oil bath with one empty dimple well away from the source droplet.
5. Repeat Steps 3 and 4 to create more capture droplets.

3.6 Translocation in Dimple Chip

1. Now the source and the capture droplets are placed parallel on the dimple chip with one empty dimple in between (Fig. 3a). Push the source droplet on this empty dimple with a clean pipette tip to make it contact with the corresponding capture droplet and form a lipid bilayer (Fig. 3b) (*see* **Note 11**).
2. Once the bilayer is formed, the size of the meniscus will stop growing (Fig. 3c). The asymmetric bilayer will drive Pep-1 to carry β-gal across the bilayer (*see* **Note 12**).

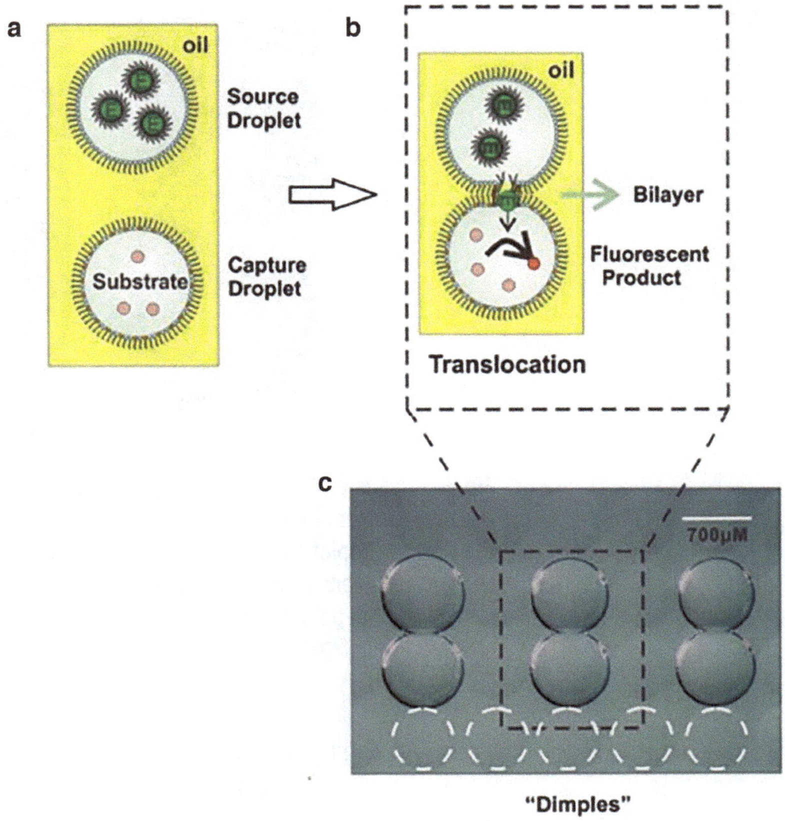

Fig. 3 Setup of the droplet interface bilayers in a dimple chip. (**a**) A source droplet containing peptide-enzyme complexes and a capture droplet containing substrate were prepared by mixing each solution with vesicles separately. (**b**) The droplets come into contact in a dimple chip under hexadecane and form a fixed bilayer based on the parameters of the dimple. (**c**) Multiple droplet interface bilayer pairs were arranged on a dimple chip for parallel translocation experiments. Lipid vesicles within each droplet supplied lipids to form monolayers that annealed to create a bilayer membrane at the contacting region. The white dashed lines highlight one row of dimples on the dimple chip. (Adapted with permission from [17]. Copyright 2017 American Chemical Society)

3. Incubate the droplet bilayers for at least 2 h. Take a dark-field image, which shows the capture droplet is bright, indicating that β-gal has translocated successfully into the capture droplet (Fig. 4) (*see* **Note 13**).
4. Use ethanol and water to rinse the chip alternatively and then allow it to air dry. Place it in a drawer to avoid dust contamination.

3.7 Calibration and Turnover Rate Analysis of Pep-1-Assisted β-Gal Translocation

1. As a control, place a capture droplet and a source droplet into the dimple chip with one empty dimple away from each other but do not make contact between the two droplets. Measure the fluorescence increase of the capture droplet over a 2 h

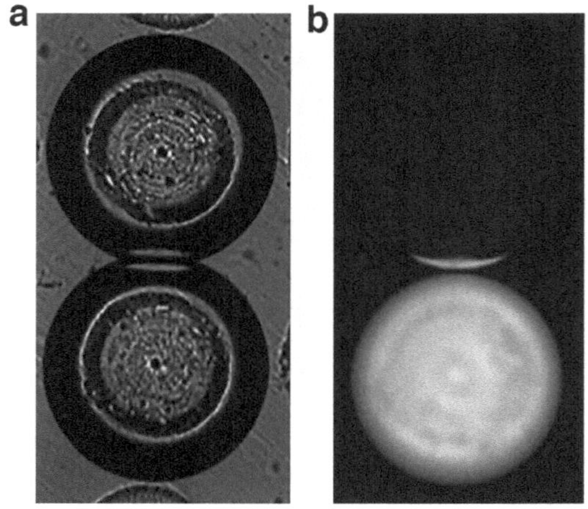

Fig. 4 Images of the DIBs taken by the microscope. (**a**) Bright field image shows that two droplets form a bilayer in the contacting region. (**b**) Dark field image of the source droplet (dark upper) containing Pep-1–β-gal complex and the capture droplet (white lower) containing the β-gal substrate, RG. After 2 h incubation, translocated β-gal hydrolyzes the substrate in the capture droplet into the fluorescent product, resorufin, which is seen as a bright white droplet

period. This baseline fluorescence represents the spontaneous breakdown of RG.

2. Calibration curve for amount of β-gal translocated: To calibrate the β-gal turnover rate, mix various concentrations of β-gal ranging from 0.1 to 2 pM, for example, with a constant concentration of 50 μM RG and 1.8 mM DphPC vesicles in LUVs buffer (*see* **Note 14**).
3. Submerge 0.3 μL of the β-gal/RG/vesicle mixture in hexadecane and image the fluorescence intensity of the droplet over a period of 2 h using a Nikon Eclipse T3i microscope with an excitation wavelength of 561 nm. The turnover rate of β-gal catalyzed release of resorufin will typically remain constant after 2 h has elapsed as the fluorescence intensity increases steadily over time.
4. Plot the fluorescence intensity of the droplets every 30 min to generate Fig. 5a. Determine the slope from Fig. 5a for the β-gal turnover rate. Then plot the slope (fluorescence intensity/time) for each β-gal concentration to generate Fig. 5b.
5. Use Fig. 5b as a standard curve for your translocation experiments and derive the resorufin turnover rate of capture droplet to quantify the translocated β-gal enzyme. Remeasure the standard curve periodically to verify the activity of the β-gal enzyme.

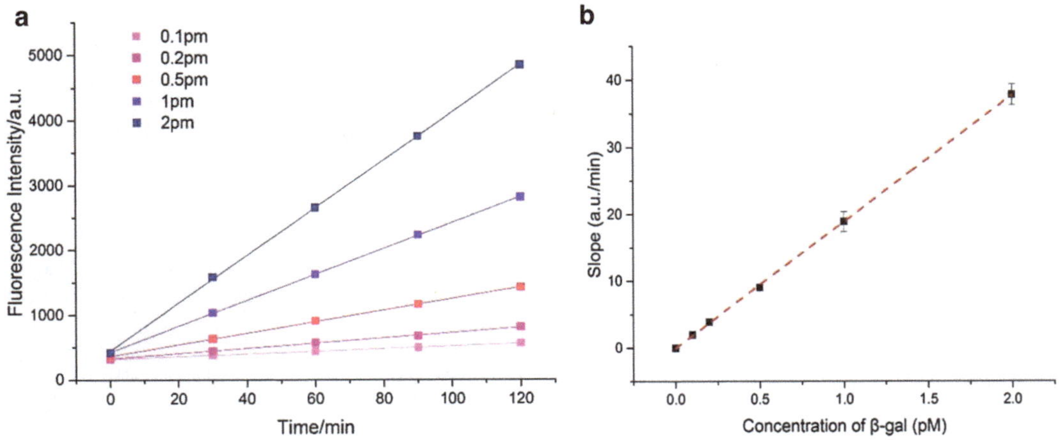

Fig. 5 Calibration curve of β-galactosidase activity. (**a**) A series of β-gal of different concentrations was mixed with 50 μM RG substrate and LUVs buffer. The fluorescence signal of the enzymatic reaction product (resorufin) was recorded over 2 h. (**b**) The β-gal turnover rates derived from the slope of fluorescence intensity vs. time in (**a**) were plotted against the β-gal concentration. The data were fitted with a linear regression model to derive the β-gal calibration curve. (Adapted with permission from [17]. Copyright 2017 American Chemical Society)

4 Notes

1. The fluorescence microscopy-based assay requires the chip to be of high transparency so that when the light is transmitted through the chip, the fluorescence intensity will not be comprised to a large degree.
2. The machine will probably create defects on the dimple chip; therefore, a silica-based toothpaste provides a gentle abrasion to smoothen any rough residues.
3. If the pH is close to 7.4 yet still a little bit lower than 7.4, further dilution of 1 M NaOH is recommended to avoid the pH being overbasic.
4. No sodium azide is added to this buffer. For long-term usage, the buffer is recommended to be stored at 4°C.
5. Pentane is volatile with a low boiling point. As the liquids exist in equilibrium between the gas and liquid states, pre-wetting the pipette tip could saturate the gas state of pentane inside the tip and thus give an accurate volume of pentane when pipetting.
6. An alternate way to dry the lipids is to load the lipid solution at the bottom of a test tube and dry the lipids under a gentle stream of pressurized air. This is an easier way and takes just a few minutes to completely dry lipids. The challenging part is to control the speed of the air and the position of the test

tube well so that the liquid does not splash out of the mouth of the test tube. It is highly recommended to wear goggles to avoid any spilled liquid.

7. The mini-extruder set includes an extruder stand, which could fix the extruder apparatus still and equilibrate the temperature with the heating block. However, the transition temperatures from gel to liquid for either DphPC or DphPG are way beyond the working temperature range [20]. We did not use the extruder stand and the heating block in the conditions mentioned here.
8. This step is not only to rinse the extrusion system with the LUVs buffer but also to check the potential extruder leakage.
9. The droplet might split into tiny droplets while moving in and out of the pipette tip; therefore, a 20–200 μL pipette is recommended for this step. Gentle pipetting skills are highly recommended.
10. Before transferring the source droplet into the oil bath, the droplet needs to be rinsed in the hexadecane to avoid small buds attached to the source droplet being transferred to the oil bath. Because these small buds might swim around in the oil bath and bring contamination to the later translocation experiments.
11. It is important to maintain the same ionic strength of both droplets to prevent the creation of a chemical potential across the bilayer. Using the buffer and reagent conditions described, the ionic strength is kept the same at 0.077 M.
12. Droplets need to be relaxed first, as indicated by the complete formationof a lipid monolayer; otherwise, there is no way to form a bilayer. Unrelaxed droplets will immediately fuse into one once they meet each other. You can prepare the droplet in a hexadecane containing lipids of the desired composition to accelerate the monolayer formation.
13. If no translocation is observed or you want to study conditions different from those mentioned in this chapter, you can alter the bilayer conditions, such as lipid concentration, lipid compositions, and the oil of the capture or the source droplet. The permeability of passive transport across the DIB was suggested to depend on membrane factors, such as phospholipid acyl chains and head groups, together with the nature of oil surrounding the droplets [21]. Also, ensure whether the activity of the β-gal enzyme is retained.
14. Prepare the droplets of different concentrations in parallel, adding each reagent to the tube one at a time, or use a multichannel pipettor. Add the RG last and image the droplets within 60 s of mixing all the reagents.

Acknowledgments

This work was supported by the US National Science Foundation under Grant No. 1253565.

References

1. Benga G, Holmes RP (1984) Interactions between components in biological membranes and their implications for membrane function. Prog Biophys Mol Biol 43:195–257
2. Bloch KE (1983) Sterol, structure and membrane function. Crit Rev Biochem 14:47–92
3. Noyhouzer T, L'Homme C, Beaulieu I, Mazurkiewicz S, Kuss S, Kraatz H-B, Canesi S, Mauzeroll J (2016) Ferrocene-modified phospholipid: an innovative precursor for redox-triggered drug delivery vesicles selective to cancer cells. Langmuir 32:4169–4178
4. Miwa A, Kamiya K (2022) Control of enzyme reaction initiation inside giant unilamellar vesicles by the cell-penetrating peptide-mediated translocation of cargo proteins. ACS Synth Biol 11:3836–3846
5. Keller H, Worch R, Schwille P (2013) Model Membrane Systems. In: Williams MA, Daviter T (eds) Protein-Ligand Interactions. Methods in Molecular Biology, vol 1008. Humana Press, Totowa, NJ. Print ISBN: 978-1-62703-397-8; Online ISBN: 978-1-62703-398-5. https://doi.org/10.1007/978-1-62703-398-5_15
6. Saksena S, Wahlman J, Teis D, Johnson AE, Emr SD (2009) Functional reconstitution of ESCRT-III assembly and disassembly. Cell 136:97–109
7. Struck DK, Hoekstra D, Pagano RE (1981) Use of resonance energy transfer to monitor membrane fusion. Biochemistry 20:4093–4099
8. Mueller P, Rudin DO, Tien HT, Wescott WC (1962) Reconstitution of excitable cell membrane structure in vitro. Circulation 26:1167–1171
9. Yeh V, Lee TY, Chen CW, Kuo PC, Shiue J, Chu LK, Yu TY (2018) Highly efficient transfer of 7TM membrane protein from native membrane to covalently circularized nanodisc. Sci Rep 8:13501
10. Denisov IG, Sligar SG (2024) Nanodiscs for the study of membrane proteins. Curr Opin Struct Biol 87:102844
11. Bayley H, Cronin B, Heron A, Holden MA, Hwang WL, Syeda R, Thompson J, Wallace M (2008) Droplet interface bilayers. Mol BioSyst 4:1191–1120
12. Syeda R, Holden MA, Hwang WL, Bayley H (2008) Screening blockers against a potassium channel with a droplet interface bilayer array. J Am Chem Soc 130:15543–15548
13. Holden MA, Needham D, Bayley H (2007) Functional bionetworks from nanoliter water droplets. J Am Chem Soc 129:8650–8655
14. Hwang WL, Chen M, Cronin B, Holden MA, Bayley H (2008) Asymmetric droplet interface bilayers. J Am Chem Soc 130:5878–5879
15. Huang J, Lein M, Gunderson C, Holden MA (2011) Direct quantitation of peptide-mediated protein transport across a droplet–interface bilayer. J Am Chem Soc 133:15818–15821
16. Morris MC, Depollier J, Mery J, Heitz F, Divita G (2001) A peptide carrier for the delivery of biologically active proteins into mammalian cells. Nat Biotechnol 19:1173–1176
17. Li X, Huang J, Holden MA, Chen M (2017) Peptide-mediated membrane transport of macromolecular cargo driven by membrane asymmetry. Anal Chem 89:12369–12374
18. Gehan P, Kulifaj S, Soule P, Bodin JB, Amoura M, Walrant A, Sagan S, Thiam AR, Ngo K, Vivier V, Cribier S, Rodriguez N (2020) Penetratin translocation mechanism through asymmetric droplet interface bilayers. Biochim Biophys Acta Biomembr 1862:183415
19. Faugeras V, Duclos O, Bazile D, Thiam AR (2022) Impact of cyclization and methylation on peptide penetration through droplet interface bilayers. Langmuir 38:5682–5691
20. Hung WC, Chen FY, Huang HW (2000) Order–disorder transition in bilayers of diphytanoyl phosphatidylcholine. Biochim Biophys Acta Biomembr 1467:198–206
21. Faugeras V, Duclos O, Bazile D, Thiam AR (2020) Membrane determinants for the passive translocation of analytes through droplet interface bilayers. Soft Matter 16:5970–5980

Check for updates

Chapter 17

Droplet Contact Method (DCM) for Stable and Reproducible Lipid Bilayer Formation and its Application to Nanopore Measurements

Nanami Takeuchi, Zugui Peng, Shoko Fujita, Mana Sato, Izumi Shibayama and Ryuji Kawano

Abstract

Droplet contact bilayers provide a versatile platform for reconstituting and measuring the activity of membrane proteins at the single-molecule level. This method enables stable lipid bilayer formation between aqueous droplets and allows the insertion of protein nanopores for electrophysiological analysis. Here, we describe the fabrication of a poly(methyl methacrylate)-based microdevice with a parylene membrane, the formation of bilayers by the droplet contact method, and the reconstitution of both protein nanopores and *de novo* designed peptide nanopores. We further outline procedures for detecting peptide fragments and performing nanopore-based sequencing using this system. This approach offers reproducible bilayer assembly and quantitative analysis of ion channel activity, supporting the development of nanopore technologies for diverse molecular sensing and sequencing applications.

Keywords Planar bilayer, Lipid bilayer, Lipid membrane, Bilayer formation, Nanopore sensing, Protein nanopores, Microfluidics, Electrophysiology

1 Introduction

Artificial cell membranes serve as controllable interfaces, enabling researchers to isolate and study membrane functions in a simplified, synthetic environment, which is crucial for understanding biological processes and developing new technologies such as targeted drug delivery and next-generation drug screening [1]. Planar lipid bilayers offer advantages over traditional patch clamping, such as the ability to study purified ion channels in a precisely controlled lipid environment and to access channels from intracellular organelles that are difficult to study in living cells [1, 2]. Classical planar bilayer techniques, such as the monolayer adhesion method developed by Montal and Müller, as well as the painting method, require operator skill. The resulting membranes are often fragile when subjected

Fanjun Li (ed.), *Nanopore Technology*, Methods in Molecular Biology, vol. 3072,
https://doi.org/10.1007/978-1-0716-5527-6_17,

to mechanical vibration or voltage switching [3, 4]. The droplet contact method (DCM) overcomes these limitations by forming a bilayer simply by touching two lipid-coated aqueous droplets in an oil phase, so that the preformed monolayers adhere to create a stable membrane, called the droplet interface bilayer(DIB) (Fig. 1a) [5–7]. First reported in 2006 by S. Takeuchi and later adopted by other groups, this approach can be implemented in microfabricated chambers that guide droplet placement (Fig. 1b, c) [5, 8–13]. With microfabrication and microfluidics, the DCM has been implemented in chip-based chambers that guide droplet placement and confine the contact area. Confining the bilayer improves mechanical stability and supports extended recordings [8, 14, 15]. Electrodes integrated at the bottom of the wells enable voltage clamp single-channel measurements (Fig. 1d) and basic membrane characterization [8, 16]. As a result, DIBs provide a reproducible and stable, yet easy-to-use platform for studies of ion channels and other membrane proteins, with potential in drug discovery and biomimetic sensing [7, 14, 17]. In this chapter, we focus on bilayer formation by the DCM and summarize procedures for protein nanopore measurements [18–21], peptide nanopore measurements [22–24], and peptide detection [20, 25, 26] and sequencing.

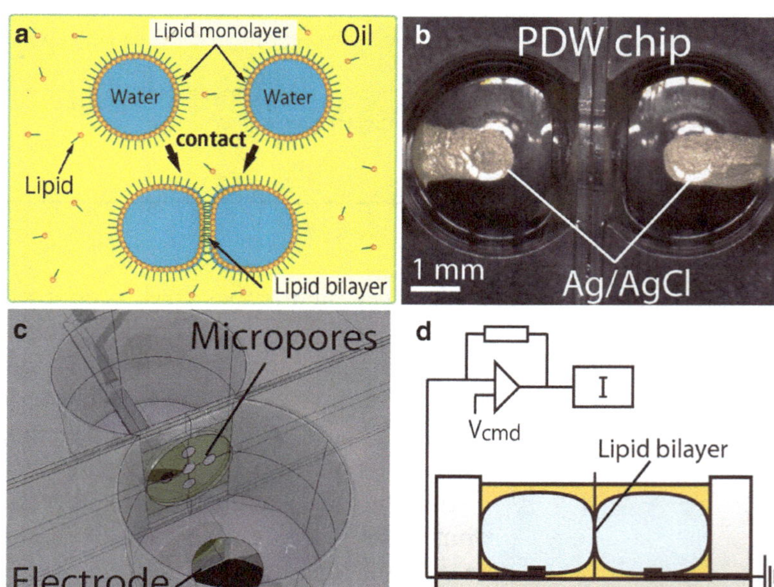

Fig. 1 Ion channel-recording system using the droplet contact method(DCM). (**a**) Formation of a bilayer lipid membrane by the DCM. Two lipid monolayers are brought into contact with one another, and spontaneously form a bilayer. (**b**) The parylene double-well (PDW) chip for the simple preparation of lipid bilayers. (**c**) Schematic illustration of the PDW chip. (**d**) Schematic diagram of the PDW chip connected to a patch-clamp amplifier [8]

2 Materials

Prepare all solutions using ultrapure water prepared by purifying deionized water to attain a resistivity of 18 MΩ·cm at 25°C.

2.1 Droplet Contact Method

1. Microdevice: A 6.0 mm thick, 10 × 10 cm poly(methyl methacrylate) (PMMA) plate, a 0.2 mm-thick, 10 × 10 cm PMMA sheet, a 5 μm-thick parylene-C film, silver wires (0.25 mm diameter, 99.99%) for electrodes, silver/silver chloride ink for preparing reference electrodes, electrical sockets/connectors, and adhesives suitable for acrylic resin and general laboratory assembly.
2. Lipid solution: 10 mg/mL 1,2-diphytanoyl-*sn*-glycero-3-phosphocholine (DPhPC) in *n*-decane. DPhPC had originally been stored as a 25 mg/mL stock solution in chloroform in a light-shielded vial at −30°C. To make 100 μL of the working solution, 40 μL of the 25 mg/mL chloroform stock, corresponding to 1.0 mg of DPhPC, was transferred to a clean vial, and the chloroform was completely evaporated under a nitrogen stream. The dried lipid was then dissolved in 100 μL of *n*-decane, and the resulting solution was stored at 4°C in a light-shielded vial.
3. 2 × electrolyte recording buffer: 2 M KCl, 20 mM MOPS, pH 7.0. To make 100 mL of this solution, 14.91 g of KCl (molecular weight 74.55 g/mol, corresponding to 0.200 mol) and 0.418 g of MOPS (molecular weight 209.3 g/mol, corresponding to 2.0 mmol) were weighed and dissolved in approximately 80 mL of ultrapure water in a beaker. The pH was adjusted to 7.0 with 1 M KOH, and the final volume was brought up to 100 mL with ultrapure water. Store at 4°C.

2.2 Protein Nanopore Formation

1. Protein nanopore solution: A 1 mg/mL protein nanopore stock solution stored at −80°C was diluted with electrolyte recording buffer and ultrapure water to the concentration appropriate for measurements. The working solution was stored at 4°C and replaced with a freshly prepared solution every 1–7 days. Examples of protein nanopores used include α-hemolysin (αHL) [27] for nucleic acid sensing and exported protein 2 (EXP2), a parasite protein whose nanopore activity was first electrophysiologically characterized by our group [20, 25, 28].

2.3 Peptide Nanopore Formation

1. About 10–200 μM peptide nanopore stock solution was aliquoted for each measurement and stored at −20°C to avoid freeze–thaw cycles. Where required, peptide nanopores were subjected to structural transition [22]. The stock solutions were diluted with recording electrolyte buffer and ultrapure water to the concentration appropriate for measurements. Working solutions should be prepared fresh for each day of recording. Examples of peptide nanopores used include

coiled-coil transmembrane (CCTM) peptides [29, 30], consensus Wza (cWza) [31], pPorA [32, 33], LEK and FFK [24] α-helical peptides, SVG28 [22], and SVG28-D2 [23] β-hairpin peptides, which associate to form α-helical bundles and β-barrels, respectively.

2.4 Nanopore Sensing/Sequencing

1. Nanopore sensing: Both nucleic acids and peptides/proteins can be detected in electrolyte-containing buffers with pH values ranging from three to nine. DNA or peptide samples stored at −20°C were diluted with electrolyte recording buffer and ultrapure water to concentrations appropriate for measurement. It is recommended that the target analyte stock solution be diluted at least 20-fold to reach the working concentration. Examples of nanopore sensing of DNA and peptides can be found in previous studies [20, 25, 34, 35].
2. Nanopore sequencing with Phi29 polymerase/Hel308 helicase: For Phi29 polymerase, the sequencing buffer contained 0.3 M KCl, 10 mM $MgCl_2$, 1 mM EDTA, 1 mM DTT, 10 mM HEPES, and 250 μM dNTPs (pH 7.9). For Hel308 helicase, the buffer contained 0.4 M KCl, 10 mM $MgCl_2$, 1 mM EDTA, 1 mM DTT, 10 mM HEPES, and 1 mM ATP (pH 8.0). Phi29 polymerase and Hel308 helicase, stored at −30°C, were diluted with electrolyte recording buffer and ultrapure water to concentrations appropriate for measurement. A typical working concentration of either enzyme was 50 nM. Examples of nanopore DNA sequencing can be found in previous studies [36–40].

3 Methods

Carry out all procedures at room temperature unless otherwise specified.

3.1 Microdevice Fabrication and Droplet Contact Method

1. Using CAD, design an outer frame for a microdevice, typically 10 × 10 mm in size, with two 2-mm-diameter, 4.5-mm-deep wells, each with a bottom through-hole. Include a PMMA sheet separator insertion groove between the wells.
2. Using CAD, design a separator with a vertical rectangle measuring 4 mm × 4.5 mm and a circular hole with a diameter of 1 mm at the bottom.
3. Mill the PMMA plate and sheet using an automated CAD/CAM modeling machine (*see* **Note 1**).
4. Form a 5-μm-thick parylene-C film by vapor deposition polymerization on a single-crystalline silicon wafer.
5. Deposit an aluminum layer onto the parylene film.

6. Pattern the aluminum using a standard photolithographic process.
7. Etch the parylene through the aluminum mask by oxygen plasma.
8. Remove the aluminum layer after etching.
9. Peel the parylene film with a single micropore (typically 100 μm in diameter) from the silicon substrate using a tweezer.
10. Sandwich the parylene film between two separators with openings to expose the micropore.
11. Insert the sandwiched parylene film between the two wells of the microdevice.
12. Fix the film and PMMA sheets in place with adhesive (*see* **Note 1**).
13. Insert silver wires into each through-hole and cover the electrode holes from the inside of the microdevice with silver chloride paste.
14. Connect the electrode to the patch-clamp amplifier via the socket.
15. Add the lipid solution into each chamber (*see* **Note 2**).
16. Add the recording solution and buffer solution into each chamber (*see* **Note 2**).
17. Apply a constant voltage of +100 mV to the recording chamber.
18. Ground the opposite chamber.
19. Detect signals, for example, using an 8-kHz low-pass filter at a sampling frequency of 40 kHz.
20. Analyze channel current signals and dwell times with Clampfit software.

3.2 Protein Nanopore Formation

1. Add the lipid solution into each chamber.
2. Add diluted protein nanopore solutions (e.g., αHL [41] or EXP2 [20]) to one side of the bilayer chamber (*see* **Note 3**).
3. Fill the opposite well with only electrolyte buffer.
4. Allow the bilayer to form and observe spontaneous insertion of nanopores into the membrane as discrete current steps within several minutes.
5. Record single-channel currents in 1 M KCl and 10 mM MOPS buffer (pH 7.0) under a constant potential of +100 mV using a patch clamp amplifier, for example, Axopatch 200B (Molecular Devices, USA) (Fig. 2a, b).
6. Use a sampling frequency of 50 kHz with a 10 kHz low-pass filter.

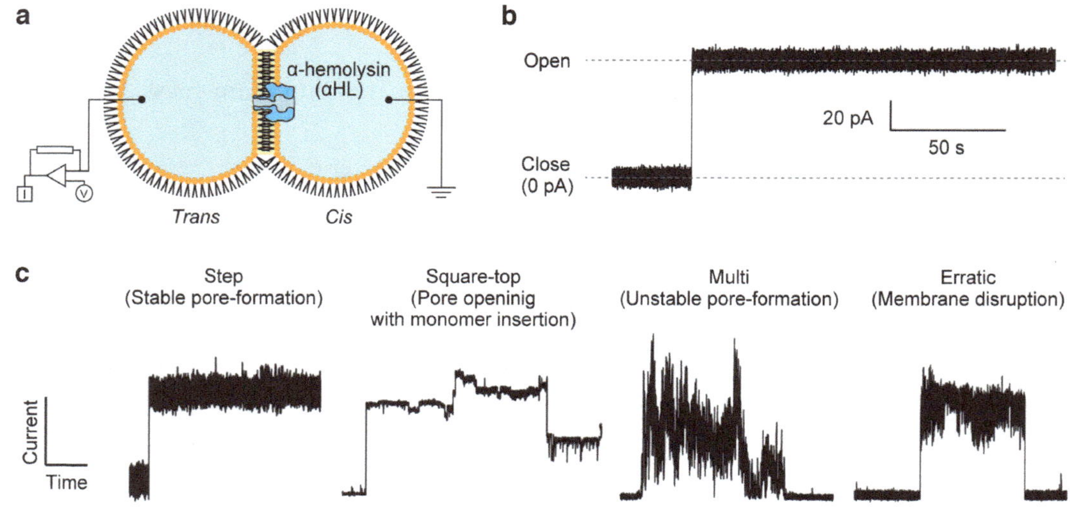

Fig. 2 Protein and peptide nanopore formation by the DCM. (**a**) Schematic of α-hemolysin (αHL) insertion into a lipid bilayer. (**b**) Typical single-pore opening of αHL recorded in 1 M KCl and 10 mM MOPS (pH 7.0) at + 100 mV. (**c**) Typical current signals of *de novo* designed pore-forming peptides, SVG28 or SVG28-mutants. To quantify pore-opening behavior, current signals were classified into four types: step and square-top (stable pores) and multilevel and erratic (unstable pores); definitions follow our previous work [23]

7. Analyze current amplitudes, conductance states, and dwell times with Clampfit software.
8. Perform additional analyses using Excel, Origin Pro, and Python software.

3.3 Peptide Nanopore Formation

1. Add the lipid solution into each chamber. In some cases, we use a lipid solution that has been co-incubated with the peptide solution as a layered (biphasic) system or as an emulsion, at 37°C for 1–24 h.
2. Add diluted peptide nanopore solution (e.g., SVG28 or SVG28-D2) to one side of the bilayer chamber (*see* **Note 4**).
3. Fill the opposite well with only electrolyte buffer.
4. Allow the bilayer to form and observe spontaneous insertion of nanopores into the membrane as discrete current steps within several minutes (*see* **Note 5**).
5. Record single-channel currents in 1 M KCl and 10 mM MOPS buffer (pH 7.0) under a constant potential of +50–200 mV (*see* **Note 6**) using a Pico patch-clamp amplifier or a Jet multi-channel amplifier (Tecella, USA) (Fig. 2c).
6. Use a sampling frequency of 40 kHz with an 8 kHz low-pass filter (Pico), or 20 kHz with a 4 kHz low-pass filter (Jet).
7. Analyze current amplitudes, conductance states, and dwell times with Clampfit software.
8. Perform additional analyses using Excel, Origin Pro, and Python software. For instance, the current signals can be

classified into four types: step, square-top, multi-level, and erratic (Fig. 2).

3.4 Nanopore Sensing/Sequencing

1. Add the lipid solution into each chamber.
2. Add the diluted protein/peptide nanopore solution to one side of the bilayer chamber and fill the other side with electrolyte buffer. The sensing or sequencing target may be premixed with either the protein solution or the electrolyte buffer and can be added to either chamber, depending on the application. For nanopore sequencing, Phi29 polymerase/Hel308 helicase should be included in the same chamber as the DNA (*see* **Note 7**).
3. Apply a constant potential, typically +100 mV, to allow the bilayer to form and to observe the spontaneous insertion of nanopores into the membrane as discrete current steps within several minutes (*see* **Note 8**).
4. Record single-channel currents under an appropriate potential for the application (*see* **Note 9**).
5. Perform recording at a typical sampling rate of 50 kHz with a 10 kHz low-pass Bessel filter (*see* **Note 10**).
6. Analyze the recording data with Clampfit software (Fig. 3).

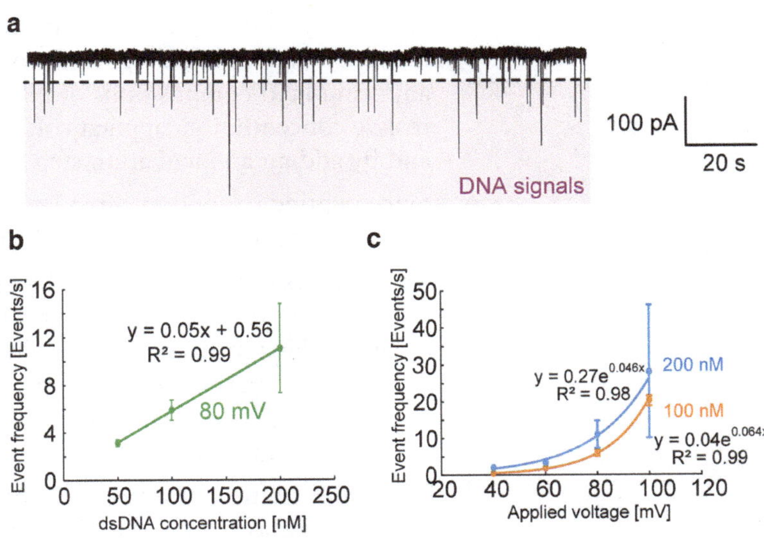

Fig. 3 The double-stranded DNA (dsDNA) translocation through a *de novo* designed SV28 peptide nanopore. (**a**) Current-time trace of the SV28 nanopore with 1 kbp dsDNA (100 nM) at +40 mV. (**b**) The event frequency of the translocation as a function of the concentration of dsDNA. (**c**) The event frequency of the dsDNA translocation as a function of the applied voltage. Blue and orange lines indicate the dsDNA concentration at 100 and 200 nM [22]

4 Notes

1. Remove all shavings thoroughly using ethanol or similar solvents. The separator adheres firmly to the bottom of the device. Residual shavings or insufficient adhesion will prevent the separator from fully inserting into the device and cause the *cis*-side and *trans*-side solutions to mix.
2. Add the lipid solution to the bottom of the device, filling the pores of the parylene film. Then, add the recording and buffer solutions. Be careful not to introduce air bubbles into the bottom of the device.
3. The optimal concentration of nanopore proteins for single-pore measurements varies by protein species. The appropriate concentration for measurement is calculated based on the duration of the observed single pore during the initial trial.
4. The optimal concentration of nanopore peptides for single-pore measurements varies by peptide species. The appropriate concentration for measurement is calculated based on the duration of the observed single pore during the initial trial. Determine the appropriate chamber for nanopore peptide addition based on the peptides' net charge and electrophoretic direction.
5. When nanopore peptides fail to form stable nanopores, current traces show noisy current steps or indistinct current fluctuations. Stable nanopore formation can be improved by adjusting lipid compositions of bilayer lipid membrane, electrolyte concentration, applied voltage, peptide concentration, and by adding an incubation step.
6. Since peptide nanopores often show voltage-dependent pore stability, explore the optimal applied voltage.
7. The analytes can also be added by pipetting the analyte solution into the recording chamber after a single nanopore has opened. This approach is technically more challenging, as the membrane sometimes ruptures after pipetting. However, it is also advantageous for comparing signals recorded before and after analyte addition.
8. The concentration of nanopores should be adjusted between experiments, as only a single nanopore is desired in typical nanopore sensing or sequencing. The threshold concentration for single-pore opening varies depending on the pore type.
9. The exact potential depends on factors, such as the translocation kinetics of the analytes, pore gating, and the stability of the lipid bilayer.

10. A higher sampling rate is beneficial when the analytes translocate through the pore with very fast kinetics.

5 Perspective and Remarks

The DCM has matured into a robust approach for reconstituting lipid bilayers and characterizing ion channel activity at the single-molecule level [5] .Compared with traditional planar bilayer techniques [3, 4], DCM offers superior reproducibility and stability, while its compatibility with microfabrication enables integration into versatile chip-based devices [8, 11, 42]. These advantages position DCM not only as a research tool for fundamental membrane biophysics but also as a practical platform for molecular sensing and sequencing.

Further development of this platform is expected to open several directions. The use of *de novo* designed peptide nanopores demonstrates that the platform can support synthetic biology approaches to generate tailor-made pores with tunable selectivity and transport properties. Integration with high-throughput microfluidics and parallel recording arrays could significantly expand measurement capacity and improve robustness. In addition, coupling with complementary readout modalities, such as optical detection or machine-learning-assisted signal classification, may enhance both sensitivity and resolution in nanopore analysis.

Beyond basic research, DCM-based nanopore systems have potential applications in biotechnology and medicine. Examples include drug screening using membrane protein targets, rapid detection of biomolecular markers in diagnostic settings, and portable sequencing platforms. By combining microdevice engineering, membrane biophysics, and synthetic nanopore design, this methodology provides a foundation for extending nanopore technologies toward increasingly complex and clinically relevant applications.

Acknowledgments

This work was supported by JSPS KAKENHI (Grant Nos. 19H00901, 21H05229, and 25H00416 to R.K.) and JST-CREST (Grant No. JPMJCR21B2 to R.K.). The authors used ChatGPT, DeepL Write, and Grammarly only for language editing and take full responsibility for the final content.

References

1. Kawano R (2015) Planar lipid bilayer formation using droplet contact method and its applications. Biophys Soc Jpn 55:77–80
2. Kageyama H et al (2022) New aspects of bilayer lipid membranes for the analysis of ion channel functions. Membranes 12(9):863

3. Montal M et al (1972) Formation of bimolecular membranes from lipid monolayers and a study of their electrical properties. Proc Natl Acad Sci USA 69:3561–3566
4. White SH et al (1976) Formation of planar bilayer membranes from lipid monolayers—Critique. Biophys J 16:481–489
5. Funakoshi K et al (2006) Lipid bilayer formation by contacting monolayers in a microfluidic device for membrane protein analysis. Anal Chem 78:8169–8174
6. Bayley H et al (2008) Droplet interface bilayers. Mol BioSyst 4:1191–1208
7. Elfaramawy MA et al (2018) Quantitative analysis of cell-free synthesized membrane proteins at the stabilized droplet interface bilayer. Chem Commun 54:12226–12229
8. Kawano R et al (2013) Automated parallel recordings of topologically identified single ion channels. Sci Rep 3: 1995
9. Kawano R et al (2014) A portable lipid bilayer system for environmental sensing with a transmembrane protein. PLoS One 9:e102427
10. Misawa N et al (2012) Efficient lipid bilayer formation by dipping lipid-loaded microperforated sheet in aqueous solution. Micromachines 12(1):53.
11. Tsuji Y et al (2013) Droplet-based lipid bilayer system integrated with microfluidic channels for solution exchange. Lab Chip 13:1476–1481
12. Tonooka T et al (2014) Lipid bilayers on a picoliter microdroplet array for rapid fluorescence detection of membrane transport. Small 10:3275–3282
13. Scott C et al (2025) Electrophysiological characterization of monoolein-fatty acid bilayers. Langmuir 41:2293–2299
14. Kamiya K et al (2018) Electrophysiological measurement of ion channels on plasma/organelle membranes using an on-chip lipid bilayer system. Sci Rep 8:17498
15. Osaki T et al (2017) Artificial cell membrane systems for biosensing applications. Anal Chem 89:216–231
16. Numaguchi Y et al (2024) Real-time monitoring of the amyloid beta(1-42) monomer-to-oligomer channel transition using a lipid bilayer system. PNAS Nexus 3:pgad437
17. Saigo N et al (2019) Electrophysiological analysis of antimicrobial peptides in diverse species. ACS Omega 4:13124–13130
18. Takeuchi N et al (2022) Pattern recognition of microRNA expression in body fluids using nanopore decoding at subfemtomolar concentrations. JACS Au 2:1829–1838
19. Takiguchi S et al (2023) Simultaneous recognition of over- and under-expressed MicroRNAs using nanopore decoding. Anal Chem 95:14675–14685
20. Miyagi M et al (2022) Single polypeptide detection using a translocon EXP2 nanopore. Proteomics 22:e2100070
21. Liu P et al (2020) Recognition of single-point mutation using a biological nanopore. Small Methods 4:2000101
22. Shimizu K et al (2022) De novo design of a nanopore for single-molecule detection that incorporates a beta-hairpin peptide. Nat Nanotechnol 17:67–75
23. Fujita S et al (2023) Cell-free expression of de novo designed peptides that form β-barrel nanopores. ACS Nano 17:3358–3367
24. Peng Z, Usami M, Nakada A, Sekiya Y, Fujita S, Mijiddorj B, Kawamura I, Kawano R (2025) De novo design of α-helical peptide nanopores for single-molecule detection using helix packing motifs. ACS Nano 19(49):41789–41802. https://doi.org/10.1021/acsnano.5c15080
25. Miyagi M et al (2025) Redesign of translocon EXP2 nanopore for detecting peptide fragments. Small Methods 9(4):2401562.
26. Yamaji M et al (2025) Complex and non-sequential current signatures of a β-hairpin peptide confined in a nanopore. Anal Chem 97:2044–2051
27. Kasianowicz JJ et al (1996) Characterization of individual polynucleotide molecules using a membrane channel. Proc Natl Acad Sci U S A 93:13770–13773
28. de Koning-Ward TF et al (2009) A newly discovered protein export machine in malaria parasites. Nature 459:945–949
29. Scott AJ et al (2021) Constructing ion channels from water-soluble α-helical barrels. Nat Chem 13:643–650
30. Niitsu A et al (2025) Rational design principles for De Novo α-helical peptide barrels with dynamic conductive channels. J Am Chem Soc 147:11741–11753
31. Mahendran KR et al (2017) A monodisperse transmembrane alpha-helical peptide barrel. Nat Chem 9:411–419
32. Krishnan RS et al (2019) Autonomously assembled synthetic transmembrane peptide pore. J Am Chem Soc 141:2949–2959
33. Krishnan RS et al (2022) Assembly of transmembrane pores from mirror-image peptides. Nat Commun 13:5377
34. Tada A et al (2023) Nanopore filter: a method for counting and extracting single DNA molecules using a biological nanopore. Anal Chem 95:9805–9812
35. Fujii S et al (2017) Pesticide vapor sensing using an aptamer, nanopore, and agarose gel on a chip. Lab Chip 17:2421–2425
36. Manrao EA et al (2012) Reading DNA at single-nucleotide resolution with a mutant MspA nanopore and phi29 DNA polymerase. Nature Biotechnol 30:349–353

37. Derrington IM et al (2015) Subangstrom single-molecule measurements of motor proteins using a nanopore. Nature Biotechnol 33:1073–1075
38. Brinkerhoff H et al (2021) Multiple rereads of single proteins at single-amino acid resolution using nanopores. Science 374:1509–1513
39. Yan SH et al (2021) Single molecule ratcheting motion of peptides in a mycobacterium smegmatis Porin A (MspA) nanopore. Nano Lett 21:6703–6710
40. Chen ZJ et al (2021) Controlled movement of ssDNA conjugated peptide through mycobacterium smegmatis porin A (MspA) nanopore by a helicase motor for peptide sequencing application. Chem Sci 12:15750–15756
41. Song LZ et al (1996) Structure of staphylococcal alpha-hemolysin, a heptameric transmembrane pore. Science 274:1859–1866
42. Le Pioufle B et al (2008) Lipid bilayer microarray for parallel recording of transmembrane ion currents. Anal Chem 80:328–332

INDEX

Fanjun Li (ed.), *Nanopore Technology*, Methods in Molecular Biology, vol. 3072,
https://doi.org/10.1007/978-1-0716-5527-6,

FSC
www.fsc.org
MIX
Papier | Fördert
gute Waldnutzung
FSC® C083411

Kolibri 360 GmbH
Ferdinand-Jühlke-Straße 7
99095 Erfurt, Deutschland
produktsicherheit@kolibri360.de